Praise for

PRESIDENTIAL A........

"You'll never breathe the phrase 'good old days' again after following Shenkman's lively excursion through two centuries of the conniving and dealing that every president except Washington had to undertake to win or hold the Grand Prize. Tough, fair-minded, and a wonderful story."
—Bernard A. Weisberger, *American Heritage* columnist

"*Presidential Ambition* is essential reading for anyone interested in the presidency. Shenkman has identified what may be the most important character trait of our presidents . . . their driving ambition . . . and he shows how that single trait has influenced the key moments and decisions in our nation's history. . . . A lively, enjoyable journey through the minds of the people who have led our nation and changed our lives."
—Professor Leonard Steinhorn, School of
Communication, American University

"A study of ambition focused primarily on the first 32 presidents . . . Washington through FDR. Shenkman applies the sort of character study Robert Caro did in his multiple volumes of Lyndon Johnson, and Richard Ben Cramer accomplished in his massive take on the 1988 election, *What It Takes*."
—*Seattle Times*

"A timely take on morality and the nation's highest office. Shenkman . . . knows how to make history entertaining. . . . Revealing. . . . Fascinating."
—*Omaha World-Herald*

"This provocative book explores the ways presidents obtained, kept, and exercised political power during periods of profound change within the country, including the most devious and underhanded acts imaginable. Readers who enjoyed his previous works on the legends, lies, and myths of American and world history will find this latest book both stimulating and entertaining."
—Robert V. Remini, author of the National Book
Award–winning *Andrew Jackson*

"Here is a fresh, frank, and entertaining study of presidential character, warts and all. No one who reads this book will ever again be surprised by what any president might do. At a deeper level, this work raises disturbing questions about the linkage of ambition, power, and corruption in the American system."

—W. J. Rorabaugh, Professor of History,
University of Washington

PRESIDENTIAL AMBITION

PRESIDENTIAL AMBITION

GAINING POWER
AT ANY COST

RICHARD SHENKMAN

HarperPerennial
A Division of HarperCollins*Publishers*

For BillyMac

A hardcover edition of this book was published in 1999 by HarperCollins Publishers.

HarperCollins books may be purchased for educational, business, or sales promotional use. For information please write: Special Markets Department, HarperCollins Publishers Inc., 10 East 53rd Street, New York, NY 10022.

First HarperPerennial edition published 2000.

Designed by Jennie Malcolm

The Library of Congress has catalogued the hardcover edition as follows:

Shenkman, Richard.
      Presidential ambition : how the American presidents gained power, kept power, and got things done / Richard Shenkman.—1st ed.
            p.   cm.
      Includes bibliographical references and index.
      ISBN 0-06-018373-X
            1. Presidents—United States—History. 2. Ambition. 3. Executive power—United States—History. 4. United States—Politics and government. I. Title.
      E176.1.S56 1999
      973—dc21   98-27045

ISBN 0-06-093054-3 (pbk.)

00 01 02 03 04 ❖/RRD 10 9 8 7 6 5 4 3 2 1

Contents

Contents

Contents

16. FDR: The Great Depression 286
How FDR triumphed during the Great Depression through charm, talent, and deviousness, several times using the IRS to go after his political enemies

17. FDR: World War II 304
How FDR waged war without asking Congress to declare it

18. The Cold War 315
How the Cold War forced Harry Truman and Dwight Eisenhower to take new and extraordinary steps to gain power and to keep it

Introduction

Given the myriad scandals in which presidents have been caught in recent decades, one of the most pressing questions of our times, to put it starkly, is surely this: Is it the presidents themselves who are at fault or the system? It was to find an answer that I wrote this book.

I became convinced that this was the right question to ask when I read Monica Crowley's *Nixon Off the Record*, which was published in 1996. Crowley's book, a pastiche of quotes gleaned from notes she took of conversations with Nixon during the last years of his life, contains many remarkable observations. But the one that stopped me cold was this: "You know," Nixon told her during a rambling reminiscence, "politics has really gone down the tubes since 1960, and frankly it's not much fun to watch. Maybe it's because the characters involved are weak on either the personal or political side—or both. I don't know. I ran in '68 and '72, so I guess that applies to me too. It just seems that good candidates aren't running anymore. It's better not to run anyone at all than a turkey. The top offices in this country require better."

Nixon always loved to shock. This was shocking. It was like hearing Jack the Ripper complain about the high murder rate. Next, I suppose, we'll discover that Lyndon Johnson believed that recent presidents told too many lies, that Ronald Reagan believed they made too many promises they couldn't keep, and that Bill Clinton believed they haven't been faithful enough to their wives.

But Nixon was always a superb analyst of American angst. Clinton felt your pain. Nixon diagnosed it.

Millions of Americans agree with his diagnosis of recent presidents. Presidents are the problem spouse: We can't live with them, we can't live without them.

The media are the detectives in this problem marriage. They provide us with up-to-the-minute reports on the presidents' daily activities. We know when Jack Kennedy plunged naked in the White House pool with secretaries Fiddle and Faddle. We know when and where Lyndon Johnson secretly rendezvoused with Alice Glass, the apple of his eye. We have been supplied with ample lists of quotes from his family showing that Ronald Reagan, the spokesman for family values, was regarded by several of his own children as a virtual stranger.

We are all gossips now. What once was considered rumor is now news. Even if George Bush didn't have an affair with that blond from the State Department, we suspect that he did, and that makes it news.

Presidents used to be able to keep their secret lives secret. They can't anymore. We know everything now. We don't like the fact that we do. We tell pollsters we wish we knew less. But the way journalism is practiced today we always know more and more. And the more we know, the less we respect our presidents. The media have turned us into a nation of valets. And as the old saying goes, no man can be a hero to his valet.

The personal is political. Who the presidents are is what counts. Even their policy failures have appeared to be personal. Johnson's refusal to withdraw from Vietnam? It was because he wasn't manly enough to admit he'd made a mistake. Nixon and Watergate? It was because he was paranoid. Reagan and Iran-contra? It was because he was too stupid and inattentive to realize that he had traded arms for hostages. And Clinton? He lacked self-control and basic morals.

This is the age of the biographer. Biography is destiny. The presidency is what the presidents make of it. If they succeed it is because they are men of great character. If they fail it is because they lack character. Character is everything, we are told. Lincoln was great because he was moral. Warren Harding was a failure because he was immoral.

In the sixties the radicals said the system was flawed. Today we insist it is our leaders who are.

Which is it?

If it is the leaders who are to blame, we have had a bad run of luck. In the ten elections since 1960 not one of the men elevated to the presidency has seemed up to it. That's an awful lot of bad luck.

Here's what I think. First, part of the problem is that we almost know too much about the presidents. Because of the media we know every time they make even the tiniest misstep; that encourages us to think they are flitting from one mistake to another. And because the media tend to affix the "gate" label on every looming scandal as if all scandals are Xeroxes of Watergate, we think presidents are nearly constantly in deep trouble. They're not. Events in Washington aren't half as interesting as they're made out to be.

System or individual flaw? This is a flaw in the system. The media simply give us a warped view of reality.

Second, if you examine closely the mistakes recent presidents have made, almost always it was because they wanted something too much. That is, they were too ambitious. Johnson in Vietnam? His fundamental error was that he was so intent on winning support for the Great Society that he was willing to conceal the build-up in Vietnam, which threatened to undermine his domestic political agenda.

Nixon in Watergate? Insecure because of his small margin of victory in 1968 (he won with just 49 percent of the vote), he unleashed the thugs in his administration to do whatever it took to sabotage his enemies and win big in 1972. When their activities threatened to become public he orchestrated a massive cover-up.

Reagan in Iran-contra? He was so committed to freeing the hostages and helping the contras that he led his subordinates to believe he didn't care if they broke the law to achieve his goals.

Clinton in the Monica scandal? He was so determined to keep the affair secret he was willing to lie under oath.

System or individual flaws? Both, actually. George Washington presumably wouldn't have made the choices these presidents did. He wouldn't have concealed a war, condoned a cover-up, encouraged lawlessness, or risked perjury. Character does count. But when president after president after president feels compelled to break the rules to gain power or to get things done, it may be a sign that the system is partly responsible.

Here's my hunch, one I think is borne out by the story I tell in this book: The system today forces politicians to sink lower and lower to gain power and wield it. Just look at what a president has to do these days. He has to exploit his family, scrounge for contributions, perhaps exploit religion and race, reinvent himself, misrepresent himself, knock off his enemies, tolerate corruption, possibly conceal malfeasance, occasionally lie, sometimes play politics with national security, pander to immigrants, and perhaps even look the other way when votes are stolen.

You may be thinking that this isn't new. Presidents have always had to compromise themselves. True. The path to power has always been difficult. In every generation there weren't just one or two people who possessed the qualifications and the desire to become president. There were dozens and dozens. In the struggle to win only the most ambitious survived.

But here's what's not understood. In every generation the struggle has demanded greater and greater moral compromises. This was because the country kept getting more complicated as it grew from a nation of four million to one of a quarter billion. When the Republic was launched, a politician anxious to gain the presidency merely had to win the support of a few thousand men. By the next generation, the time of Jackson, a president had to seek out the votes of hundreds of thousands. By the Civil War presidents had to fight for the votes of millions, among them Irish Catholics, urban workers, and Wall Street financiers. Within another generation they had to seek out the support of industrialists and machine party bosses. By the turn of the century they had to contend with millions of immigrants, Populists, and unions. In the early part of the twentieth century they had to learn how to attract the votes of women while mastering radio—and later, television. After World War II they had to attract the votes of blacks, white ethnics, and Hispanics.

And all this was merely to obtain power. Once they found themselves in the White House they had to find ways to get things done. This, too, became an ever-more-complicated task as the country became more complicated, requiring even the most honest presidents to compromise their integrity.

Power corrupts, it is said, and absolute power corrupts absolutely. But in our system it is the lack of power that corrupts. It is when a politician lacks the power to achieve his objectives that he is tempted to compromise democratic values. And in a system in which power is shared by three branches, two parties, and fifty states, politicians are always discovering how little power they possess. Even Abraham Lincoln, at the height of the Civil War, when he held the powers of a virtual dictator, complained that he didn't have enough power. "I claim not to have controlled events," he remarked in a letter to a supporter, "but confess plainly that events have controlled me." Or as Harry Truman put it, "Being a president is like riding a tiger. A man has to keep on riding or be swallowed."

In the modern era politicians have had to grab power under the watchful eyes of the cynical and powerful media. This has had two consequences. First, it has made acquiring power even more difficult, which has tempted politicians to make even more compromises. Second, it has made our generation of politicians, their flaws on display on the front pages of the morning paper and on television, seem ever more corrupt and immoral.

This is a disturbing book. It's meant to be. The history's disturbing. It's disillusioning to see the depths to which otherwise honest men will sink when the holy grail of power is held out before them. But I came away from the book feeling good about our presidents. I discovered, for instance, that nearly all of them had been deeply influenced by idealists. (Five presidents had been raised to be ministers, another five were descended from ministers.) And none, I am delighted to report, ever purposefully tried to wreck the system to advance his own ambition. Moreover, most tried to redeem themselves by doing good works once they obtained power.

But I was left feeling uneasy about the message the book seemed to deliver. That is, that the system itself is corrupting. But that seems to be the case however much I wish it weren't.

Not every politician who seeks the presidency is corrupted by the process. Several, like Dwight Eisenhower, who came into the office with a reputation as heroes, escaped largely unblemished. But most did not. That seems to be the disturbing truth.

The book tells two interrelated stories. The main one is about the changes the country went through that increasingly made politics more and more complicated. The other is about the individuals who confronted these changes and in confronting them themselves changed, making compromises with which they were uncomfortable and then, from time to time, redeeming themselves through good works.

I decided not to tell each president's story in detail. To do so would have been tedious and in any case unnecessary, since their stories are similar in so many ways. In deciding finally which ones to dwell on, I chose those that seemed to illustrate most dramatically the interplay of history and ambition. I deliberately passed quickly over the best-known stories. The longest story in the book is about Lincoln. I believe he warrants extended treatment.

The book focuses on the presidents elected before 1960, those who served before the media began casting such a harsh beam on the presidents that all we could see were their warts. Our presidents we know well: We know who they are and what they did. So the focus here is on the presidents who came before them, those on the far side of the imaginary line that divides American history in our minds between the presidents of the past who supposedly demonstrated strong character and our own, who did not.

The historical record demonstrates amply that circumstances forced all of the presidents to bend, weave and compromise. With a single exception. The exception was George Washington. He never made inflated promises, cut deals, or told lies to get elected or to get things done. In part that was because he was special: He was the only president who was handed the office on a platter. But it was also because the country then wasn't the complicated place it's now become.

Prelude

Things *were* different at the beginning. To understand how different—and we should if we are ever going to be able to comprehend the changes that were to make governing and politics so complicated—it may be helpful to see just what life was like then, before things got so complicated.

So: what *was* life like then? Or, more precisely, what was governing like then?

As they do today, politicians faced great problems. In the 1780s the government under the Articles of Confederation had failed so miserably to provide financial stability that the farmers in Massachusetts revolted. An army had to be dispatched to put down their rebellion. When Washington took office the economy was in shambles: the currency inflated and worthless, credit in short supply.

So there were very real, very frightening problems.

But great as their problems may have been the times were very different. For life was far far simpler. . . . The country was smaller—there were after all just thirteen states. . . . The population was smaller—there were in all of America in 1790 just four million people, the great majority of whom were white Anglo-Saxon Protestants. . . . The federal government's functions were limited.

And governing was relatively easier.

To begin with, the pace of life was slow, so officials had plenty of time to think things through. Where news cycles today are counted in hours, forcing officials to react with lightning speed to events, back then the cycles lasted months. News that happened in New York might take two or three months to reach the interior of Georgia. When Congress, sitting in New York, officially announced that George Washington had been elected president of the United States the news took two weeks to reach Washington at his home in Mount Vernon. And whereas today presidents have to contend with the prattling tongues of the three thousand national reporters assigned to cover them, in Washington's day there were . . . none. In the whole country there were hardly any journalists at all. And there were hardly any daily newspapers—just eight actually. Because of the dearth of newspapers Washington was able to set the public's agenda with relative ease. Until Hamilton and Jefferson started feuding, leading to their famous newspaper war, what Washington decided was news *was* news. He controlled the news agenda.

Washington did have to worry about his image, especially after Jeffersonian partisans began openly taking him on. But he never had to worry too much. There simply weren't any other national figures with the clout to challenge his control. At the beginning of his presidency there weren't any other truly national figures period. Only Washington was known by all, north and south, east and west. And even Washington wasn't known as presidents are today. Prior to the invention of photography and the tape machine Americans didn't know what their presidents looked like or how they sounded. (One day while on a journey into the back country Washington was approached by a farmer. *Who are you?* the farmer asked. *George Washington,* the president answered. *Well, I'll be,* the man responded.)

The government Washington headed was modest. The army at the time numbered barely over five hundred men, about the size of the police force needed today to protect Des Moines, Iowa. And there was, as yet, no navy. (The navy would be established in 1798, thanks to the efforts of John Adams. It would consist of fewer than half a dozen ships.)[1]

The bureaucracy was almost nonexistent. No Commerce Department. No Interior Department. No Agriculture Department. Not much of anything really. There wasn't even a Justice Department; the attorney general had so little to do he wasn't even given an office and for decades wasn't required to live in or near the capital.

What few departments there were—State, War, Treasury, and the Post Office—hardly seemed like departments as we would understand them. If, say, you happened in at the State Department, chances were you would find about half a dozen people at work—at the War Department, even fewer. Only the Treasury Department would look like a fully functioning government bureaucracy. Go there and you'd find about a hundred and fifty people. The biggest department was the Post Office, but of course its employees were spread across the country. In all there were barely a thousand people on the federal payroll.

There was, of course, no White House staff because there was as yet no White House. Washington in fact made due with just a single assistant: a secretary.

Washington would complain bitterly that the presidency was killing his health and in his first year in office he nearly died from a bout of pneumonia. "My late change," he complained, "from [physically] active scenes, to which I had been accustomed and in which the mind has been agreeably amused, to the one of [physical] inactivity which I now lead and where the thoughts are continually on the stretch, has been the cause of more illness and severe attacks of my Constitution within the last twelve months than I had undergone in thirty years preceding put together." But much of his workload consisted of tedious paperwork: signing commissions, payrolls, and purchase orders—the kind of work future presidents would barely deign to do. (He did not have to review his departments' budgets. There were none. For decades the departments simply spent what Congress appropriated. If the money ran out they asked for more.)

The government was small under Washington because the government in the early years didn't do much and wasn't expected to. Washington twice during his administration felt so confident the government could get along without him that he left for months at a time to travel the country to see and be seen. (A chief goal of his trips was to give the federal government a human face.) During those months he remained largely out of touch with government affairs, leaving any decisions that had to be made up to the cabinet.

The fact is, heavy as his responsibilities were, and there's no denying they were heavy, the job simply required less time than it subsequently would. During the summers Washington was able to leave the capital for July and August to relax at Mount Vernon. John Adams during his presidency would leave so often to go home to Massachusetts that some people joked that he had abdicated. One year, when his wife, Abigail, fell ill, he would stay home with her for seven long months. The government was so easy to run in the early 1800s that even during the War of 1812 James Madison would return home for the summer to Montpelier to rest. There really was no reason to stay;

almost everybody deserted the swamp-infested capital as soon as the weather turned warm.[2]

Not only was the president's workload lighter, so was his cabinet's—and it would remain lighter for decades. James Monroe, inaugurated in March 1817, was able to get by without a secretary of state until September. The secretary of war didn't arrive until October. During Monroe's eight years in office there constantly were vacancies in the major cabinet posts. At these times the president simply had one of his remaining secretaries cover two departments simultaneously. In the 1840s James Polk would be able to run the government almost single-handedly. In the summer when the cabinet secretaries left to escape Washington's humid weather Polk would stay behind and run their departments for them. He would personally answer the mail, fill out forms, sign commissions, and issue purchase orders.[3]

Nothing better symbolizes the vastly simpler state of things in Washington's day than his own accessibility. While he insisted on standing on a little platform when greeting guests at public levees, anybody could come and meet him as long as they were dressed well. At the beginning he even kept an open door day and night. When that proved a little too burdensome—"From the time I had done breakfast and thence till dinner and afterwards till bedtime," he complained, "I could not get relieved from the ceremony of one visit before I had to attend another"—he limited the public to two open meetings a week. On those occasions, Tuesdays between three and four and Fridays in the evening, people crowded the mansion for a chance to see him.[4]

Personally fearless, Washington never traveled with a guard. Anybody could come up to speak with him while he was out and about. On the mornings when he went horseback riding, which was almost every day—when the capital was in New York he would often start his day by riding completely around Manhattan Island—he often rode alone. If a friend came along, Jefferson perhaps, it would be just the two of them out and about, just the president and the secretary of state taking a nice little ride. More shocking still, even his house on Broadway went unguarded. If someone wanted to knock off the president there was nobody assigned to stop them. But, of course, the thought had not yet occurred to anybody that they might want to knock him off. Kill a president? Why for heaven's sake? (And not for a long time would the thought occur to anybody. The White House would remain unguarded until the Civil War. Presidents wouldn't begin receiving full-time secret service protection until 1901, after McKinley was assassinated.)*

So life *was* simpler. And politics *was* simpler. Washington didn't even need to campaign. The office was given to him.

And because things were simpler governing was simpler. Not simple. Simpler.

---

*The first president to be seriously attacked was Jackson, on January 30, 1835. Visiting the Capitol one day to attend a state funeral for a congressman from South Carolina, Jackson was accosted by a madman armed with two pistols. Both misfired. the gunman, a deranged Englishman who believed he was an heir to the British throne, was apprehended and committed to an insane asylum. Marquis James, *The Life of Andrew Jackson* (1938), vol. 2, pp.390–91.

1. In the Garden of Eden

How George Washington, alone among the presidents, was able to gain power and get things done without compromising himself or his principles

There was a touching scene at the outset of his presidency that was almost too good to be true. When Charles Thomson, the secretary of the Congress, arrived at Mount Vernon to tell Washington he had been elected president, the two men withdrew to the main room of the house and delivered little speeches to each other. Thomson told Washington that the Congress was delighted he had agreed "to sacrifice domestic ease and private enjoyments to preserve the happiness of your country." Washington responded that he had accepted in deference to the public's desires. He couldn't promise to be a great president, he added, but "I can promise . . . to accomplish that which can be done by honest zeal." It was almost comically stilted, like one of those scenes out of a 1930s Frank Capra movie in which Jimmy Stewart stands up and delivers a sincere and selfless sermon on patriotism. But it happened. And Washington came off looking exceptionally decent.[1]

He hadn't always seemed so decent.

As a young man there was a certain crassness about him that was almost palpable. Though he was a born aristocrat he was very much a man on the make. Land was everything in Washington's youth, the symbol of wealth and prestige, and he had set out to acquire as much of it as he could. Through inheritance he had received Mount Vernon and about two thousand acres. But that hadn't come nearly close to satisfying his appetite. He didn't want just a lot of land. He wanted more land than anybody else. Which was, apparently, the prime factor in his decision to court Martha. She was, even though youthful, neither particularly pretty nor particularly socially adept. And Washington didn't love her (not at first anyway); as he admitted in a letter at the time of his engagement, he was actually in love with Sally Fairfax, his best friend's wife. "You have drawn me," he wrote Sally, "into an honest confession of a simple Fact." But keep it a secret: "The world has no business to know the object of my Love, declared in this manner to you, when I want to conceal it."

But Martha was not without her attractions. One of the richest widows in North America, she possessed thousands of acres of land. Under the laws then in effect, her land became his upon marriage, instantly turning Washington into one of the richest men in America. Through Martha he received a hundred slaves, another six thousand acres, and enough money to buy thousands and thousands of acres more.

Not even all that was enough to satisfy him. In 1767, eight years after his marriage, he made a grab for land expressly set aside for the Indians. It happened to be illegal under laws promulgated by the Crown. Washington didn't care. He told his surveyor "to keep this whole matter a profound secret." If anybody asked the surveyor what he was up to, Washington instructed, the surveyor was to lie. Over the next few years he was to acquire another twenty thousand acres from the British government in return for his service as a colonel in the Virginia militia. He wasn't really entitled to the land; in fact when he had signed up the government had made it clear that the property was supposed to go to soldiers, not officers. But Washington had dextrously arranged for the officers to receive land, too. As the leader of his regiment, Washington had the responsibility of deciding who received which parcels (two hundred thousand acres of land were to be handed out). Washington saw to it that he received the best, "the cream of the Country," as he subsequently boasted.[2]

But then the Revolution had come. It changed Washington as it did many of the leading figures in the colonies. Suddenly Washington, the "inveterate land grabber," as historian John Clark called him, became Washington, the enlightened revolutionary. Acquiring land was no longer enough. Being rich was no longer enough. Believing himself to be in a position to affect history, Washington lifted his sights and became something no one had any right to expect he would. Now, instead of acquiring land, he would seek to acquire what people in the eighteenth century called fame.

In our time fame has taken on a pejorative meaning. But in his day fame was far more sublime. To be famous was to be immortal. It was believed at the time that there were many ways to gain fame. But the most honorable way of all, it was felt, was to found a commonwealth. Thus did Washington, as fired by ambition as ever, decide to dedicate himself to the patriot cause, inspiring his fellow Americans as no one else did.[3]

He didn't prove to be a brilliant general. In fact, he never won any major battles. But he kept the army together during awful times, and by strength of character was able to command the people's respect. At the end of the war he was held in such high esteem that he might very well have been able to crown himself king—as a lot of people wanted. But Washington refused, wouldn't even consider the subject. All he wanted to do was return to Mount Vernon. When friends in the army demanded that he make himself dictator after Congress refused to pay the soldiers their back wages, he looked upon the proposal with sheer horror.

When the war finally ended and the British evacuated the country Washington, like Cincinnatus, his Roman hero, laid down his sword and went back to his plow. It was his intention to remain at his plow for the remainder of his life. But events intervened: The Confederation collapsed, the Constitution was adopted, and Washington was drafted for president.

The kind of trust Washington inspired is difficult for us to comprehend because no figure today is comparable. Nobody since Washington has been comparable to Washington. He is sui generis. It is not just that he led the country through the Revolution and then at the end willingly surrendered his sword and returned home to his plow. It is not even that he was the Father of His Country, and a country can have just one father. It is that Washington, singly among our presidents, did not want to be president. In his case, and his case alone, the office actually did seek the man. Not only did he not lift a finger to be made president. He actually preferred *not* to be president. He accepted in the end only because he had no choice. If he failed to accept the presidency there was a good chance that the Republic he had worked so hard to bring about would collapse. All others after him would try to claim that they, too, had no choice, that duty required them to run; for to admit ambition for the office was to prove oneself instantly unworthy of it. But only he really was motivated solely by duty. For what need did he have of the presidency? He already was the Father of His

Country. All the presidency could do was possibly put that reputation at risk.

It could be argued, of course, that he agreed to run for president because he feared that that achievement was in jeopardy; if the commonwealth collapsed, his efforts would have been for naught. He would be the founder of nothing. Thus, it could be said, he ran for the same reason all the others would run afterward—because he was ambitious for fame. And yet—there would seem to be a vast difference between someone who eagerly seeks an office and one who, as it were, has had it thrust upon him, as the presidency was indeed thrust upon Washington.

The chief object his first term was to put the country's financial house in order. The federal government was loaded down with debt. The states were loaded down with debt. And the people were hostile to taxes. Raise taxes, and the people might rebel. Try to ignore the debt, as both state and national governments had been doing for years, and the government would appear weak. Appear weak, and the states would begin to go their own separate ways, and rich people would become alarmed at the instability. Washington himself didn't have any idea how to solve the problem; he wasn't very bright, and he knew nothing about economics. But he was terrific at spotting talent, and long ago he had spotted Alexander Hamilton. Hamilton, put in as secretary of the treasury, would know how to fix things.

It was a brilliant choice. Hamilton was a genius at economics as he was at almost everything he put his mind to. And he quickly came up with a brilliant plan. He would have the federal government assume the states' debts, thereby helping cement the Union together. And he would establish a national bank to help provide the government with money to pay off the debts. Because the bank would be backed by the government, rich people would lend the bank money; as creditors they would then have a stake in the survival of both the bank and the government.

Having come up with its program, the administration still had to sell it to Congress. And that was exceptionally difficult. Difficult because it was a complicated plan and in order to work, all of its various features had to be kept intact; remove one in the course of bargaining and the whole edifice would collapse: Eliminate the bank, and the government couldn't pay its debts. Refuse to assume the states' debts, and the government would lose the ability to command the support of the states. And yet every feature was controversial. Strict constructionists argued that the Constitution did not give the government the right to establish a bank. States that had worked hard to pay off their debts didn't want the states that hadn't to get a free ride. Further complicating the matter was that most of the states that *had* paid off their debts were located in the South, the others in the North, which exposed a sharp cleavage between the sections—a cleavage everybody had feared and wanted to hide. If there was one thing that could tear the country asunder, it was sectionalism.

Washington worried—and did nothing. It was his belief that he shouldn't become involved in political controversies, even controversies that threatened the very foundation of his government. He felt he had to remain aloof. Get involved in the nitty-gritty of politics, and he would begin to look like a politician. And if he looked like a politician the country would no longer rally around him as a symbol of national unity. He would be tainted. Undoubtedly he could win in Congress this time and at countless other times if he put his prestige on the line, but with each victory he would make more enemies, thereby diminishing his stature as a statesman.

Washington could afford to remain above the fray because the other members of the administration, particularly Hamilton and Secretary of State Jefferson, were willing to become deeply engaged in politics. It was Jefferson who dreamed up the compromise that brought the North and South together by creatively joining the debt issue to a second one that also divided the sections: the location of the country's capital. New York wanted the capital to remain where it was, in New York City. Pennsylvania wanted it moved to Philadelphia. Virginia wanted it moved to the Potomac. Jefferson's ingenious solution, worked out one night at dinner with Hamilton, was to give the South the capital in exchange for its support of the federal assumption of state debts, the measure the North wanted. To mollify Pennsylvania, the capital would be moved to Philadelphia for ten years.

The wisdom of Washington's decision to remain aloof from politics became clear when the financial plan suddenly became enmeshed in scandal, the first scandal in the history of the United States government.

What happened was that the plan to pay off the government's debts made a few people very, very rich.

Much of the money the federal government owed in 1790—more than forty-two million dollars in all—was owed to average people of modest means, including thousands of Revolutionary War veterans whose IOUs had never been redeemed by the government under the old Articles of Confederation. The war certificates they held had fallen dramatically in value, most fetching just fifteen or twenty cents on the dollar in the open market. By any standard of justice the veterans should have been the ones to benefit from Hamilton's plan, under which the certificates would be paid off at par; after all, the securities they held had literally been purchased with the blood of patriots—their blood. But the veterans by and large weren't the ones who were to receive the benefit: Rank speculators were.

Just as the debt bill was about to be passed, speculators in New York, Philadelphia, and Boston raced to the countryside in advance of the news and bought up the long-devalued Revolutionary War certificates owned by the veterans. A few congressmen even got into the act, hiring ships to send agents deep into the remote areas of the country. Before the veterans figured out what was happening, the speculators had managed to scoop up the bulk of the once worthless but now valuable certificates.

Hamilton, who never cared for money—he was nearly bankrupt when he was killed in a duel with Aaron Burr—wasn't personally implicated in the speculation. But he didn't mind it, either. Eminently practical, he reasoned that the speculation, involving some of the nation's most important bankers and merchants, would help win them over to the support of the federal government. And anyway he believed that he couldn't have prevented what happened. It would have been a bureaucratic nightmare, he alleged, to try and find the original certificate holders.

Perhaps it would have been, though it's been suggested he easily could have avoided the trouble simply by drawing up in advance a list of the people who were owed money. Of greater concern was the effect the scandal had on the administration. Of course, the administration was pilloried, as any administration today would be. But Washington himself escaped largely unscathed, Hamilton bearing the brunt of the criticism. In part that was simply because Washington was such a towering figure that no one wanted to risk taking him on. But mostly it was because of the manner in which he had conducted himself. After approving in secret the broad outlines of the financial plan, he had left its fate up to the cabinet and Congress, neither publicly endorsing it nor lobbying for it. On just a single occasion did he even express his opinion; it was in a letter to a politician—that was as far as he was willing to involve himself in anything as sordid as politics.*

A president taking so passive a role today on so important a piece of legislation is unimaginable. But Washington could because nobody expected him to be involved in politics—nobody even wanted him to be. He hadn't been elected because he was a politician but precisely because he wasn't one. If he'd done anything more than he did, he would have risked damaging the very thing that sustained his authority, his image as a man above the fray.

Washington was no mere figurehead. As Henry Graff has pungently observed, he actually called the shots in the government. It was he who picked the cabinet, of course. He who decided to back Hamilton's financial plan. He who decided, without consulting Jefferson, whom he would send to France and England to represent us. He who decided where the White House would be built.

It was Washington who decided he would not consult with the Senate personally about treaties, a decision that had a profound influence on the presidency. The Founding Fathers had anticipated that the president would use the Senate as a kind of parliamentary sounding board for executive decisions, giving the legislators vast influence over the operation of the government. But Washington had found it impossible to consult with the Senate. He had personally gone to the legislature to brief the body about a treaty being negotiated with the Creek Indians. But things had immediately gone sour. First he had had to sit through two readings of the proposed treaty because the noise outside made it difficult to hear the first reading. Then he had had to face questions about the treaty, an arrangement he found objectionable. Finally, Sen. William Maclay had moved that the Senate should submit the treaty to a committee for additional study. That had so angered Washington that he stood up and stamped out of the chamber, complaining, "This defeats the very purpose of my coming here!" He was to return to the Senate once more, but never again after that, setting

---

*Americans then did not make the assumption that the president automatically would support a measure backed by a cabinet secretary, especially the secretary of the treasury. By law the treasury secretary reported to Congress, not to the president! Washington happened, of course, to support almost all of Hamilton's plans, but not *all*. He specifically objected to Hamilton's proposal to subsidize business and told Jefferson he would veto any measure that included such a subsidy. (None was ever passed.)

the presidency on a decisively different course than anybody had expected.

In another precedent-setting matter, it was Washington who decided that the president should have the right to fire employees of the federal government without having to ask the Congress for permission. Had he not insisted on this right, the presidency would have become wholly subject to the will of the legislature. (A bill giving Congress this right came to a tie vote in the Senate; Vice President John Adams broke the tie by voting against the measure.)

And finally, it was Washington who made the decision that Congress is entitled to see any and all executive documents, even those that might prove embarrassing, a hugely important precedent.

But while Washington was in overall charge, it was Hamilton and Jefferson who did most of the interesting work. It was they who wheeled and dealed and made things happen. In a way it was almost as if there were two administrations, the apolitical administration, headed by Washington, and the political administration, led by Hamilton and Jefferson.

As long as Jefferson and Hamilton worked together, the line between the two administrations, the apolitical and the political, remained firm and clear. But when Jefferson and Hamilton split—Jefferson reaching the opinion that Hamilton wanted to set up some kind of military dictatorship, and Hamilton becoming convinced that Jefferson was out to sabotage him in Congress—Washington found himself increasingly drawn into the political maelstrom. Somebody had to decide ultimately who was to prevail, and that somebody had to be Washington. As might be expected, he resisted taking sides and pleaded with both Jefferson and Hamilton to stop attacking each other. But by then the attacks had become so personal that they almost couldn't.

Hamilton had gone so far in his battle with Jefferson that he had helped set up a newspaper to carry his arguments to the public. Jefferson, in response, had then set up a newspaper of his own to serve as a vehicle for *his* arguments. Each then used their position in the government to help finance their own side's organ. Hamilton gave his newspaper hugely lucrative contracts to print Treasury Department documents. Jefferson arranged to hire his newspaper editor as a translator in the State Department, then made sure the man had so little work to do that he could spend most of his time writing editorials against Hamilton.

Fortunately, just as the battle heated up between Jefferson and Hamilton an event intervened that helped Washington avoid having to choose between them. It was the election of 1792. Both were so eager for Washington to run again that they (temporarily) agreed to set aside their differences for the good of the country, leaving Washington free to remain apolitical. This sounds strange to our ears. Elections in our day incline presidents to be more political, not less. But that's because presidents in our day want to be reelected. Washington didn't.[4]

Washington hadn't wanted to run the second time any more than he had the first. Four years and out; that was the plan. He'd even had Madison write up a farewell address. But that was before the panic. In quick succession two bubbles had burst. First there had been the problem with the BUS—the Bank of the United States. Everything at first had gone as expected—better than expected. Bank stock had gone on sale the year before on July 4—that was a good ploy!—and sold out in an hour. Eight million dollars in stock. *In an hour!* Now things didn't look so good, however. Speculators had stepped in and driven up the price of scrip that could be exchanged for bank stock. Then, after they'd made their bundle, they'd gotten out of the market and scrip prices had plummeted. Scrip that had sold for $325 fell to $160, then to $100, then to $50. Hundreds of people had been wiped out. Hundreds of *rich* people! The financial system was tottering—the very system that Washington had lent his prestige to create.

And then a second bubble had burst. This one was even worse, affecting the entire economy. It had begun when a group led by Robert Livingston had cornered the gold and silver markets in New York. That, in turn, had forced the banks, short of hard specie, to call in their loans. And that had led to the collapse of thousands of businesses dependent on easy credit. By the spring of 1792, the year of the second presidential election, the economy was in tatters.[5]

An additional factor putting pressure on Washington to run for reelection was the looming division of the country into two warring factions, one led by Jefferson, the other by Hamilton. Atty. Gen. Edmund Randolph was so concerned that he warned Washington that the country was on the verge of civil war.

Washington resisted making a decision about running until the very last possible moment—a shrewd move. By delaying he forced Hamilton and Jefferson to hold off on their attacks, neither wanting to alienate him and risk his not running. (Both wanted him to run for the sake of the country; they realized that without Washington the country might implode.) Finally, a month before the electoral college was due to vote, Washington let it be known that he would be available for reelection.

In the spring of 1793 Washington settled in to begin his second term. It would not, however, be as productive as his first. History was against him this time. The conditions that seemed to operate in his favor the first time around—the absence of parties, the isolation of the country from foreign conflicts, the goodwill of Jefferson and Hamilton—seemed to reverse themselves. Jefferson lasted another year in office, then left, and after a time helped form the Democratic-Republican Party.* Hamilton quit a year after Jefferson to take control of the Federalists. Worst of all, the United States was drawn into conflicts with the world's two leading powers, France and Great Britain, and these conflicts in turn divided the country. Jeffersonians took the side of France, which was then in the throes of revolution. Federalists took Britain's side. Washington persisted in standing above it all, but fewer and fewer people appreciated his stance. Federalists felt he naturally should have taken their side and resented it when he didn't; Republicans felt that was precisely what he had done. In the press Washington was vilified, "in such exaggerated and indecent terms," he complained, "as could scarcely be applied to a Nero, a notorious defaulter, or even a common pickpocket."[6]

From the first Washington disapproved of the changes taking place in the country. He especially disapproved of the birth of the inchoate party system. Parties were bad, in his opinion. In his Farewell Address he was to warn against them. For what was a party anyway? It was a formal institution whose chief purpose seemed to be to divide the country. Washington wasn't alone in his distrust of parties. Virtually all the founders felt the same. Even Hamilton and Jefferson had originally inveighed against parties. But under the pressure of events they had come to a new opinion. Washington, however, hadn't. The stronger the parties became, the more discontented he became. And yet there was nothing he could do about them. That he knew for sure, considering that it was his own two trusted lieutenants, Hamilton and

---

*Variously known as either the Democratic-Republican Party or, simply, the Republican Party. Eventually this Republican Party would evolve into the modern Democratic Party. The Republican Party of today can be traced back to the Whigs, a party that developed in the late 1820s.

Jefferson, who had taken the lead in creating them. If *they* could be drawn into the party system, then other Americans would be, too.[7]

History did indeed seem to be against him. But unlike so many of his successors, Washington did not react by being devious. He may have been desperate to stop the rise of parties, but he was never so desperate that he was willing to sell out his principles to do so. He would stand like a wall against parties, even if in the end he had to stand alone, which he very nearly did. The difference between Washington and his successors was not that he was strong and they were weak. Many of the men who came to sit in his place were equally as strong—Jackson, Lincoln, the Roosevelts—these men were as strong as men anywhere have ever been. The difference was that Washington could afford to stand against history, and they could not. Indeed, the more he leaned against history the stronger he became. Americans expected Washington to do the right thing as he saw it, even if they disagreed with him. His power derived from their sense that in the end he would always do what he thought best, not what he thought was expedient. And that was true of none of his successors. They became weaker when they stood against change. Only he became stronger. Of course, it was sometimes difficult for people caught up in the battles of the day to admit that Washington in following his own star was doing what he should. Often they were too angry at the moment to see it. When, for instance, Washington decided to accept the Jay Treaty, which settled our differences with Great Britain on terms largely favorable to Britain, the country went wild, burning both Jay and Washington in effigy. Jefferson, who knew Washington so well, condemned him bitterly. But by the time his term was up Americans had already begun to return to their earlier opinion. When he died he was nearly universally praised as the hero who was "first in war, first in peace, first in the hearts of his countrymen."

He was not just the first person to be president. He was for all intents and purposes the inventor of the presidency. Nobody knew what a president was when the framers created the office. The framers themselves didn't know. But they knew what Washington was, and he was what they had in mind. Only because they were confident that he would be the first president were they willing to make the position as powerful as they did. The president is the commander in chief because Washington was the framers' idea of a commander in chief. They gave him primary responsibility in foreign affairs because they were confident he knew how to conduct foreign affairs. The presidency, in short, is what it is largely because of George Washington. All the extensions of presidential power that were to come later would have been impossible without him.

All the presidents since Washington have lived in his shadow. It was he who gave them—and us—our idea of a president. He was the model. Politicians pretend they are unambitious for the office because he was unambitious for it. They pretend to be apolitical because he was apolitical. All politicians in the eighteenth century paid lip service to the ideals Washington came to symbolize, but it was because he actually seemed to live up to them that his successors often appeared to be so inferior. In a way, he is his successors' great curse. For every president is judged by how he compares with Washington, and yet none ever truly can. Washington gained power by denying he wanted it; the denial proved he was worthy of it. But who after him could gain power by sincerely denying he wanted it? No one. Unlike Washington, they would not be given the power, they would have to compete for it. Every generation there would be dozens of people competing for the office that just a handful of them could have. Washington not only didn't have to compete, there was no one to compete with. He was the only national figure whom all Americans, north and south, east and west, knew and trusted. Each region had its heroes; the nation had just one.[8]

Because of Washington, Americans would lean toward military heroes; such men seemed above politics. These were to include Andrew Jackson, William Henry Harrison, Zachary Taylor, Franklin Pierce, U. S. Grant, Rutherford B. Hayes, Teddy Roosevelt, Dwight Eisenhower, and John Kennedy.* But not one of these military heroes was in the favorable position that Washington was. Each one of them had to campaign for the office.

* * *

---

*Candidates who couldn't truthfully be described as war heroes were, if possible, packaged as military veterans. These included: James Garfield, Chester Arthur, William McKinley, Harry Truman, Richard Nixon, Gerald Ford, Jimmy Carter, and George Bush. (Both Lyndon Johnson and Ronald Reagan claimed to be military veterans, but weren't in the common sense view of the matter. LBJ flew one mission under fire as a Congressional observer of operations in the Pacific, then spent most of the rest of the war in Washington; Reagan made movies for the army but never left Hollywood during the war.)

Politics *was* simpler then. People hadn't thought of killing a president yet. They hadn't thought of lots of things yet. It wouldn't occur to Washington, for instance, to exploit his family to win election to the presidency—he had it won without trying and anyway, what would his family have to do with his running for office? And how could they possibly be of help to him? Martha help George? How, pray tell? By giving her opinions, perhaps? But nobody would think of asking Martha for her opinion about anything. Nor would anyone think to ask her to tell about her marriage as a way of humanizing her husband. That would strike people as utterly ridiculous. So limited was her role that nobody even thought to give her a title after Washington was elected. She was, simply, Lady Washington. (Julia Tyler, wife of John Tyler, would be the first presidential spouse to have a title. She insisted on being called "Lovely Lady Presidentress." Aptly, she went around town in an imported Italian carriage pulled by six white Arabian horses. The first spouse to be referred to as "First Lady" was Dólley Madison, on the occasion of her death in 1849. For some reason, perhaps its attractive simplicity, the title stuck.[9])

Nor, for that matter, would it occur to Washington to scrounge for contributions, reinvent himself, steal votes, or do any of the other things candidates do nowadays to win election. Again, he didn't have to. He had the election won without trying. And anyway, he wouldn't have had to scrounge for donations because presidents as yet didn't run campaigns, and therefore didn't need donations. And even if they did campaign, there wouldn't be much to spend their campaign funds on. All the money in the world couldn't buy a spot on radio or TV because radio and TV as yet didn't exist. Money couldn't even be used to send out photos of the smiling candidate: Photography didn't exist.

Reinvent himself? If being the Father of His Country couldn't get him reelected, what could?

Steal votes? How? While he might have cut deals with the members of Congress or other political bigwigs there weren't any political machines yet in existence capable of wholesale vote stealing. Nor would it be clear whom the votes would be stolen from. "The people" didn't vote for president. "The people" by and large didn't vote at all. Blacks didn't vote, women didn't vote, adult males lacking property didn't vote.

Exploit race? Why? Whites then didn't fear blacks; most blacks were firmly shackled in slavery. They weren't marching for rights, demanding to be freed, or openly threatening the social order.

Exploit religion? Nearly all Americans were Protestants. The great tides of immigration that were to sweep millions of Catholics and Jews into America hadn't yet arrived. The Catholic wave wouldn't begin for another half century. The Jews wouldn't arrive in meaningful numbers for nearly a full century.

Once in office President Washington *would* feel many of the same pressures presidents do today to succeed. Like them he had to deal with forces outside his control, such as the rise of the party system. But unlike them, he gained power by holding to a straight line, not by bending, weaving, or ducking. He succeeded not by doing, but simply by being. Besides, he was relatively lucky. He got through eight years in office without a war, and therefore was never tempted to lie about how well one may have been going. Nor did he have to worry about patronage. He had hardly any jobs to give out, and anyway he didn't have a political party dependent on patronage to keep its supporters content. There was corruption, to be sure; Hamilton's number two man at the Treasury was caught using inside information to make a killing in the bond market. Hamilton himself was accused of corruption by the Jeffersonians. To clear himself of the charge that he had consorted with a known speculator he had to admit that his actual crime was consorting with the man's wife, Mrs. Reynolds.* But would Washington need to cover up these scandals? Hardly. Being Washington, he could take the hits, no matter how hard. He had a lock on the presidency. As long as he himself remained clean, any number of scandals could swirl beneath him and hardly soil his shoes. And Washington of course would remain clean. Although he was always pressed for cash—who wasn't back then in cash-poor America?—and he had to borrow money to be able to attend his own inauguration, he was one of the richest men in the entire country. And rich men don't need to steal.†

---

*Hamilton had owned up to the affair at a private meeting with Republican James Monroe, in an attempt to stop the Republicans from casting doubts on his integrity as a public official. The gesture was in vain. Monroe promptly leaked the story of Hamilton's adultery to James Callender, one of Jefferson's favorite journalists. This would prove to be ironic. A few years later, after Callender and Jefferson had had a falling out, Callender would expose the rumor of Jefferson's affair with Sally Hemings.

†In agrarian America hardly anybody had cash. The economy ran on credit and barter.

Exploiting their family, exploiting religion and race, covering up corruption—all that lay in the dismal distant future.

Washington died in 1799. With him died the founders' hope that the presidency would be above politics. It couldn't be. Only Washington could be above politics—above ambition—and there was only one Washington. After Washington would come the politicians.

2. The Birth of the Two-Party System

How John Adams played politics with foreign policy as a means of
keeping power and Thomas Jefferson's supporters bargained for the
presidency as a means of gaining power

The United States that elected John Adams was as simple as the
United States that elected George Washington, with one difference:
Adams's United States was neatly divided into two parties. And that
changed everything. For eight years Americans had lived with the
myth that theirs was not a system based on parties because Washington had been above parties. But with Adams's election the myth was
shattered. Adams was a party man pure and simple.

Adams hadn't wanted to be a party man. Like Washington he had
wanted to stand above it all. But he couldn't. He wasn't Washington.
And anyway, the world had changed. There would be a party system
whether he liked it or not. The country was changing, and nothing and
nobody could stop it—not even the president of the United States.
Adams would either have to accommodate the change or be run over
by it. A president could not ignore the party system if he hoped to
accomplish something; the only way to get legislation through Congress would be by working with the party in power. Ambitious to succeed, Adams chose accommodation. *He* would play politics.

Unlike Washington, Adams was very bright—brilliant, actually. In his
spare time he would read the classics—in Latin and Greek. But he
was an easy man to dislike. He was vain, quick to rage, and cursed
with a streak of bold independence. In 1768, already a critic of the
British, he turned down a position with the Crown as an attorney with
the Court of Admiralty because he felt that he was being bought off.
On the eve of the Revolution, at substantial cost to his reputation in

the patriot community, he volunteered to defend the redcoats indicted for murder in connection with the Boston Massacre because he felt that they were entitled to legal representation.

At the time of the Constitutional Convention, he worked up in a matter of months a two-thousand-page tome on the history of self-government that proved utterly indispensable. But he continued to remain controversial, Jeffersonians detesting him for his relentless Anglophilia. When the first Congress met, Adams, presiding over the Senate in his role as vice president, had been obsessed with the subject of titles, especially the title to be used to address the president. Washington himself simply preferred to be called "President Washington." Adams, imbued with English ideas formed during his service as the American minister to Great Britain, preferred more involved nomenclature: "His Highness the President of the United States and Protector of the Rights of the Same." The Senate went along, but the House would not. It would be, simply, "President Washington." (Adams, it was decided, would be addressed as "Vice President Adams." But behind his back his enemies would call him "His Rotundity"—in view of his fat waist.)

In 1796 he ran against Jefferson for president, fully expecting to win. "I am Heir apparent you know," he wrote his wife, Abigail. As the election approached, Federalists lined up behind Adams for president and Gen. Charles Cotesworth Pinckney for vice president. But then an impediment was encountered. Secretly Alexander Hamilton began maneuvering to make Pinckney president and Adams vice president again. (Hamilton figured Pinckney would be easier to control.) Adams was crushed. He wrote Abigail that while he thought he had "firmness of mind enough to bear it like a man," he would, if passed over, "groan like Achilles, and roll from side to side abed sometimes, at the ignorance, folly, injustice and ingratitude of the world."

He fought for the office with great passion but honestly. Although his supporters played mean—Federalists claiming that Jefferson was an atheist, destroyer of religion, and a Francophile—Adams himself refused to join in the attacks. He didn't even campaign. (Nor did Jefferson; not until the 1830s would a presidential candidate take to the hustings.)

He won, of course, but he still seemed to believe that people were ungrateful—they had merely elected him president. They hadn't said that they liked him. And anyway, they were tardy in electing him president now. After all, he felt, he should have been elected president in 1789 instead of Washington. What had Washington known of government? He was nothing more than a soldier.[1]

His first two years he tried to govern like Washington, acting as though he were indifferent to politics, with the president operating in one world and rank politicians in another. When France, then in a life-and-death struggle with Great Britain, began raiding American ships, eventually confiscating more than three hundred vessels in retaliation for our support of Great Britain, Adams resisted the demands of the politicians in his own party to make war on the French, though he had every reason to feel that the French were out to get him. During the election of 1796 the French minister to the United States, Pierre Adet, had openly supported Jefferson for president over Adams. Adet had even gone on the road to help rally Jefferson's supporters. But Adams felt that the United States could not afford to take France on, and he was right to think so. We didn't even have a navy yet. Instead of war Adams chose diplomacy, dispatching three diplomats to Europe to meet with Talleyrand, the French foreign minister, to negotiate a peace settlement.

It was a great act of courage—a Washingtonian act, Adams behaving as Americans expected their presidents should. But Adams was not Washington. He lacked Washington's charisma—lacked Washington's balanced judgment, lacked Washington's name. Because he lacked these, he found it difficult—and then, in 1798, impossible—to govern apolitically. In that year events careened out of his control.

The French, instead of seizing the diplomatic initiative as a way out of a messy situation, made things worse by refusing even to give the American diplomats an official audience. Three French agents—whom the Americans immediately dubbed X, Y, and Z, to conceal their identities—revealed that Talleyrand would not deign to open negotiations unless he was paid a bribe of a quarter of a million dollars. That pushed Adams over the edge. Until then he had been willing to put up with the French tribulations in order to avoid war. But war began to seem unavoidable.

It was at this delicate point in the history of the country that Adams chose to do what Washington had strenuously avoided doing for eight long years: He politicized foreign policy. Instead of working with the Jeffersonians to create a united front against France, he played a dirty trick on them. He virtually begged them to demand to see the diplomatic correspondence that laid out in detail the French demands for a bribe in what was to become known as the XYZ Affair. Aware of what was in the correspondence, Adams knew that the papers' release would instantly lead to public revulsion against France. But he let the Jeffersonians think otherwise. The trap laid, they fell into it. When Adams, at the insistence of the Jeffersonians, sent the papers over to the Senate, they were completely humiliated. In a flash the country could see that they were wrong and Adams was right. The Senate immediately voted to arm American merchantmen and to beef up national defense.[2]

But it was a costly victory. In his zeal to sabotage his enemies and win public support for a tough policy against France, Adams had made the loyal opposition look disloyal, associating Jeffersonians in the public mind with France at a time when France was becoming identified as public enemy number one. In effect, what he did was to make Jeffersonians seem unpatriotic. Perhaps at the time Adams didn't see where that might lead. American politics was still inchoate; no one yet really understood its patterns. No one yet realized what happens to the civil rights of minorities when Americans get caught in the grip of patriotic fury. Red scares and Communist witch hunts lay far in the future. But if Adams did not at first see what dark forces he had unleashed, he did not have long to wait. Within weeks it was evident. As soon as the public had absorbed what had happened people began to demand that the Jeffersonians be punished in some way. What happened next was not too great a surprise. As war fever mounted, the people demanded that the *government* punish the Jeffersonians. Four months later, in July 1798, the Congress, dominated by Federalists—altogether eager to deal their enemies a crippling blow before the elections that fall—passed the Alien and Sedition Acts, which gave government officials the right to lock up anybody who criticized them.

Adams promptly signed the measures, forever changing history. After that there never again was any real hope that the American presidency could live up to the standards set by Washington. Adams laid that dream out dead on the ground as effectively as if he had used his fists to score a bloody knockout blow. Ever after, presidents would be seen to be operating in the very same world as other elected officials. Where before there had been two worlds, there now was to be just one. In the opinion of most Americans—then as well as now—the wrong one had triumphed. Americans didn't want presidents playing politics (they didn't even want *ordinary* politicians playing politics). But the world that triumphed was the one that had to. A president had to dirty his hands in the grimy world that was. He couldn't wash it away.

It didn't matter that just fourteen Americans would actually be punished under the Sedition Act, or that Adams himself was to be personally involved in only two of the cases. Nor did it matter that he hadn't lobbied for the legislation or even given his approval in advance. He signed it. And that put him officially on record in favor of laws expressly designed to punish his political enemies by tossing them in jail.

It was ludicrous legislation in a supposedly free country—as was almost instantly and comically to be revealed. After Congress adjourned, Adams left Philadelphia to journey home to Quincy. Along the way he happened to stop in Newark, New Jersey, where local Federalists gave a parade in his honor. A citizen by the name of Luther Baldwin, standing by a tavern, heard a customer remark as Adams went by, "There goes the President and they are firing at his ass." Baldwin, a Jeffersonian, retorted, "I don't care if they fire through his ass!" For that remark he was promptly arrested, charged with violating the Sedition Act, and put behind bars. Other cases were more serious. A U.S. represenative, Matthew Lyon, was prosecuted and sent to jail for four months.

In school, when students are taught about the Alien and Sedition Acts, the emphasis is almost always put on the latter. But the Alien Acts were equally reprehensible—and potentially more damaging to the prospects of the Republicans. The Sedition Act was like a blunt knife against their party, frustrating criticism of the government but not really preventing it. In contrast, the Alien Acts were like a sharp knife pointed directly at the party's heart. For what the Alien Acts did was restrict the party's ability to enlist new voters. In one swift stroke they extended the term of residency required to become a citizen from a brief five years to an onerous fourteen. The restriction applied to all immigrants, but that hardly hurt the Federalists: New immigrants by and large weren't becoming Federalists, they were becoming Jeffersonians. And now those immigrants, tens of thousands of them, would have to wait years and years longer before they could vote. One group of immigrants in particular would be deprived of the vote for many years: the twenty thousand white French immigrants who had recently arrived from the Caribbean island of Saint Domingue, then in the midst of the slave revolt led by Toussaint L'Ouverture, virtually all of whom were expected to join the Jeffersonians.[3]

Jeffersonians took the Alien and Sedition Acts seriously, so seriously that in response they began contemplating the abolition of the experiment in federal government. Jefferson and Madison refused to go that far, but they, too, were alarmed. Out of their concern they developed in secrecy what came to be known as the Virginia and Kentucky Resolutions, which introduced the radical idea of nullification: If the Congress and the president approved legislation that the states believed was unconstitutional—as Jefferson and Madison believed the Alien and Sedition Acts to be—then the states had the right to declare such legislation null and void within their borders. Through deft political maneuvering, Jeffersonian agents were able to persuade the state legislatures of both Virginia and Kentucky to approve the resolutions. Ever after, Jeffersonians would pair two dates as turning points in the history of freedom: 1776 and 1798. In '76 a blow had been leveled against British tyranny, in '98 against federal tyranny. "Remember '98," they would say.[4]

That fall the Federalists ran a vicious campaign in the off-year election for Congress, exploiting the passions Adams had raised in the imbroglio with France to tar their opponents as traitors. Some may actually have believed what they were saying. Most clearly did not. They'd have said anything to win: Winning seemed vital.

While the Federalists then controlled all three branches of the federal government, there was every indication that their control was slipping, that they were becoming identified with extreme conservatives—the moneyed elite—while Republicans were proving adept at winning over the young, the poor, and the working class. In the last election Adams had beaten Jefferson by just three electoral votes. As likely as not the next time round Jefferson would beat Adams, who no longer could bask in Washington's glow. The Federalists could survive the defeat of Adams, but they might not survive the loss of Congress. And that was just what they were beginning to worry might happen. While they held a commanding majority in the Senate—twenty Federalists to twelve Democratic-Republicans—in the House of Representatives their majority was much smaller, fifty-eight to forty-eight. A shift of just six votes and the Jeffersonians would control the House. The more Federalists dwelt on that number six, the more depressed they became. *Just six votes!* It had haunted them from the moment the election results had come in two years earlier.

Winning by demagoguery was embarrassing, but Federalists found themselves unable to resist the temptation. Running against France and the Jeffersonians was a lot more exciting and profitable than running against Jeffersonians alone. By bringing France into the campaign, the Federalists could make people forget what they didn't like about the Federalists. In effect, they could use the crisis with France to continue discrediting their opposition.

By the time the campaign was in full swing, actual war broke out, armed American merchantmen firing on armed French merchantmen. This raised passions even further, giving the Federalists ever greater opportunities to reap political benefits. The hotter the war turned, the better their prospects.

There was nothing unique in the Federalists' strategy to use war to score political points. Throughout American history many politicians would do the same. But the Federalists were unique in one respect: They were the first to do it—and the first to figure out how.

There were some Federalists, to be sure, who objected to the party's machinations, who held true to the ideals of the Republic as symbolized by George Washington. But then the election returns came in. The Federalists triumphed. In the House, which the Federalists had worried about losing, they won a stupendous victory, winning sixty-four seats—twenty-two seats more than the opposition. After that the politicians agreed that playing politics with national security was essential.

Luckily for Adams, the war (which was undeclared) went well. Talleyrand, impressed by American resolve and eager to focus on the fight with Great Britain, signaled that he was prepared to open negotiations without payment of a bribe, and an agreement to end the war was reached. A grateful nation sang Adams's praises—and then, as so often happens in American politics, abruptly changed its tune. In 1800, when Adams ran for reelection, the voters threw him and the Federalists in Congress out of power, voting in the Jeffersonians. In a way the peace settlement that had been Adams's triumph was also his undoing. By settling the war with France he took foreign policy issues, where he had an advantage, off the table. That gave the voters time to refocus on domestic issues, where Federalists were weakest. As war fever had abated Jeffersonians had even discovered that they were able to rally public opinion against the Alien and Sedition Acts.

And that was a reminder of a basic lesson: In politics victory is ephemeral.

Before he left office Adams pulled off one last political trick. In the waning weeks of the lame-duck Congress—the Federalist-dominated Congress—the party passed a sweeping reorganization of the federal judiciary, doubling the number of circuit courts and creating twenty-three new federal judgeships. Adams promptly signed the legislation and then nominated Federalists to fill each of the newly created positions. As his administration wound down, he labored feverishly to have his nominees approved by the Senate before the Republicans took over. On his last day in office he was still filling out their commissions, giving rise to the criticism that he had created "Midnight Judges." His apologists later derided the term as a Jeffersonian myth. But on his last evening as president, Adams, his boxes packed, was still signing away. At five the next morning he slipped out of Washington unnoticed and went home to Massachusetts, not caring to see his longtime opponent installed as his successor. Not many were sad to see him go.[5]

When the Jeffersonians came to power they repealed the Alien and Sedition Acts. The Virginia and Kentucky Resolutions, in consequence, became moot, no court ever ruling on their constitutionality. But nobody could repeal history. The change John Adams had wrought in the presidency was permanent.

Adams is usually remembered, when he is remembered at all, for the amazing dynasty he started. For generations, defying the odds, Adamses prevailed in both politics and academe. After Abigail and John came John Quincy Adams, himself a president; then Charles Francis Adams, congressman, minister to England, biographer, and historian; and finally Henry Adams, the dyspeptic author of *The Education of Henry Adams*, who, out of family pique, wrote a scathing history of the Jefferson and Madison administrations. But John Adams's effect on the presidency, though forgotten, was decisive—so decisive that he can be said to have reinvented the office.

Until 1798, however misguided Presidents Washington and Adams had seemed, it was assumed that they were simply misguided; they were almost never accused of acting out of base political motives. After '98, presidents would find it almost impossible to escape the criticism that their actions were often politically motivated. The change was owing to Adams's handling of things: Everyone knew that *his* actions had in fact been political. And once the genie had gotten out of the bottle, there was no way to put it back inside. Fewer than ten years had passed since Washington had been elected. But a hundred years might as well have elapsed, so great was the effect of the change in the office.

There was an irony in the fact that it was Adams who was the agent of change—of *this* change—for if anybody had wanted to operate above politics, it was John Adams. But a force beyond Adams's control—beyond, really, anybody's control—was taking over. However much presidents might wish to ignore politics, as many would profess they did, they would find that they actually could not, now that the two-party system was taking hold. Not even the most iron-willed of men could—not Jackson, not Lincoln, not either of the Roosevelts. If they wished success, wished to survive in office and to control events, then they had to play the game—and play it better than their enemies. That, too, was a lesson of the Adams administration.

While it was clear that Adams had lost the election, it was unclear who won it. Not unclear in some metaphorical way, but in a very real, very practical way. Jefferson supposedly was the winner; he after all was his party's choice as president. But when the electoral votes were counted it suddenly became evident that Aaron Burr, the party's vice presidential candidate, might in fact become president in what would amount to a stolen election.

The presidency once again was embroiled in politics.

Through a quirk in the election laws then in effect, presidential and vice presidential candidates did not run together as they do now on a joint ticket. Instead all of the candidates, whether intending to seek the presidency or the vice presidency, ran for president. The person who received the most votes took the top position, while the person receiving the next most became vice president. The system had worked well initially because the candidates had run as individuals. But the system was unworkable once the candidates began to run as representatives of a party. For then the candidates for president and vice president of the winning party would receive the same number of electoral votes, and that would result in a tie in the electoral college. The obvious solution was to amend the election laws to allow electors to note which candidate they wanted for president and which for vice president. But that wasn't a simple thing to do. The election rules were embedded in the Constitution itself; to change the procedure the Constitution would have to be changed. Instead, therefore, the parties had settled on a simple expedient measure. To avoid a tie they made arrangements with several of their own electors to vote for someone other than the party's designated vice presidential candidate, which would leave the presidential candidate a few votes ahead.

Nothing could be simpler, really. All the party had to do was contact a couple of its electors, as both parties had done in 1796. And contacting a couple was just to be on the safe side. In reality just one elector had to be contacted—for the presidential candidate had to receive just one more vote than the vice presidential candidate to win. *Just one more vote!*

And yet somehow this simple thing that needed to be done was not done. In the hubbub of the election nobody bothered to make the necessary arrangements, though Jefferson himself had asked party leaders to be sure to take care of this single, extremely important, detail. When the electoral ballots were counted Thomas Jefferson and Aaron Burr both received exactly the same number: seventy-three. (Adams received sixty-five.)

If there had been any doubt that the Constitution needed to be amended there was no doubt any longer. But that would have to take place in the future.* In the here and now, the House of Representatives, under the procedures outlined in the Constitution, would pick the next president. And that created a genuine political crisis. For Jefferson to win, Burr would have to tell his friends in the House to vote for Jefferson for president. But when the time came to do so, he refused. A rabidly ambitious man, he hungered for the presidency and was willing to do anything, it seemed, to obtain it. The Republicans had selected him for the vice presidency in part because he was ambitious. Sharp, smart, and powerful, and powerful in particular in New York (which was critical; the Republicans needed New York to win), in the 1790s he had even managed to outmaneuver Alexander Hamilton and take control of the local government in Manhattan. But not until now did party leaders realize quite how ambitious he was. When the House met in February to choose the next president Burr secretly connived with the Federalists to make himself president instead of Jefferson. There is no evidence that Burr himself ever actually lobbied any of the members. But then, he didn't

---

*The Twelfth Amendment provided for the necessary change. It was passed in 1804, in time for the next election.

have to. All he had to do was have his friends do the lobbying. They would do his dirty work.

The Federalists held the key to the election, for they were in control of the House. Although they'd lost their majority in the recent election, it was not the newly elected Republican Congress that would decide the presidential contest, it was the old lame-duck Federalist Congress. Because they held the balance of power, the Federalists could, if they had been brazen enough, even have made Adams president again. Actually, there was some talk of them doing so. But that would have amounted to felony theft—an outright stolen election—and not even the most manipulative Federalist politicians thought they could get away with that. But because Burr had put himself into play in the presidential contest they thought they would be able to justify *his* election. And they much preferred Burr to Jefferson. In part that was because they simply loathed the Virginian, but it was also because they hoped to be able to strike a bargain with Burr: They would make him president and in exchange he had to promise not to fire Federalists on the government payroll and to retain the Hamiltonian system of finance. (The Federalists seemed to have had no hope of striking such a bargain with Jefferson; he expressly refused to negotiate.)

With the presidency up for sale, the House began voting. The first vote took place on Wednesday, February 11, 1801, just after Congress had formally opened the ballots of the electoral college. Jefferson received the votes of eight states, Burr six, with two dividing. Since there were sixteen states in the Union, Jefferson needed nine to win a majority.* Getting that majority would prove difficult. That afternoon more ballots were taken, each time with the same result. So there were more ballots, one after another, on into the night. Members sent home for blankets and pillows. And still there was the same result. At sunrise on Thursday, after twenty-seven ballots, they finally took a break. That afternoon they voted again, on Friday two more times, and on Saturday three more times. By then they had been at it four days and gone through thirty-three ballots—with no winner. But they would need to take just three more ballots. For that evening Federalist James Bayard, the sole representative from Delaware, announced that he was

---

*The Constitution prescribed that each state had one vote, and a majority of the states was necessary for victory.

switching to Jefferson. On Tuesday, on the thirty-sixth ballot, Jefferson was finally elected president. "Thus has ended," commented Republican Albert Gallatin, "the most wicked absurd attempt ever tried by the Federalists," which was a little unfair. It hadn't been the Federalists alone who were responsible for the mess. Aaron Burr was, too.

Why had Bayard switched? While Jefferson had refused to bargain for the presidency, his supporters had decided they had to. Under the pressure of events they promised Bayard that Jefferson would retain both the Hamiltonian system *and* Federalist officeholders—all that the Federalists had hoped to gain by supporting Burr. Hamilton, ironically, played a decisive role in the drama. He came out in favor of Jefferson. It was not because of a new-found love of his old enemy. It was because he feared Burr and only hated Jefferson.[6]

It had not been Jefferson who had made the ultimate partisan deal, trading the presidency for some promises. It was his supporters. But that didn't really matter. The deal had been struck in his name. And Jefferson would be honor-bound to deliver on the promises his supporters had made. There would, therefore, be no dismantling of the Hamiltonian financial system. And there would be no wholesale removals of Federalist employees from the public payroll.

Jefferson actually had had no intention of destroying the Hamiltonian edifice. But he certainly had expected to replace Federalist bureaucrats. For the only way Jeffersonians could hope to build up their party quickly was by offering key supporters jobs. As long as Federalists held those jobs, the Federalists benefited instead of Republicans. Jefferson before long was to complain bitterly about his inability to rid the government of the Federalist holdovers: "How are vacancies to be obtained? Those by death are few, by resignation none." But he never in fact did get rid of them. At the end of his two terms some two-thirds remained in office.*

The presidency had in fact been sold. And it had been sold for the reason that it *had* to be. For by then the politicians had become desperate. The Jeffersonians had tried to win by appealing to reason and patriotism and that had gotten them dozens of ballots and no winner. So they had

---

*Jefferson, it should be noted, made one exception to his promise not to fire Federalists wholesale. He decided he had every right to remove Federalists appointed to office after the election results were made known in December—and did.

decided to win any way they could. In the end the man who should have won, won. But victory had come at the cost of principle.

Washington's election in 1789 seemed farther away than ever.*

A single great force was driving politicians to take actions they otherwise wouldn't have dared to take, and that force was, simply, the emergence of a full-blown two-party system. That same force would now lead Jefferson to do things as president he would have preferred not to do.

Just two weeks after the tumult in the House, it seemed that everything was back to normal, back to the way it had been before the election, before politics had become so intensely partisan. It was Jefferson who brought about the change. At his inaugural he famously announced, "We are all federalists, we are all republicans." They were just eight little words, but already presidents were so powerful that eight little words could affect the country's whole outlook on politics. They were so powerful that they were to become an emblem of nonpartisanship through the ages. In the 1970s scholar Kenneth Stampp, speaking to a group of hundreds of historians at a contentious conference on slavery at the University of Mississippi, would try to quiet things down by beginning, "We are all federalists, we are all republicans." The audience, immediately grasping Stampp's reference, hooted in delight.

The funny thing was that Jefferson did not mean to be apolitical.

---

*Burr would remain a source of terrible trouble. While still vice president he would run against Hamilton's handpicked candidate for governor of New York and lose. Exasperated, he would challenge Hamilton to a duel and kill him. Insane with ambition Burr then would conspire with the head of the army, General James Wilkinson, to stage a coup against Jefferson, in hopes of establishing a new country to be carved out of the southwestern United States. At the last minute Wilkinson would expose the plot and turn Burr in to the authorities, leading to the trial of the former vice president of the United States on charges he had attempted to make war on the United States. Never before or since would there be a trial to match and Americans followed the proceedings avidly. But the result would disappoint. Because the prosecutors could only prove that Burr had raised an army and not that he had planned to use it to make war against the government, he was acquitted. Subsequently he would be indicted by several states but before he could be tried again he escaped to England.

He was only pretending to be apolitical.† To *be* apolitical would mean governing as Washington had, and Jefferson had no intention of trying to be another Washington. In fact, he was to be the first avowedly political president in American history. Adams had been political in his actions but had not admitted he was being political. Jefferson was political and admitted it. In the whole history of the presidency few presidents have been more political. The longer he remained in office the more political he became. Indeed, one of his chief goals as president was political. It was to bury the Federalist Party forever. Jefferson despised the Federalists, doubting they should even be considered a legitimate group. Republicans were legitimate; they represented the voice of the people. Federalists represented special interests: bankers, merchants, businessmen. In Jefferson's view what was good for the Republican Party was good for the country. The best thing that could happen to the Federalist Party was for it to expire.[7]

That it was Thomas Jefferson who first institutionalized party politics in the office of the presidency is something of a paradox. For no one was more reluctant to play the politician than he. His battles with Hamilton in the Washington administration had so worn him down that he had retreated to Monticello to take up the life of a virtual hermit. He even stopped reading newspapers for a time. He returned to politics in 1796 only at the insistence of his friends, who feared that the "monocrats" were taking over. By 1800, fearful of the direction in which the Federalists were taking the country and alarmed by their attempt to use the government to silence their critics, he had become passionate about running for president and had even orchestrated a nationwide campaign to elect Republican majorities in the state legislatures to help bring about his own election. But Jefferson remained in many ways a paradox. It was almost as if he thrived on contradiction (as many creative people do). A slaveholder, he believed that "all men

---

†Significantly, in the handwritten copy of his inaugural address the words "federalists" and "republicans" were written in lowercase; that spoke volumes. All Americans *were* federalists. All Americans *were* republicans; the Constitution established a federal system based on republican principles. Jefferson expressly did not say and did not mean "We are all Federalists, we are all Republicans," though the official printed draft of the speech contained the capitalized spellings.

are created equal." A true believer in agrarianism, he started a nailery at Monticello that presaged the beginning of the Industrial Revolution in northern Virginia. An opponent of government debt, he ran up personal debts in the hundreds of thousands of dollars, bankrupting his estate. A lover of simplicity, he added so many stories and rooms to the main house at Monticello that the building took twenty-five years to finish.

Some of the compromises he made were so small as to be almost attractive. When some urban workers complained about an essay he had written years before in which he criticized "the mobs of the great cities who add just so much to the support of pure government, as sores to the strength of the human body," he temporized. "I had in mind," he sheepishly wrote, "the manufacturers [manual workers] of the old country," not the fine, upstanding workers found in American cities. Other compromises were less attractive. But they were always necessary—or seemed so at the time.

One of the first in this class was his order concerning the deposit of government funds. Whenever possible, he instructed his treasury secretary, government funds should be deposited in banks favorable to the administration.[8]

Another was his decision to ignore a scandal involving his postmaster general, Gideon Granger, who had become caught up in the most fantastic swindle in American history: the Yazoo land fraud. In this one deal, named for a river, Georgia agreed to sell speculators thirty-five million state-owned acres out west at a cut-rate price, one and a half cents an acre. In return for their largess the legislators demanded a boodle in kickbacks: stock in the four companies that bought the land, out-and-out cash payments, and in many cases, title to some of the land itself. The sale of the land was completed in 1795, but a new legislature elected in Georgia after the scandal was exposed rescinded the deal, leaving investors in the companies in the lurch. After years of controversy the investors eventually petitioned Congress to make them whole on the grounds that they had been illegally deprived of their property. Granger joined the petitioners in their action, but then, according to John Randolph, went a step further, resorting to bribery to try to get Congress to act.

Because the allegations came from Randolph, a Jefferson friend turned foe, Jefferson had every reason to believe that they had been manufactured. But he also had a duty as president to find out if they actually had been. This he refused to do, and the reason was politics.

Granger was the only New Englander in the cabinet at the time, and Jefferson didn't want to do anything to undermine him; Jefferson's support was weakest in New England. Investigate Granger, and New Englanders would be sure to rise up in anger. So Jefferson let him remain.[9]

Jefferson was very good on civil liberties. One of his first acts after taking office was to order the release of people imprisoned for violating the Alien and Sedition Acts. But Jefferson didn't like press criticism, it turned out, any better than Adams or Washington had. It hurt to be called incompetent, ambitious, and atheistic, and it was troubling that so much of the press coverage of the administration, at least the coverage provided by the Federalist papers, was plainly untrue. It especially hurt to find himself depicted as a fornicator, as journalist James Callender was claiming. And it was extraordinarily painful to have it said that the person with whom he was fornicating was his very own slave, Sally Hemings. But for a good long while Jefferson simply put up with the situation.

Finally, understandably perhaps, he blew. After years and years of abuse, he decided secretly to encourage state prosecutors to crack down on the presses owned by his enemies in Pennsylvania, a center of opposition to the administration. "I have," he wrote the Republican governor of the state, "long thought that a few prosecutions of the most prominent offenders would have a wholesome effect in restoring integrity of the presses." At the time press freedom was not understood to mean the same thing it means today; in Jefferson's day most Americans believed the press had the right to print only what was true. It wasn't until later that the First Amendment was construed to give the press the right to print virtually anything. (And not until the *Sullivan* decision in 1964 did the courts recognize the press's right to make erroneous statements about public officials; and not even *Sullivan* shields journalists from libel suits if their reporting is found to have been malicious or reckless.) But that it was Jefferson who aggressively (and secretly) connived to take on the press was out of character. It was Jefferson after all who had said, "Were it left to me to decide whether we should have a government without newspapers, or newspapers without a government, I should not hesitate to prefer the latter." But then that was before he was president. Things looked different once he was.

There is no telling how far Jefferson might have been willing to carry his secret campaign against the press, because before long it came to a sudden and dramatic end. At yet another trial of yet another editor on the charge of seditious libel, this time at the federal level, the defense decided to take the offense. As the case against the editor went to trial his supporters quietly let it be known that they had evidence against Jefferson that they planned to release if the prosecution moved forward. In effect, what they were trying to do was blackmail the president of the United States. Jefferson was incensed, but he did not want the public to know what the blackmailers promised to reveal: that as a young man he had, as he put it, "offered love to a handsome lady." At the time he was a bachelor, but the lady in question was his neighbor's wife, Mrs. John Walker, whom Jefferson was supposed to be looking after while Mr. Walker was away. The affair actually had never been consummated. Mrs. Walker, loyal to her husband, had refused Jefferson's offer. But Jefferson ever after felt extremely guilty about the affair. And the last thing he wanted was to see it written up in the papers. Beside the personal fallout, there would be the political embarrassment. So when he found out what the defense had on him, he told the prosecutor to drop the case.[10]

Jefferson liked to pretend that he was always straightforward, though in practice he often wasn't. But he was commendably straightforward in dealing with one of the great issues of his administration, the Louisiana Purchase. Louisiana had been a source of great trouble for the United States for years. France had owned it originally, then lost it to Spain after the Seven Years' War (known in the United States as the French and Indian War), then won it back after Napoleon came to power. And that had made Americans hugely uneasy. Americans didn't much care about most of Louisiana. At the time they felt they had more room than they knew what to do with. But they didn't approve of the swapping of the territory by European powers. It brought European conflicts too close to home. And they especially didn't approve of European control of the Mississippi Delta. It was the delta they cared about, not the rest of Louisiana. For it was the delta, New Orleans in particular, through which the commerce of the western states passed after traveling down the Mississippi. Thus it became a great object of American diplomacy to try to secure New Orleans. To the European powers it was an insignificant place. In 1800 there were barely seven thousand people living there, most living in wooden shacks. Americans, therefore, were hopeful.

Napoleon at first seemed uninterested in a deal, even though the Americans offered a princely sum: ten million dollars, an amount about equal to the entire annual budget of the United States.* But then one day, completely out of the blue, Talleyrand, the French foreign minister, who for months had given the American minister handling the negotiations little encouragement, asked, "What will you give for the whole?" Taken aback, the American said the United States didn't want the whole thing, just New Orleans, and for several days insisted that was all the United States wanted. Talleyrand practically had to force Louisiana on the diplomat. Eventually a deal was struck, and in 1803 for fifteen million dollars we got New Orleans and everything north to the Canadian border and west as far as . . . well, nobody could be sure how far west, but as far west as anybody had ever traveled. (The western border came to be set in the Rocky Mountains.)[11]

It was one of the great bargains of all time and a better bargain than anybody had a right to expect. Napoleon had agreed to sell only because he had lost his foothold in Saint Domingue after the slave revolt led by Toussaint L'Ouverture; without Saint Domingue there was no hope for a French empire in the New World. But there was one difficulty, and it was a great one for Jefferson. The Constitution nowhere expressly authorized either Congress or the president to make purchases of foreign territory. And Jefferson had repeatedly vowed as president to read the Constitution literally. One of the chief reasons for his break with Hamilton in the 1790s had been Hamilton's insistence that the Constitution be interpreted loosely.

But the fact was the Louisiana Purchase was too good a bargain to be passed up. Jefferson knew it, and so did everyone in the country. In his message to Congress requesting the approval of the Louisiana treaty Jefferson did, to be sure, sidestep the constitutional questions. But he did not lie about them, did not make up some elaborate sophistical argument to befuddle the issue and rationalize his conduct. He simply did not need to. As he explained to Sen. John Breckenridge, he and the members of Congress must simply "throw themselves on the country for doing unauthorized, what we know they would have done for themselves had they been in a situation to do it."[12]

The truth is that Jefferson acted as forthrightly as he did because he

---

*The money was also to be used to cover the purchase of West Florida.

knew that he could. But it was a unique moment. Never again would a president be able to make a grab for vast territory in the knowledge that his actions would be approved wholeheartedly by the general public. In the future every acquisition would be fought over fiercely. And in each of those fights presidents would lie.* It was the price the country would have to pay for western expansion.

That Jefferson was willing to resort to subterfuge when he needed to to gain what he wanted in foreign affairs is abundantly clear. He even contemplated the use of bribery. The issue of bribery came up in connection with the effort to secure Spanish Florida, which for years had proved to be a source of terrible trouble to southern slaveholders. Indians and fugitive slaves used Florida as a haven from which to launch daring raids against the slave owners. Jefferson, a slave owner himself, was sympathetic and vowed to put a stop to the depredations. But when he went to Congress with a plan for action, he lied. He told Congress and the public that he expected to be able to resolve the situation through the use of military force. But his real intention was to offer the French a two-million-dollar bribe to put pressure on Spain to relinquish the area. (Spain was then virtually under French control.) Jefferson lied for the same reason other presidents were to lie when they made a grab for land: He had to. No president could admit that he wanted to pay a bribe to acquire territory. That was what Europeans did. We were supposed to behave better.[13]

Jefferson could be like that—practical—when he had to be, and in office he had to be quite often; it was the only way to keep power and to get things done. A sign of just how far he had come from the idealistic days of his youth when he had dreamed of a life studying botany, law, and architecture, a life of the mind, was how he handled the selection of his second vice president when he ran for reelection in 1804. It might have been thought, after the disastrous experience he'd had the first time, when the Republicans picked Burr, that he might want to see to it that this time they got someone who had been better vetted, someone who at least would be a worthy holder of the office. The last time, after all, they had gotten themselves a traitor. They owed it to the country, if not to themselves, to find somebody this time who would do the office proud. But when the time came to choose, they did not

*Except in the case of Alaska, Seward's Folly.

make their choice on the basis of competence or rectitude. They made it on the same basis on which they had made their choice of Burr four years earlier: for its political effect. Jefferson went along.

That was how the party got stuck with George Clinton.

George Clinton had very little going for him in 1804. He was old, sixty-five at the time. He had no interest in national legislation. He did not even want to move to Washington. And he had a past. In 1792, when he had run for reelection as governor of New York, he had lost fair and square. But because he controlled the political machinery in the state he had managed to rig the results and get himself reelected. He did this by getting the election commissioner to disqualify the votes cast by constituents in three counties. The pretext for doing so was flimsy; in one case the ballots were tossed out because they had been delivered to the election commission by a person appointed by a county sheriff and not by the sheriff himself. But the real reason was political. Clinton's support was weak in the three counties; eliminating their votes was vital for victory.

There was only one thing in Clinton's favor, only one thing that made him an attractive candidate for vice president of the United States on the Republican ticket—and that was his place of residence. He was from New York. And New York was critical to the election. The only way the Republicans could win was by taking New York. They had taken Burr because he had been from New York. And now they took Clinton because he was.

After the party caucus settled on Clinton, John Quincy Adams commented that "a worse choice could hardly have been made." And he was right. In office Clinton proved to be a disaster. Not, to be sure, quite the disaster that Burr had been. But bad enough. Too old by then to be effective, he took hardly any interest in his post as president of the Senate, his only constitutionally mandated responsibility, often declining even to show up for meetings. For long stretches he remained at home in New York. Even when he did happen to stir himself and visit the Senate, he made a mess of things, at one point creating three committees to do the work of one. While in the presiding officer's chair he frequently forgot points of order and miscounted votes.

(Shockingly, in 1808 Clinton would be renominated by the Republicans, this time to serve with James Madison. Madison quickly regretted the selection. Clinton began openly criticizing the president and in 1811 used his power to break a tie vote on a bill to recharter the

Bank of the United States, voting no, against the administration. He died soon after, the first vice president to die in office.)[14]

America was still a simple place when Jefferson left office after eight years, but it wasn't as naive a place as it had been under Washington. Adams had played politics with national security. Jeffersonians had bargained for the presidency, and Jefferson himself had used the power of the government to go after his enemies. He had also tolerated corruption and compromised his principles.

Both Adams and Jefferson were strong men. But neither could stand up to the forces that were quickly transforming the small agrarian Republic into a thriving little empire. The Adams presidency had been twisted and turned by the force of European war and the birth of the two-party system. Jefferson, in turn, had had to make adjustments as the two-party system began to mature.

3. The Revolution in the Suffrage

How three presidents in a row tried to gain power by unprecedented means: John Quincy Adams by personally bargaining for the presidency, Andrew Jackson by countenancing personal attacks on his opponents, and William Henry Harrison by reinventing himself

---

During the presidencies of James Madison and James Monroe the country keeps changing. Madison and Monroe, in turn, have to change with it to remain in control.

Madison, against his every wish, decides to lead a divided country into war against the British.

During the war Federalists foolishly let themselves become identified with radical antiwar extremists in New England, some of whom demand that the region secede from the Union. As a result, the Federalists, already weak, become thoroughly discredited, leading to the collapse of the two-party system.

When Monroe takes office after the war he discovers that without a party behind him he cannot control Congress. In response he begins to strike innumerable bargains with individual members in an ambitious attempt to curry their favor. Inevitably, he finds that to keep their support he must overlook their transgressions, turning a blind eye on corruption and malfeasance.

But important as these changes are in making politics more complicated, they are as nothing compared to what follows: the revolutionary expansion of the suffrage.

As the curtain rises on this next act of American history, John Quincy Adams and Andrew Jackson are fighting fiercely for the presidency. It is 1824.

He was, as even his own supporters had to admit, the least likely of candidates. Andrew Jackson, candidate for the presidency of the United States? It was almost laughable. This was, after all, a time of great statesmen: of John C. Calhoun, of John Quincy Adams, of Daniel Webster, of Henry Clay. These were the men to watch, these were the go-getters, the best of the best, the kind of men America had turned to in the past and would undoubtedly turn to in the future. For these were the men who ran things currently. Calhoun was Monroe's secretary of war. Adams was secretary of state. Webster was chairman of the House Judiciary Committee. Clay was Speaker of the House of Representatives. And these were just some of the men with the right kind of résumés who were ambitious to be president. There were others, equally capable, like William Crawford, who had served as secretary of the treasury and secretary of war.

Jackson, to be sure, had achieved national prominence. In the War of 1812 he had thrilled the country with his victory over the British at New Orleans (British casualties: 2,000; American: 71). In the Seminole War he had seized Florida, whipped both the Spanish and the Indians, and put to death two British subjects accused of inciting the Indians. But he wasn't any kind of a statesman. In fact, he was downright unstatesmanlike in almost every conceivable way. At the time of the Seminole War the Monroe cabinet had all but unanimously condemned his behavior, saying he had far exceeded his orders. He had been given authority to put an end to Indian depredations, not to take over Florida. Besides, he had hardly any knowledge of political philosophy or political science. Though he had served briefly in both the House and the Senate, he had distinguished himself in neither office. As a young congressman he had embarrassingly suggested that George Washington should possibly be impeached for authorizing John Jay to negotiate the Jay treaty without first obtaining the advice and consent of the Senate, a "[d]aring infringement of constitutional rights." Surely a man such as Jackson would not have a prayer of becoming president.[1]

At any earlier time he would not have. Jackson as president in 1800, 1804, 1808, 1812, 1816 or even 1820 *was* inconceivable. But 1824 was different. And it was different because politics was different. The old order had become discredited. Jackson, the outsider, the renegade, was the beneficiary.

In 1824 that was not, however, immediately apparent. His candidacy had come about because an old friend, John Overton, who owned a bank in Nashville, was hoping to use Jackson's popularity to undermine the governor of Tennessee, who was trying to establish a rival bank in the state. Overton never imagined that Jackson's candidacy would go anywhere. Jackson himself had been astonished by his political popularity. Although he was an exceedingly ambitious person—born poor, he had become an outstanding general and a rich businessman, rich enough that by the 1820s he owned much of what was to become downtown Memphis—he had never had an ambition to be president. In 1821, in response to suggestions that he run for president, he had said, "Do they think I am such a damn fool! No sir; I know what I am fit for. I can command a body of men in a rough way: but I am not fit to be president." Of course, he changed his mind.[2]

Two developments accounted for the change in Jackson's political prospects, one inevitable, the other an accident of history. The accident was the Panic of 1819. The American economy had been growing stupendously, growing so fast and so powerfully that people had come to expect it would keep on growing. Nobody had yet heard of the business cycle. Then, suddenly, as a result of overspeculation, it stopped growing and started slipping into decline. It was a terrible time for people. Credit, which had been easier to get than ever before, became exceedingly tight. In Kentucky alone forty banks closed. The price of goods collapsed. Tobacco that had been going for twelve dollars a bale now went for four. The price of cotton fell from thirty-three cents a pound to eighteen. The price of rice fell by half. All regions of the country were affected, even Philadelphia, which had been more prosperous than in its entire history. The Philadelphia Society for the Promotion of National Industry was so worried by the "calamitous situation of our agriculture, manufactures, trade, and commerce," that it demanded that the president convene a special session of Congress to increase the tariff to protect American jobs.

The net effect of the panic was to shatter people's confidence. Suddenly, it seemed, the people who ran things, the bright people who had set up the system of paper money and banks, didn't really know what they were doing despite their fancy explanations and complicated machinations. Under the circumstances logic dictated that these people should be replaced in power by people who did know what they were doing, people like Jackson, who were successful and who were proven leaders, but who weren't part of the existing political establishment and didn't share the establishment's enthusiasm for paper money and banks.[3]

The other development, an inevitable one, was people power. The founders hadn't wanted the people to have power. They had in fact designed the government to run with very little interference from the people. Of the three branches just one was to be subject to popular will, the legislature, and at that just one half of the legislature, the House of Representatives; senators were elected by the state legislatures. All that had been part of a deliberate plan to keep power in the hands of an elite that would rule in the name of the people but not be controlled by them. Washington, Adams, Jefferson—all had been part of that elite establishment. And then that elite establishment had begun—slowly, to be sure—to lose its grip on power.

The loss of the elite's control could be seen most obviously in the change in the election of presidential electors. In 1800 just five states permitted the people to select electors directly, the rest leaving the power in the hands of the state legislatures. By 1816 ten states did. By 1824 nineteen did (out of a total of twenty-four). By 1832 all but one state did.* While that change was taking place, another, of equal moment, was also having an effect. Restrictions on voting were being dropped. In 1800 in almost every state the suffrage had been restricted to white adult males with property. Now in almost every state all white adult males were allowed to vote, with or without property. Almost immediately the newly enfranchised began taking advantage of their newfound power. In 1824 three hundred thousand turned out to vote for president. In 1828 1.1 million. In 1832 1.2 million. In 1836 1.5 million. In 1840 2.4 million. It wasn't exactly a revolution. The change took place over twenty years. But it was a real change. In 1824 just a quarter of the eligible voters voted. By 1840 more than three-quarters did.

Jackson slowly began to recognize that—given the way things were changing—a man such as he might actually have a shot at the presidency. But one thing stood in his way: King Caucus. And it was a very very big thing. For King Caucus virtually monopolized the power to pick the presidential ticket. King Caucus was in actuality nothing more than a congressional caucus composed of members of the

---

*The one holdout was South Carolina, which was controlled by an oligarchy led by Calhoun. Not until after the Civil War did South Carolina reform its election laws to permit voters to select presidential electors directly.

Democratic-Republican party, who met every four years to endorse a candidate for president. But because the parties at this time didn't hold national conventions as they do today to pick the presidential ticket, King Caucus functioned as a kind of party nominating convention. If King Caucus liked you, you got to be the party's presidential nominee. If it didn't like you, well—you wouldn't, and there was nothing you could do about it.

As long as King Caucus remained powerful, the reforms that were taking place, while beneficial, could never reach their full potential, could never truly transform the politics of the country. For with King Caucus the people were left out of the presidential selection process. While more and more people could vote, and they could vote directly for electors, they couldn't decide the name of the person on the ballot. In effect, then, the elite continued to pick the president, the same old elite that always had. If anything, and everybody could see this, the old elite really had more power than it had had years ago.

In the 1790s the elite had been divided into two parties, the Federalists and the Democratic-Republicans, which gave the voters a choice between two slates of candidates. But as the years had gone by the Federalists had become less and less of a factor in politics. In the War of 1812 they became completely marginalized. While most Americans were enthusiastically behind the war effort, Federalists by and large opposed the war and in doing so killed their party. In 1816 they didn't even bother putting up a presidential candidate. By 1820 the United States for all intents and purposes was a one-party nation. Monroe, up for reelection, ran unopposed and nearly won the unanimous vote of the electoral college. (Just one elector voted against him.) Under these circumstances King Caucus truly was king. Whoever the caucus picked became president. The only votes that really counted, therefore, were those cast by the little elite who got to vote in the caucus.

Andrew Jackson hated King Caucus, not just because it was undemocratic, not even primarily because it was undemocratic, but because it was an obstacle in his path. That made the fight personal. For Jackson fights were always personal. In the War of 1812 it wasn't the United States Army versus the British, it was Andrew Jackson versus the British. During the Seminole War it was Andrew Jackson versus the Indians. It was always Andrew Jackson versus somebody. (In a few years it would be Andrew Jackson versus the Bank of the United States. The bank "is trying to kill me," Jackson would say at the time, "but *I will kill it*.")

Now it was Andrew Jackson versus King Caucus. And that made the fight, for Jackson, a no-holds-barred, fight-to-the-finish kind of fight. And that was just the kind of fight Andrew Jackson loved. He was, more than anything else, a born fighter. There never was a time in his life when he wasn't fighting somebody. His body was a testament to that. On his forehead was a scar left by a British soldier with whom he got into a brawl during the Revolution, when he was just a boy. Next to his left lung was a bullet he took during a duel in 1806. In his left arm was a second bullet he took during a bar fight with the Benton brothers (Thomas Hart and Jesse) in 1813. Some of his friends thought that one reason Jackson was so feisty was because he was so full of holes; certainly his wounds put him in a great deal of pain. The bullet next to his lung continually caused him to cough up blood. The bullet in his arm continually abscessed; as president he finally would have to have it removed in an exceedingly painful operation conducted without anesthesia.

But he was ornery even before he'd been shot up. It was just part of his nature. Once when he and another white man on their way home from Natchez, Mississippi, were taking a group of twenty-six slaves into Indian territory, a posse of twenty armed men looking for runaways had stopped them at the border. *Got a passport?* Jackson was asked. *No,* said Jackson, *and I don't need one. My American face is my passport.* Given the situation, it would have behooved Jackson to act nice. After all, he was outnumbered, twenty to two. But that wasn't Jackson's way. When the posse leader demanded that Jackson leave his slaves with them, he exploded in rage. *Leave my slaves? Hell I will.* He then went back to his slaves, who were still on the Mississippi side, removed their chains, and armed them with guns. Then he marched them like an army across the border. Needless to say the posse let this oddest of little armies pass by unmolested. As soon as his little group was out of danger Jackson collected the weapons and put the slaves back in chains. *They* weren't about to challenge Andy Jackson either.[4]

Jackson knew that King Caucus would never select him to be president, and when it met, it didn't. Instead it picked William Crawford, Monroe's treasury secretary, who was considered part of the old-boy network. Crawford, however, was a terrible choice. Just recently he had suffered a stroke that left him half paralyzed, partially blind, and unable to work or even to speak. If ever there was a chance to defeat the caucus, it was now. The stroke was a secret outside Washington, but once the country learned of his condition the caucus's selection of Crawford would undoubtedly be seen as an astonishing act of arrogance, almost as if the members believed that they could foist anybody they wanted on the people.* And it gave Jackson a great issue. Naturally he made the most of it, pillorying Crawford and the caucus as antidemocratic.

The caucus's selection of Crawford was almost an act of political suicide. But it wasn't just Jackson who saw opportunity in Crawford's choice. So did his rivals Henry Clay and John Quincy Adams turning the election into a four-way contest as all joined the fray.

King Caucus *was* in its death throes. Its selection of Crawford was evidence of that. Moreover, the meeting had been poorly attended. In the past nearly the entire contingent of Republicans in Congress had participated. This year just 66 out of 231 had. Proof that it had finally outlived its usefulness, that it was finally irrelevant, came with the results of the November election. Crawford came in third, behind Jackson and Adams. Clay came in fourth. The king—King Caucus—was dead.

Jackson, although he won more electoral and popular votes than anybody else, was not made president however. Because so many candidates had competed he failed to receive a majority of the electoral votes. That meant that the House of Representatives would once again have the responsibility of choosing the winner. And that, in turn, meant that the old elite would have the chance, again, to pick the president of the United States. Under the rules prescribed in the Twelfth Amendment, the House had to pick from the top three candidates: Jackson, Adams, and Crawford. Crawford's hopes quickly faded, leaving the contest between Jackson and Adams. On February 9, 1825, the House picked Adams.

It was later said, by those who took a narrow view of the matter, that Adams had been elected by pure chance. The election in the end

---

*In fact, the members simply hadn't realized how ill Crawford was when they voted. New York political boss Martin Van Buren had falsely led them to believe that Crawford had staged a full recovery. Van Buren's push for Crawford was part of a long-term strategy designed to create a new national party based on an alliance between New York and Virginia (Crawford, a Georgian, was enthusiastically backed by Virginians). Van Buren felt he could only win the trust of Virginia if he stood by Crawford no matter what. See Donald B. Cole, *Martin Van Buren and the American Political System* (1984), pp. 126–41.

had come down to the vote of New York, and the vote of New York had come down to a single representative, Stephen Van Rensselaer, a doddering and weak-willed old man who at different times had promised *both* Adams and Jackson supporters that he would vote with them. When the time came for him actually to cast his vote, he'd closed his eyes and bowed his head in prayer. When he finally raised his head he happened to catch a glimpse of an Adams ballot lying on the floor beneath his seat. Believing this to be a sign from God, he promptly voted for Adams. "In this way," recalled Martin Van Buren, "it was that Mr. Adams was made President."

In truth the matter was far more complicated. Adams won because over the preceding weeks, through winks and nods, he'd struck a series of bargains with various key politicians to win them over to his side. He got Maryland's vote by vowing to keep old Federalists on the federal payroll. He helped win Missouri by pledging to keep its congressman's brother on the federal bench, even though the judge had killed a colleague in a duel. He gained Webster's support by hinting he'd consider appointing Webster minister to Great Britain in the future. And, most famously, at a secret meeting with Henry Clay, Adams left the impression that he'd appoint Clay secretary of state.

Of all the deals this was the most vital. Clay, as Speaker of the House, was hugely influential. With Clay's help victory was assured. Afterward it was said that Adams never actually made any specific promises to Clay, nor Clay to Adams. But they didn't need to do anything so crude as to say *you do this and I'll do that*. In the parlance of the day, they merely had to reach "an understanding." And this they did. Clay delivered the states of Kentucky and Ohio. Adams was elected president. A few weeks after, Adams appointed Clay secretary of state.[5]

Adams always insisted that there was no bargain between him and Clay. Maybe he even believed it. He almost had to. Adams had a very strong sense of tradition, a very strong sense of family, and both family and tradition were against bargaining for the presidency. The son of John Adams didn't seek office, it sought him. "If my country wants my services," Adams had confided a few years earlier, "she must ask for them." When his friends pointed out that "the others are not so scrupulous" and that he had to make an effort to win the office because "you won't be on an equal footing with them," Adams had insisted firmly, "That's not my fault. My business is to serve the public to the best of my abilities in the station assigned to me, and not to intrigue for further advancement. I have never, by the most distant hint to anyone, expressed a wish for any public office, and I shall not begin to ask for that which of all others ought to be most freely and spontaneously bestowed."

Of course Adams very much did want to be president, even if he felt he couldn't admit it; having his father be president made it that much more important for *him* to be president, too. Indeed, his parents groomed him for the White House, expected him to be president. But he sincerely detested the idea of bargaining for the office.

That a man with such beliefs in the end found he had no choice *but* to bagain was a cruel turn of events. It was especially cruel because Adams prided himself on his superiority to Jackson, and Jackson—Jackson the Barbarian, as he was known in Adams's circles—had *not* stooped to bargaining. It was as if the world had been turned upside down. The man who wouldn't have been expected to bargain, had; the man who would be expected to, hadn't. Looked at strictly in terms of the two individuals involved the situation *was* totally mystifying. But if one took into account their different circumstances, it was entirely understandable.

The difference between them wasn't their age. Both happened by chance to have been born in exactly the same year, 1767, just a few months apart. It wasn't even that one was a born aristocrat and the other wasn't. Both by this time were on an even footing financially and even in a way socially. While Jackson certainly didn't rank as high in eastern circles as Adams, in western circles he was considered to be at the height of the social pyramid. Jackson was in fact a member of the country's elite. He lived on a huge plantation. He was friends with the best-bred people. Easterners when they met him for the first time were always surprised. From the stories that made the rounds they half expected to find a tomahawk-carrying savage, "but little advanced in civilization over the Indians with whom he made war," as one of them put it. Instead they encountered a true gentleman.

The real difference between them, when you dug deep enough, was that Adams was a man of the past, Jackson a man of the future. For Adams to win he had to fight the future—hold it off, in effect. For he lacked popular appeal. He read books, knew the classics, was well educated (went to Harvard), knew several foreign languages, was well traveled; in short, he was everything that the average person wasn't at a time when the average person was beginning to become politically powerful. Jackson, in contrast, *was* the future. He did have popular appeal. Unlike Adams he seemed like a regular fellow, *one of us*, not *one of them*. He wasn't well educated, hadn't traveled, hadn't even gone to college, didn't know how to spell—and didn't even care. He was quoted as saying, though no one knows if he actually did, that he never thought much of a man who could only think of one way to spell a word.

Jackson by no means had a lock on the presidency. But under the new circumstances, the new politics, the emergence of a nascent popular democracy, Adams clearly was more seriously disadvantaged. The only way he could become president was by cheating. And Adams very much wanted to be president, expected to be president. He was so desperate that at one point he even considered asking Jackson to become his vice president. His friends were aghast; Adams's response was that the "vice-presidency was a station in which the General could harm no one and in which he would need to quarrel with no one."[6]

The election of 1824 was historically important not because of who was elected, but because of how he was elected. It was the old way. The same way Jefferson had been elected in 1800. A bunch of politicians sitting in Washington had connived and cut deals and finally settled on a winner. But it was the last time things would be done this way. The country was changing again, and as a result politics was changing. The control the politicians had exercised for decades over the presidential election process was slipping. From now on they would have to take the people into account—and that made things messier. The elite no longer would rule with quite the same degree of power as they had until then. The brightest lights of the political firmament couldn't sit in a room and among themselves decide what was right for the country as they had for so many years. They had had their day and their day was over.

It was, of course, an inevitable development the way things were going, with the demise of King Caucus and the reform of the election laws. But because the elite had overreached, because they had not just cut deals as they had in 1800 but had actually bluntly thwarted the people's will, depriving them of the candidate they obviously favored—Jackson, after all, had won more popular and electoral votes than any of the other candidates (Jackson: 153,544 votes; Adams: 108,740)—the new age of democracy was ushered in a lot more quickly than it otherwise would have been.

The power in the country was shifting. For a generation, since the Constitution was drafted, power had been moving centripetally, toward Washington, away from the states, away from the people, which was kind of curious (curious because the Constitution today is thought of as a democratic instrument and yet it initially took power away from the people). Now it would move centrifugally, away from Washington and back toward the states, back toward the people.

The day of the old elite was not over because they had performed badly. Actually they had governed admirably, providing the country with an exceptional succession of leaders: Washington, Adams, Jefferson, Madison, Monroe—these were all truly exceptional men. Even John Quincy Adams, the last of the men chosen the old way, was exceptional.

Under their leadership the country had largely flowered. The institutions of popular government had been firmly planted. And politics, by and large, had attracted a mostly honest class of officials and employees. While some were greedy, like the politicians who had swindled the Revolutionary veterans out of their government bonds, and some were dastardly, like the traitors who had taken part in the Burr conspiracy, most were dedicated public servants. In all the history of the country there never would be another time so largely scandal-free. Not until the War of 1812 was there a noticeable rise in government corruption, and that was because of the War of 1812. As was to happen during nearly all our conflicts, the mass acquisition of military supplies and expensive armaments provided opportunities for corruption that some found irresistible. In a typical case after the war, the government agents responsible for disbursing pensions to army veterans received payments in hard specie, then paid out the claims in devalued paper currency, and pocketed the difference.

Of all the old-style presidents, there was only one whose integrity was questionable: Monroe. Always hard-pressed financially, he accepted a five-thousand-dollar loan from fur trader John Jacob Astor and subsequently rewarded Astor with a change in government policy that proved to be of immense help to Astor's business. After Monroe left the White House he billed the government for tens of thousands of dollars in false claims stemming from his years of service as a diplomat. One claim involved the ten-thousand-dollar loss he'd suffered on the sale of his house in Paris. Although he had no right to expect the government to reimburse him for the loss, he insisted he was owed not just the ten thousand dollars but the interest he would have earned on the money had it been paid to him at the time of the sale. That amounted to another twenty thousand dollars. On the eve of his death in 1831 Congress, without looking deeply into the merits of the claims, paid him off.[7]

That the old-style presidents had been honest mattered a great deal.

Americans always demanded honesty in their presidents, and at the beginning valued honesty more than almost anything else. Being honest and being apolitical seemed the same thing. But honesty was not enough in the new climate in which politicians were operating. Talent and experience were not enough. Good policies were not enough. A government *for* the people was not enough. Now the government had to be *of* the people. And that meant the president had to be *of* the people.

Adams, of course, was not. He was an elitist. And that was crippling given the change in politics and the unseemly way he had come to power. And all of that doomed his presidency. Adams, aware that his legitimacy as a leader was in question, and perhaps feeling guilty over the dubious means he had employed to gain office, tried to govern like a Washington, to be president of all the people, to operate above the battle, to be truly nonpartisan. He appointed a nonpartisan cabinet and kept in office the people appointed by his predecessors, refusing to clean out his enemies even when his enemies used their position in the government to undermine his authority and power. But it was no use. The times had changed. The people no longer wanted a Washington, no longer wanted a man who stood apart from them, as Washington had. Washington was an anachronism now. Washington had gained power by doing what was right regardless of popular opinion. Adams, attempting the same, lost power.

Americans did not admit that the Washington model was irrelevant. But it was, and Adams was the proof. That he was the wrong man for the presidency in the 1820s was obvious from the moment he took office. In part the problem was that he happened to believe in big government and big projects—a national university, a naval academy, an astronomical observatory, national highways and canals—at a time when the people were growing suspicious of big government. Mostly, though, it was his very mood, his outlook. He held the public in contempt. Worse, he virtually admitted it. In his very first message to Congress, after outlining his controversial program of public improvements, he insisted that the government not be "palsied by the will of our constituents." The statement had shocked. Not even Hamilton had been so bold.

If he hadn't come to power with the Clay cloud over his head, it's possible he might have done a little better. But with the people against him, and he against the people, he could do nothing. Not even the old elitists in Congress stood by him. They had created him. He was their monster, in effect. But they wanted nothing to do with him. Clearly one wasn't going to improve one's political position by siding with John Quincy Adams.

His father, faced with a changing polity, had changed with it, in an attempt to remain influential, to remain in control. But the son did not. There would be no deal making, no accommodating, no politicking. The result was that the son became irrelevant. Ambitious as he had been to gain power, he was not ambitious enough, or ambitious in the right way, to use it to get anything done. In four years in office he managed to persuade Congress to pass just a single one of his visionary initiatives, a bill to fund a diplomatic mission to the Congress of Panama, a Latin American forum created to resolve regional conflicts. And not even this minor achievement could be said to have been in any way a success. Congress took so long in approving the measure that by the time the funding came through the conference was over.[8]

History was on Jackson's side as it had not been on Adams's. But Jackson was not the kind of man to leave things to chance. There was a fight to be won. Adams and the politicians had stolen the election of 1824. Jackson would see to it that he won the next one, won it so big that no one would dare try to deprive him of victory. Whenever he was asked about the loss, his eyes would grow hot with anger and he would repeat what he told everybody, that the people had had the election stolen from them. But that wasn't what drove him. This was personal. *Andrew Jackson* had had the election stolen from him. And that was intolerable.

The Jackson strategy was not to try new ways to win. He wouldn't give speeches, wouldn't campaign, wouldn't make promises. But whatever candidates had done in the past to win, he would do more of. Lots more. He wouldn't break tradition so much as stretch it. Instead of leaving presidential politicking to the final year before the election, as candidates had been accustomed to do, Jackson would politick for all four years. He would write letters. He would meet with constituents. He would meet with politicians. Nobody would doubt that Andrew Jackson was running for president again. From the moment of his loss until the day he won he would spend virtually all his waking hours working to be elected president. Running would be, for him, as it had been for no other person before him, a full-time job.

When it was suggested that it might be worthwhile to have a newspaper to control, as Jefferson had had, Jackson agreed, but went a step further. He personally put up the money needed to buy the paper:

three thousand dollars. When Martin Van Buren, by now a fervid Jackson loyalist, began pressuring Jackson to create a political party as Jefferson had, Jackson agreed and then went on to do what Jefferson never did. He actually began seeing to it that his supporters built Jackson political machines across the country. Of all the changes, the rebirth of the two-party system was easily the most dramatic and the most important of his contributions. In 1824 there had been just a single party in the country, the old Republican Party. All four candidates for president—Adams, Crawford, Clay, and Jackson—had run as Republicans. In 1828, directly as a result of Jackson's efforts, there would be two parties: the National Republicans and the Democrats.[9]

The outcome of the election of 1828 was never seriously in doubt. Of course Jackson would win. And aside from his winning, it is seldom remembered. But one thing happened during the campaign that was significant. It was the first election in which a candidate's wife became an issue. Until then elections had mostly been about real issues, about whether the defense budget should be increased, about the threat of war, about hard, substantive things—things that mattered. Even the smear campaigns against John Adams and Thomas Jefferson had mostly been about real things. It mattered if Adams really was a monocrat who had sought to make himself king. It mattered if Jefferson really was a Francophile. Just once had sex been dragged into presidential politics: Jefferson had been accused of having had an affair with Sally Hemings. And on that one occasion sex had remained an issue of minor concern to the electorate. In the election of 1828 sex became a chief issue. Not *the* issue, by any means—more newspaper ink was spilled over the charge that Clay and Adams had struck a corrupt bargain than over sex—but sex loomed large.

Of Adams it was said that he had had premarital relations with his wife, that she was illegitimate, and that when he had served as minister to Russia he had pimped for Czar Alexander I. Of Jackson, that his mother had been a whore, that he was the son of a lascivious mulatto named Jack (hence "Jack's son"), that he had seduced his wife, Rachel, wrecking her marriage to another man, and then lived with her in sin.

Rachel had in fact been married when she and Jackson met. But she and her husband were no longer living together, he had filed for divorce, and both she and Jackson believed that the divorce had gone through. It had not, but they didn't realize it. When, two years later, they found out that the divorce hadn't actually been final when they married, they promptly got married again.[10]

Given that the candidates were Adams and Jackson, two of the most straitlaced gentlemen to sit in the White House, it was an astonishing development, an awful sign, a portent of things to come. And it was inevitable. Once the people became part of the political equation, the politicians had to figure out ways to reach them, ways to get their adrenaline pumping, to motivate them, and to get them to the polls. The corrupt bargain was an important issue, but people hanging over the backyard fence couldn't gossip about it as they could about sex. Sex was special—a real motivator. Besides, sex was easy to understand.

Adams never personally approved of the attacks on Jackson, but his supporters were responsible for starting the whole sex smear campaign and Adams never tried to stop it. He never even chided those of his friends who carried on the attacks, though they carried them on in the *Daily National Journal*, a paper he controlled through its editor, a paper that lived off the big federal printing contracts he made sure it received, a paper everybody in Washington regarded as *his* paper. He wasn't the one making the attacks. But because they were appearing in the *Daily National Journal*, it was as if he were. And that, too, was an astonishing development.

For nearly four years John Quincy Adams had taken the high road, had tried to be like George Washington. Now, suddenly, faced again with an election, he reverted to the old Adams, the Adams of 1824 and 1825, the Adams who did everything to win, even at the cost of principle. In fact, if anything, Adams's behavior was even more egregious this time. Cutting a deal was one thing; it was what politicians at all times have always done. Attacking an opponent because of his sex life was another; that had never been done, not on the scale it was being done now, at least. It was almost as if Adams was more desperate to win this time around than the last. And in a sense, he was. The last time he had merely wanted to be elected president. This time he wanted to get back his honor. And the only way he could do that was by winning more popular votes than Jackson. That would in a way wash away the smell from the last election. It would say to the world, in effect, that it had not just been the politicians who thought Adams should be president, but the people. It would legitimize the Adams presidency.

Jackson was appalled and angered by the attacks. He opposed this kind of politics. But Jackson was also tough, tremendously tough, and he believed in the justice of the Old Testament, an eye for an eye. So when Adams's supporters persisted in smearing him, he allowed as how his supporters had a right to smear Adams. "Should the administration continue their systematic course of slander," he wrote Duff Green, the editor of *his* newspaper, "it will be well now and then to throw a fire brand into their camp by the statement of a few facts." "Female character," he went on, "should never be introduced or touched by my friends, unless a continuation of attacks should be made against Mrs. J." Jackson felt bad about the attacks on his wife because no husband wants to see his wife's name dragged through the mud. But he felt especially bad about the attacks on Rachel because she didn't like politics, didn't want him running for president, and didn't really want him to win.

Every time he left the Hermitage, their plantation home in Nashville, to go to Washington on political business, there'd be a scene, Rachel hysterical. Jackson tried his best to protect her from the harsh side of election politics, keeping her in in the dark about the attacks being made against her, and for the most part was successful. Which was good because Rachel had a weak heart and collapsed under pressure. If she ever found out what had been said about her—well, Jackson tried not to think about it. Then in December, just after he won the election, she did find out about it. Tired after a walk around Nashville while doing some shopping, she happened to stop for a rest at the office of one of Jackson's friends. There, by accident, she came across a pamphlet containing the worst of the vicious smears. Rachel, always weak, and especially weak at the moment because of the recent death of her adopted Indian son, suffered a complete breakdown, right there in the office. A few weeks later she died.[11]

Jackson was inconsolable. At her funeral he thundered, "I can and do forgive all my enemies. But those vile wretches who have slandered her must look to God for mercy." The simple truth was that Jackson himself was to blame, in part, for her death. If he hadn't run she'd be alive. But the simple truth was also misleading. She really died not because Jackson ran, but because of what happened to candidates who ran for president in the new environment created by people power. Now that the people were directly involved in the political process, the candidate's family was open to political abuse. And this was new.

In the past a candidate's family had been almost irrelevant. Of all the president's wives thus far, only one had achieved any political visibility, Dolley Madison, and that was simply because she was such a powerful presence (and her husband was such a weak one). But now, suddenly, a president's family was to be highly visible. And that greatly complicated matters, for it meant that now when a man decided he wanted to be president, he had to take into account the toll a race would take on his family.

Of course, politics had always taken its toll on families. The pressure, the long absences, the long hours, all that had always been a part of politics and all that had been tough on the families involved. But now the ordeal they would have to go through would be far, far worse. This was a fundamental change. It meant that the moment a man decided to run for president he was putting his family at risk. A man who ran now was deciding, in effect, to put his ambition ahead of his family's welfare, not because he was any more ambitious than candidates in the past or less sensitive, but simply because politics had changed. It was a sad development and, in Jackson's case, an ironic one. The very thing that was responsible for his accession to the White House—people power—was also, indirectly, the thing that caused him to suffer his greatest loss, the death of his dearly beloved wife, Rachel.

On March 4, 1829, Jackson was sworn into office as the seventh president of the United States. That night thousands of people—thousands of average people, people with dirty fingers and dirty boots—crowded inside the White House, the first time such people had been permitted inside the mansion. So many came that Jackson, tiring, had to escape through a window, the party continuing without him until four in the morning.

The next day the place would look as if it had been struck by a tornado—the furniture torn, the plates broken, the rugs soiled—everything in tumult.

In a way it *had* been struck by a tornado—the Jackson tornado.

A Jackson presidency was guaranteed to be different from any other because Jackson was different. He would dare to do things other presidents wouldn't have dreamed of doing. When he discovered that the wives of some members of the cabinet were refusing to socialize with the wife of the secretary of war, John Eaton, a fellow Tennessean and his very good friend, because Eaton's wife, Peggy, had a checkered past—it was rumored that she'd slept with the visitors to the inn where she worked before marrying Eaton—he called a cabinet meeting to hear the charges against her. Satisfied that she had been wrongly accused—as had Rachel—he demanded that the cabinet wives give Mrs. Eaton the respect they gave one another. When they continued to snub her he grew irate and finally decided to demand the resignations of the entire cabinet so he could make a fresh start.

Jackson was to go through cabinets the way other presidents go through suits of clothes. In his eight years in office he went through four secretaries of state, five secretaries of the treasury, three secretaries of war and three of the navy, two postmasters general, and three attorneys general. He even drove his vice president, John C. Calhoun, to resign.

Calhoun and Jackson were almost destined to have a falling-out, in part because they were such strong personalities, in part because Calhoun was so rabidly ambitious. What triggered the resignation was Calhoun's insistence on blackballing Peggy Eaton. But what may have been an additional factor was Jackson's discovery that in 1818 Calhoun, then secretary of war, had opined in secret discussions held by the cabinet that Jackson had exceeded his authority in the Seminole War. That helped finish Calhoun in Jackson's eyes. Once Calhoun realized that he was finished, realized that Jackson would never help him achieve his goal of becoming president, he turned against Jackson and threw himself into the arms of the nullifiers of South Carolina, whom the president was battling over the Tariff of 1828, the so-called Tariff of Abominations.

The big battle of the administration wasn't over Peggy Eaton or John C. Calhoun, however, but over the Bank of the United States. This was a quintessentially Jackson controversy. Most of his own cabinet was against his decision to declare war on the bank. But Jackson decided a war was what he wanted; the bank symbolized corrupt power, loans going to the institution's political friends, such as Webster, who weren't required to pay them back. So he went to war and he won—just up and broke the bank. Despite all its power. Despite its deep roots in the American economy. What Jackson wanted Jackson got.

It was people power that made his presidency possible and people power that saw him through the succession of crises he created. With the people behind him, a president could do almost anything, even almost control history. Jackson was so confident he had the people behind him that he felt he could even dare to name his own successor and install him in office. His idea was to put Van Buren in as his second vice president and then, after the election, resign and let Van Buren become president. It was a crazy idea, and Van Buren had the sense to scotch it. But it showed just how supremely confident Jackson was that the people would back him up. Because Jackson had the people behind him, he rarely felt desperate even in the midst of crisis. Being confident, he almost never felt it necessary to stoop to unprincipled means to achieve his objectives.

But he wasn't perfect. One of his chief goals was to build up the Democratic Party, and the only way he could do that was by giving his supporters jobs in the government. That required taking jobs away from current officeholders—experienced officeholders, whose only crime was that they weren't Jacksonians—and giving them to inexperienced workers, many of whom were nothing more than party hacks. Jackson gave the practice a fancy name. "Rotation in office," he called it. Which sounded good. The way he explained it, it *was* good. Workers would be rotated in and out of office to keep the government in touch with the people, to close the gap between the governed and the governing. Very democratic! But the purpose was plainly political: to use the government to help build up the party—his party. Jackson couldn't see anything wrong with that. And since the extent of his rotations was relatively small—only twenty percent of federal workers were replaced during his two terms—there really wasn't much wrong with it. And anyway, he hadn't invented the spoils system. It was as old as the Republic. But over time, as the political parties grew and the government grew, the spoils system would become totally corrupt. The Jackson machine wasn't corrupt, it was just a machine. The corruption of machine politics, the padding of payrolls with the totally incompetent, the forced collection of kickbacks from government workers to the party, all that would come later.

Jackson in fact had been determined to run a government free of corruption. One of his claims against Adams was that the government had become bloated and venal and needed to be cleaned out. His claim was an exaggeration. When he finally took over after Adams, his minions, expecting to find wide evidence of corruption, scoured the bureaucracy for proof but came up with little. The irony was that it would be a Jackson appointee, Samuel Swartwout, collector of the Port of New York, who was to prove to be truly venal on a grand scale, becoming the first person in American history to swindle the government out of a million dollars. But he was the exception. The Jackson government on the whole was run honestly, though Jackson ignored warnings about Swartwout.

He had no choice really but to be a spoilsman. If he was going to succeed in the age of people power, he had to have a party. It was the party that gave him his power, the party that let him accomplish all that he wished to in the name of the people. It was through the party that he communicated with the people and through the party that they communicated with him. And to keep the party content he had to be able to give party workers jobs. Putting party workers on the government payroll wasn't, therefore, a luxury, an indulgence. It was a necessity—the only way to keep this new democratic instrument vital and percolating—the only way to keep control.[12]

Because Jackson had been so popular, more hugely popular than any other figure of his generation, so popular he had been able to mold events to his liking, and so powerful a presence that he seemed larger than life, a force of nature almost, the tremendous effect of people power on politics—its tremendous negative effect—had been largely overlooked while he was in office. What people noticed more than anything was Andrew Jackson. Jackson the dragon slayer. Jackson the war hero. And most of all, Jackson the man who caused a titanic shift in power. Before Jackson the great power of the federal government resided in Congress. Now it resided in the presidency. That was a remarkable development. It turned the Constitution on its head—at least the Constitution as the founders had understood it.

The Congress still was powerful. With men in it like Thomas Hart Benton, Henry Clay, and Daniel Webster, it could hardly be otherwise. But Congress no longer spoke for the people. Congress represented the special interests. Webster represented the Bank of the United States, which kept him on retainer. Clay represented the bank and the emerging class of go-getter businessmen, people who wanted to build roads and canals. Benton represented the West. Only the president spoke for the people as a whole. And that made the presidency superpowerful. For the people were the source of power now, the mother lode of power. And Jackson owned the mine, lock, stock, and barrel. Congress could pass legislation, but if he didn't like it, he'd unpass it, through the use of the veto. Nearly all the presidents had employed the veto, but they had done so rarely and only on the grounds that the legislation was, in their eyes, unconstitutional. Jackson vetoed legislation simply because he didn't like it. And he vetoed and vetoed and vetoed again, more times than all previous presidents combined.

"King Andrew," his enemies called him. The president who felt he could do as he pleased. A cartoon captured the essence of the image. It showed Jackson standing by a throne, a royal scepter in one hand, a copy of his veto message in the other. The caption read: "Born to Command."

If Jackson's presence overwhelmed, obscuring the developments that were taking place in the culture because of people power, it would be telling who was to come after Jackson. The choice would say much about what politics had become. It would have to be a party man, of course, either a Democrat like Jackson or a Whig like Henry Clay.* But who?

The Whigs were so unpopular that they couldn't choose. Not one of their candidates, not even Henry Clay, was deemed attractive enough to win in a head-to-head contest with a Democrat riding on Jackson's coattails. So the Whigs put up four candidates, each one strong in a particular region of the country. The inability to settle on a single choice was not in itself indicative of the changing state of politics in the country. But the strategy behind the selection of the four candidates was. The hope was that by putting up four candidates the Whigs could throw the election into the House of Representatives, and then possibly wheel and deal themselves into the White House.

This was as sure a sign as any that the new politics would inspire manipulation and deceit. In the past a candidate would have been too embarrassed to try such a stunt. It would reflect badly on his reputation. But candidates didn't represent just themselves anymore. They represented parties. And parties didn't worry about their reputations too much. Parties were organizations, and what organizations cared about was winning. Now that politics was to be dominated by parties it would become much more ruthless. For much more was at stake. If a candidate lost it was more now than just he and his family who suffered. It was the whole organization behind him. It was all the people who worked hard to bring his election about. All the people who hoped to gain by his victory. The Whig strategy was the first real indication of this.

The Democrats' choice was Martin Van Buren. The Little Magi-

---

*The opponents of Jackson, the old National Republicans, had taken the name Whig as a shrewd way of reminding people of Jackson's kingly pretensions; it was the Whigs in England who had rallied in opposition to Charles II.

cian, the Sly Fox of Kinderhook.* The most political of all the extremely political leaders then operating in American politics. *Question*: Why does Martin Van Buren's bread always land butter-side up when it falls to the ground? *Answer*: Because Van Buren always butters both sides at once.

Nobody could point to a single incident that was especially damning, one that put him in a special category. It was that there was an accumulation of incidents. Like the time he had arranged for Federalist Henry Fellows to be kicked out of the New York General Assembly because some of the ballots in Fellows's close election had been marked "Hen. Fellows," not "Henry Fellows." (With Fellows out, Van Buren's forces were able to take control, giving them the power to name the next speaker.) Or like the time he switched from firm opposition to the Erie Canal to firm support after Gov. DeWitt Clinton, the builder of the canal, won reelection in a landslide. Or when he agreed to nominate Clinton to yet another term as governor and then, at the end of his address, suddenly came out in favor of somebody else whom Van Buren believed would be more popular. And those were just the stories that trailed him from his days as a state politician.

After he became a national figure there were more stories. More decisions that seemed to be driven by pure politics and pure ambition: his decision to accept Jackson's offer to become secretary of state just weeks after he had taken office as the governor of New York. His offer to resign from the cabinet to become the minister to Great Britain just as the controversies over the bank and the tariff were breaking, saving him from having to commit himself on the issues. It had been considered a testament to his political skills that he and Jackson had even become friends because in the 1824 election Van Buren had been the leader of the Crawford forces, the very forces that were thwarting Jackson's elevation to the presidency. But his switch to Jackson hadn't been a surprise. Jackson was obviously going to be president so of course Van Buren switched.

Naturally the Democrats would have preferred to have had another Jackson at the helm. But every generation there is only one Jackson. So they had to turn instead to a politician, a politician's politician,

---

*Van Buren was from Kinderhook, New York, sometimes referred to as Old Kinderhook, or OK for short, which some believe was, somehow or other, the linguistic root of the expression "okay."

somebody who, although he wasn't a hero and wasn't personally popular, could by wheeling and dealing gain control of the process. Once
party politics had become predominant Van Buren's rise was almost
inevitable. He wasn't just good at party politics. He'd invented it. It
was Van Buren who single-handedly had revived party politics in New
York after the Era of Good Feelings and then, afterward, helped nur-
ture the development of a two-party system throughout the country.
More to the point, Van Buren wasn't just on good terms with the party
bosses. He himself *was* a party boss.[13]

Keeping control of the system had been easy for Jackson. For Van
Buren it would be hard. Jackson was so popular he rarely had to make
fundamental compromises; in fact about the only compromise he made
was to invigorate and justify the spoils system. He could do just about
as he pleased, as the opposition Whigs charged. Van Buren would have
to make compromise after compromise.

His most serious compromise of all was his selection of a vice
president. It was a very, very important choice. Who was picked could
very well determine the outcome of the election, could decide
whether Jackson's victory had been a fluke of personality or the triumph of a party. Jackson made it clear that he wanted Richard Johnson of Kentucky, a longtime Jackson supporter, who also had the
advantage that he was not William C. Rives. Rives was the choice of
the Virginia wing of the party and for that reason alone was unaccept-
able to Jackson, who long ago had decided to make sure that the party
remained anchored to the West. It was a vital concern to Jackson. Virginia to him represented the old elite. Until he came along the country
had been dominated by the politicians in Virginia and New York, and
one of his aims was to see to it that they did not resume their domina-
tion after he left the scene. Van Buren had spent a lifetime actually
trying to cement the relationship between Virginia and New York.
Even had a name for it: the Richmond-Albany axis. But Van Buren
couldn't defy Jackson. He owed his nomination to Jackson. So he
caved.

Picking Johnson was not, as it happened, a bad political move. It
made strategic political sense. He was a popular figure, popular in the
West, where he was known as a winning Indian fighter (he claimed to
have killed Tecumseh), popular in the Northeast, where he had a good
reputation among workingmen grateful for his twenty-year campaign

to end the imprisonment of debtors. But while choosing Johnson may have been good politics, it was also an indubitably appalling choice, worse even than the selections made in the past for vice president, worse than the selection of Aaron Burr, worse than George Clinton. For in both of those cases the party hadn't known in advance just how bad these men actually were. About Richard Johnson they knew.

It wasn't just that people doubted that he actually had killed Tecumseh. Westerners were known for their tall tales. Nor that he ran a tavern. It was that there was always about Johnson an unsavory smell. As a politician he was utterly dishonest. His whole family seemed dishonest. During the Monroe administration, his brother had been given a contract to explore the Yellowstone. He had gotten a hundred thousand dollars for it. A gigantic sum. But after delaying more than a year, when he finally got under way, he cried that he needed more money. Secretary of War Calhoun hadn't wanted to give him any more. A hundred thousand dollars had seemed quite enough. But Monroe overruled him, personally authorizing another eighty-five thousand. Monroe was never any good with money, his own or the government's, and the money he spent on the Yellowstone expedition was a huge waste. The expedition never got around to doing much exploring and never even made it to the headwaters of the Yellowstone, its primary goal. But Monroe had kept the spigot open because Johnson happened to be the chairman of the House Military Affairs Committee. It should have made for a big juicy scandal. But because Johnson was powerful people left him alone. The only one who might have blown the whistle on him was Jackson, who in the 1820s campaigned against government corruption, but because Johnson had been an early Jackson supporter he managed to escape criticism.

The real complaint about Johnson, anyway, wasn't political, it was personal. His love life was a shambles, a series of affairs with black mistresses who bore him several illegitimate children. That, given the times, was dismaying enough, but Johnson also had earned a notorious reputation for mistreating his mistresses. When one ran away with an Indian, Johnson, furious, had had her captured and sold into slavery.

But the decision to choose Johnson wasn't about morality. It was about pleasing Jackson and about winning. The irony is that Johnson, after all, did not really help the ticket and may have even hurt it a bit. Virginians, distraught over Johnson's personal life and angry that he had been selected over Rives, refused to give Johnson their support, depriving him of an electoral majority. That, under the Constitution, threw the election of the vice president into the Senate, the only time in our history that that was to happen. The Senate duly approved Johnson by a narrow vote. In office he proved to be as appalling as the Democrats had had every right to expect. He continued seeing his mistresses and when he grew bored with public life he retreated to his tavern. There on summer nights, dressed in country clothes, the vice president of the United States could be found serving up food and ale alongside his employees.[14]

Van Buren came into office under terrible circumstances. Not only was he not Andrew Jackson, which people unfairly held against him. The economy was against him, too. Within weeks of his inauguration, the country succumbed to a panic that lasted his entire term. Van Buren was stymied, desperate, and utterly incapable of fixing things.

The sad fact was nobody knew how to fix things. When the economy collapsed the prevailing wisdom was that the government could and should do nothing. And not for nearly a hundred years would people think differently. Worse, the politicians hadn't yet even figured out how to instill hope in the people in the midst of trouble. As people lost hope, Van Buren did, too; without the people he'd lose the next election.

The fact is he'd barely achieved the White House the last time, when the economy was going well. Out of nearly a million and a half votes he'd won by just twenty-five thousand. And he'd only won by getting people to think that a vote for him was a vote for Jackson. How many times could the Little Magician pull off that trick? Jackson in effect had been run for president four times: in 1824, 1828, 1832, and 1836. Could they run him yet again? Run him symbolically a *fifth* time? It was a politician's trick to get people to think they were voting for one man when another was on the ballot. A trick that would be tried over and over again in American politics, now that the people had the vote. For the people could be swayed by emotion. Van Buren, the trickster's trickster, was anxious to try it again, for he was very, very desperate.

Out of his desperation he came up with a wild plan. As Jackson was the party's chief drawing card, he would have Jackson return to New Orleans in 1840, on the eve of the presidential election, to celebrate the twenty-fifth anniversary of the Battle of New Orleans, the great battle of the century, the victory that made Jackson a national hero. That would help revive the loyalty of Jacksonians and perhaps revive Van Buren's sagging fortunes as well. By getting Jackson back into the news people would be thinking about him when they voted.

Jackson was old now and deathly sick, suffering from hemorrhages and daily diarrhea. Just getting to New Orleans would be an ordeal. He'd have to travel over rough roads for four days, then take a steamer down the Mississippi through dangerous ice-clogged waters in chilly winter weather. Then he'd have to put himself through endless days of ceremonial events, all of which could prove fatally taxing. And then, at the end, he'd have to retrace his trip back home. It would be a terrible struggle. Van Buren was asking Jackson to put his life at risk for the sake of the election. Old Hickory, however, promptly agreed. He was, he wrote Van Buren, "determined to go through or fail in the struggle."[15]

Jackson went and survived. But Van Buren wasn't the only one willing to play tricks on the public in an attempt to win the election. So were the Whigs. And they had a better trick to play. Instead of running a stand-in for a hero as the Democrats were doing, they would run a hero himself: William Henry Harrison. The other great hero of the War of 1812, the man who won the Battle of the Thames in 1813, the man who defeated the British in the Northwest and in defeating them defeated the Indians there as well, the man who made the Northwest safe for the white man.* The Whigs did not settle on Harrison because they thought he would make a great president. If anything they expected he would be a rather mediocre one. So mediocre that the plan was to let him be president in name only. The people who really would be in charge would be the people who deserved to be president: Clay and Webster. Clay would run things from the Senate, Webster from the cabinet. Harrison wasn't even to have the final say on issues of importance; the cabinet would decide, by majority vote. Clay, who had already made two runs for the White House, in 1824 and 1832, had wanted to make another, but the party bosses had turned him down. Clay was a loser, they felt. And they were tired of losing.

When in December 1839, at the Whigs' first national convention, the bosses settled on Harrison, no one was surprised. Harrison would be the Whigs' answer to Old Hickory. Hell, he even had a nickname like Old

---

*Both Harrison and Richard Johnson made their name in the Battle of the Thames. Their dueling claims became a subtheme in the election of 1840. Harrison boasted of defeating Tecumseh in the battle, Johnson of killing him. (Johnson again ran as Van Buren's vice president. But he was so unpopular among Democrats that the party refused to endorse him officially.)

Hickory: Tippecanoe, which he had earned in 1811 in a grisly battle with the Shawnee near Tippecanoe Creek (in present-day Indiana).

Unlike Jackson, Harrison did not seem to hold any firm views about government. Nobody at the convention knew what he believed. Harrison himself didn't seem to know what Harrison believed. But nobody cared. He wasn't chosen because of his views. He was chosen because he could win. In fact, the convention managers were happy nobody knew where he stood on the issues. It would make him a better candidate. A candidate with strong views alienated people, as Clay had. The campaign wouldn't be about issues at all. It would be about a personality, a celebrity. The party wouldn't even publish a platform. A platform could alienate people, even if it were written in the most general way.

John Tyler, the candidate the Whigs settled on for vice president, did hold strong views. He was a firm believer in states' rights, which happened to be what most Democrats believed in. Whigs, by and large, believed in an activist, centralized federal government: high tariffs, subsidies to business, the Bank of the United States. But because Tyler would be running for vice president his views wouldn't matter. Nobody would pay any attention to them. Anyway, he wasn't put on the ticket because of his views but because he was expected to help the party in the South (and because, curiously enough, he had been a Clay supporter).[16]

It was that kind of convention. Cynical. In which nothing counted but winning. And it would be that kind of campaign.

The year 1840 was the year politics in America would change for good. The year people power, in all its glory and all its insidiousness, would triumph. Van Buren, sensing defeat but unable to think creatively, unable to respond anew, conducted himself during the election the way Jackson, and Jackson's predecessors, had. He stayed away from the limelight, refused to campaign, and ran on the issues, spending his days and nights writing long letters in which he outlined his positions on all the important topics of the day, as if nothing had changed in American politics. Van Buren, as much as any other single individual, had helped change politics, but now, in 1840, when yet more change was needed, he seemed incapable of further change. Harrison and the Whigs, in contrast, seemed to be eager to change, eager to break tradition, eager to break old taboos.

One of the things they did was, simply, to make campaigns fun. The

Democrats had experimented a little with this, beginning the practice of campaign parades. The Whigs took things to the next level. They made up slogans (Tippecanoe and Tyler, too!). They made up songs and jingles ("Van, Van is a used up man"). And on a whim that would do Madison Avenue proud, they got up a big old ball of paper and rolled it from Kentucky to Baltimore, all the while chanting, "What has caused this great commotion, motion/Our Country through?/It is the ball a rolling on,/For Tippecanoe and Tyler, too." What the Whigs were doing wasn't uplifting, but it was exciting. And people responded. It was theater, of course, but politics *is* theater. And it was perfectly in tune with the theme the Whigs came up with, the need for change. And this was a theme that really resonated with people. The Democrats after all had been in power for twelve long years. People were tired of them. Tired (a little) of Jackson, even. Tired (a lot) of bad times. Ready to try something new.

It was a very people-oriented campaign. Very democratic. There was just one problem. Harrison wasn't a man of the people as Jackson was. Unlike Jackson, who could honestly claim to have been born poor, Harrison was a rich fellow who came from rich folk. His great-great-great-grandfather had been one of the largest landowners in Virginia. His great-great-grandfather had served in the Virginia House of Burgesses. His great-grandfather had been speaker of the House of Burgesses. His grandfather had been a colonel in the colonial militia. His father had been a signer of the Declaration of Independence, speaker of the Virginia House of Delegates, and the governor of Virginia. In short he was, by anybody's definition, one of the elite. He even married rich. His wife's father had been chief justice of the New Jersey Supreme Court.

The Whigs were very much in danger of being severely hampered by their man's aristocratic past when a Van Buren newspaper came to their rescue shortly after Harrison's nomination. Anxious to prove that he was an empty shell of a man, the paper played up a damning quote from one of Clay's disappointed supporters, who supposedly had said that Harrison would no doubt be delighted to give up the presidency for a two-thousand-dollar pension and a log cabin on the banks of the Ohio. The Democrats thought it was a great jab and made the most of it. But their little plan backfired. The Whigs loved the image of their guy sitting on the front porch of a log cabin, having a drink out of a barrel. It was perfect. Just exactly the kind of image the people would find appealing. Overnight the log cabin became the symbol of the Whig Party and the campaign. Whenever the Whigs held a rally they'd drag out a little log cabin as a reminder of their party's humble candidate. Log cabin mania swept the country. Things got so out of hand that at one point Daniel Webster felt compelled to apologize for not having been born in a log cabin (though, he immediately added, his siblings had been).

Harrison himself hadn't been born in a log cabin either. He'd been born in a two-story palatial brick mansion on the banks of the James River. Currently he lived in a sixteen-room mansion in Ohio. The only log cabin he ever had anything to do with was the little log cabin his Ohio mansion had been built around years before. But truth didn't count for anything in a campaign anymore. It was all just theater. Image. Slogans. Whatever the public would buy. Politics would never be the same.

The Democrats struck back hard with the only weapon they had: an accusation with just enough truth in it to hurt, that Harrison was old. Too old. A washup. Feeble. The fact was Harrison *was* old. Sixty-seven, the oldest man who had ever run for president, and a full decade older than Van Buren. And he was, by 1840, something of a has-been, his glory days long ago behind him. It had been nearly thirty years, after all, since he had triumphed at the Battle of the Thames. Ancient history for most Americans, many of whom weren't even alive then. At the time he'd been a big shot, so big that his superior, an ambitious social climber who hoped to become president, had sabotaged Harrison's career, fearful that Harrison would eclipse him. Harrison, in response, had resigned from the service. Afterward he'd served in a variety of posts, as the governor of the territory of Indiana, as a congresman, senator, and diplomat.

In 1836 he had been selected by the Whigs as one of their four regional candidates for president. But nothing he'd done after the Thames had won him much attention; in Congress he'd worked on the unexciting issue of army pensions. And most recently he had served as clerk of the court of common pleas in Hamilton County, Ohio, where he now lived. Not exactly an impressive landing from which to jump into the presidential waters.

Nothing could be done about his unimpressive record. He was what he was. But there *was* a way he could disprove the charge that he was too old to be president, that he was feeble. He could go out among the people and give speeches, make appearances, let himself be seen and heard.

Harrison wasn't comfortable with the idea of personal campaigning. It rubbed against the bedrock principle laid down by George Washington, that the office should seek the man, not the man the office, a principle still quite full of life. Even Jackson had been against personal campaigning. Told that he could win the presidency in 1828 if he made a tour of the New England states Jackson had sworn he never would. "If I was to travel to Boston where I have been invited that would insure my election—But I would feel degraded the balance of my life." Harrison turned out to be a little less stubborn than Jackson. In June he hit the road for a three-week tour. In September he campaigned for the whole month. Only one candidate for president had ever done anything like it before: Harrison in 1836.[17]

Of course he wasn't supposed to say anything. That would undermine one of the campaign's key strategies. "Let him . . . say nothing—promise nothing," Nicholas Biddle, the former head of the Bank of the United States, had advised. "Let the use of pen and ink be wholly forbidden as if he were a mad poet in Bedlam." So Harrison went out and campaigned, hours on end, day after day, and said—virtually nothing. He even tried to have it both ways, to campaign at the same time that he denounced campaigning, just in case anybody was tempted to accuse him of office seeking. "I am not with you today, Fellow Citizens, in accordance with my own sense of propriety," he told a crowd at Chillicothe, Ohio. "Much more consonant would it be with my feelings to remain at the domestic fireside. . . . Indeed I sometimes fear that upon me will fall the responsibility of establishing a dangerous precedent." But, really, what choice did he have? "Appearing among [his] fellow citizens" was the "*only way* to disprove" the charge that he was superannuated. "You must have already perceived that I am *not* Caged, and that I am *not* the old man on crutches . . . they accuse me of being."[18]

Van Buren's only hope had been that he could persuade the people that a vote for him was really a vote for Jackson. But what he found out was that the people no longer were susceptible to that particular ploy. A vote for Van Buren in 1840 was just a vote for Van Buren. And in 1840 there weren't enough votes for Van Buren to elect him president. Harrison won in a landslide, by nearly 150,000 votes. His electoral college victory was even more stunning: 234 to 60.

It was a short-lived victory. At his inauguration Harrison caught a cold and a month later died. Tyler, the man whose views nobody had paid any attention to, became president and abruptly reversed the administration's course, adopting a states' rights program, leading to the resignation of nearly the entire Whig cabinet and an attempt to impeach him.

But the lesson the politicians learned from their experience was not that they should be more careful about who they selected as vice president. They would go on picking vice presidents strictly on the basis of their political appeal, whether the candidate was qualified or not, whether he held similar views to the man at the top of the ticket. For all that they could focus on was winning, not governing. As Hamilton had said in the *Federalist*, man by nature seizes on the immediate, often losing sight of his own true interests. The lesson, they learned, was that to win, a party had to be willing to manipulate popular opinion anyway it could—by songs, celebrity, razzle-dazzle—whatever.

Thus 1840 had been a watershed year. The year the politicians finally came to grips with people power. The year they finally gave in to it. Jackson, a creature of people power, had been able to achieve his aims without fundamentally compromising himself because of his enormous popular appeal. But he was virtually alone. All the other pols, unsure of their power, unsure the people would back them, weaved, bobbed, and caved. It had been a frightening time for politicians. Frightening because it was all so new. Adams had been so frightened he betrayed himself, resorting to corrupt bargaining to win because he lacked the means to win honestly. Eventually, however, they had figured it out, seen what worked and what didn't, and learned how to regain control. In the process they transformed American politics. The election season lengthened considerably. Campaigns grew rowdier and coarser. Party bosses assumed a place at the center of events. Compromising became endemic. Trickery became common.

Of all the many changes that took place, one stood out. Never before 1840 had a candidate put up for the presidency been run the way Harrison was. Never again would one be run any other way. After Harrison they would all be packaged, the only difference being how.

Packaging candidates was a direct result of people power. To reach the people the politicians learned they had to give them what they wanted. Give them a hero if that's what they wanted. Give them a man like themselves, a log cabin kind of candidate, if that's what they wanted. That it was unlikely that a candidate could ever be found who actually fit all the people's wants was found to be unimportant. For the politicians learned that image counted more than reality, and they could control image. Could remake virtually any candidate into the kind of man the people wanted.

It was an awesome power. It was inevitable that it would be abused.

4. Manifest Destiny

How James K. Polk lied the country into a war with Mexico in order to achieve his—and the public's—ambition to expand westward

America by midcentury was a vastly different place than it had been fifty years before. In 1800 barely five million called themselves Americans. By 1850 nearly twenty-five million did, almost five times as many. In 1800 Americans had mostly lived on the east coast. By 1850 they had begun spreading out over the entire continent. The very borders of the country had expanded immensely. In 1800 the country covered fewer than a million square miles. By 1850 nearly three times as much. So much land was acquired so quickly that, for a time, a new state came into the union every year. In 1816 Indiana, in 1817 Mississippi, in 1818 Illinois, in 1819 Alabama, in 1820 Maine, in 1821 Missouri—and on and on. In 1800 there had been just sixteen states. By the end of 1850 there were thirty-one, nearly double.

Acquiring much of the new land Americans were now living on was easy. In 1803 Jefferson had gotten the Louisiana Territory for fifteen million. In 1819 Monroe had obtained Florida for about five million. Even gaining Texas had been easy: Texans had done all the fighting. All the United States had had to do was annex the place.

Acquiring the rest of the land, however, had been hard. Britain had claimed Oregon. Mexico had possessed California and the Southwest. And neither had wanted to surrender their claims.

And then, in 1844, James Knox Polk was elected president. [1]

James Knox *Who*? was the question most Americans asked themselves when they awoke to find Polk president. For he had seemingly come from out of nowhere to capture the presidency. Until the Democratic convention, it had been assumed that the Democrats would pick

Van Buren again and the Whigs Henry Clay.* But then Van Buren had committed political suicide, coming out against the annexation of Texas, saying it would stir up sectional tensions, and make Northerners fear the South was becoming too powerful. And that had killed him in the South, of course. Leaderless, the Democrats had turned to Polk, a Jackson protégé. And Polk had gone on to beat Clay. It had been a close race. The switch of just thirty-five thousand votes would have made Clay president. But Clay had made an awful run, the worst of his career, and one of the worst in American presidential history. He just couldn't shut up. Every day it seemed he had a new announcement to make, and with every announcement he alienated more and more voters. His statements about the tariff were comically incoherent, each one contradicting the last one. Polk, shrewder, had remained quiet, breaking his silence just once to issue an ambiguous statement of his own on the tariff, a typically wishy-washy thing that let Northerners think he was sympathetic to their concerns (for protection) and let Southerners feel he was sympathetic to theirs (for free trade).

That Polk had won really should have surprised no one. As a young man he'd been a real comer. A man dedicated to politics. A man who had no interests outside of politics. At twenty-eight he'd got himself elected to the Tennessee state legislature. At thirty to a safe seat in Congress. At thirty-four to Speaker of the United States House of Representatives. Then, when it appeared the Democrats were going to lose control of the House, he went home and ran for governor. This was a gutsy move. Tennessee, despite being Jackson territory, was overwhelmingly Whig. To give up a safe seat in Congress to run for governor was, therefore, a huge gamble. But Polk had taken it. And then he had run the most sensational race any candidate in Tennessee ever had.

In a first for Tennessee, he decided to visit dozens and dozens of villages and crossroads all across the state, to bring the campaign to the people directly, so he could be seen by tens of thousands of voters. It had been a terrible ordeal. He'd had to travel by horseback. Some two thousand miles by horseback. So many miles he wore out his horse and had to get another. And at every stop, after traveling for hours, he'd had to get up and give a speech, a long speech, some lasting two or three hours. Several times he gave speeches lasting four

---

*Whig Party leaders still feared that Clay was a loser, but after the imbroglio over Tyler they needed a solid party man around whom to rally.

hours. Once he gave a speech during a driving rainstorm. He continued the speech even after the platform he was standing on collapsed beneath him. And he did all this for four straight months, taking just a single day off to rest. For a man in good health it was an incredibly arduous schedule. For Polk, whose health was always precarious—he suffered from gallstones and diarrhea—it was a killing schedule. But in the end it turned out to have been necessary: He won by fewer than twenty-five hundred votes.

Until then Polk had lived a charmed life. Everything he'd wanted, he'd gotten. But then he entered into an awful dark period, where nothing seemed to go right. When he ran for reelection two years later, he lost. Then he ran for the vice presidential nomination. And lost. Then he ran again for governor. And lost. On the eve of the Democratic convention in '44 he was all but dead politically. In worse shape than Nixon was to be when *he* made what was said to be the miracle comeback of American politics in 1968. (Nixon had lost only one race for governor.) But through grit and determination—and luck—Polk finally had triumphed, becoming at age fifty the youngest president ever.

Polk had willed himself back from the dead, but there was no guarantee he could will himself Oregon, the Southwest, and California, which he very much wanted. Great Britain and Mexico stood in his way. And they could not be controlled. In time they might be persuaded to give up their claims, for Americans were steadily moving into the area. But that could take years and years, and Polk didn't *have* years and years. Because he had promised to serve just a single term in return for the nomination, he had just four years. Desperate, he decided he had to take an extreme measure. Do what no previous president had ever dared to do—lie brazenly to the American people. (Jefferson, too, had lied—in a bid for Florida—but his lie was as nothing compared to the ones Polk was to tell.)

To gain Oregon he told a whopping lie. Said that the United States was entitled to all of it. Every last inch. Which was flatly untrue. Britain and the United States had equal claims to the territory. In recognition of this the two countries for ten years had shared ownership of the area. Polk knew it was untrue and didn't even want all of Oregon. He was willing to settle for the part below the forty-ninth parallel. But to put pressure on Britain he felt compelled to demand it all, even if in the process he had to lie to the American public about his true aims and the justice of the American demand. Predictably, Britain caved. Polk felt vindicated.

To gain the Southwest and Mexico he had to tell another whopping lie. Actually, a series of whopping lies. Lies that would involve the country in war.

In his first year in office he didn't tell any lies about Mexico. Preoccupied with filling patronage jobs—as a Democrat following a Whig he had lots of jobs to fill—he barely had time even to think about Mexico. But he did send an emissary to offer twenty-five million dollars for the land.

Twenty-five million wasn't a bad price for land that the Mexicans might eventually have to give up for nothing. But Mexico wasn't in the mood to sell. Wasn't even in the mood to receive Polk's emissary. That left Polk with just one option if he was to achieve his objective. And that was to go to war.

The trouble with the war option was that it wouldn't be unanimously popular. Far from it, in fact. Polk couldn't even be sure Congress would let him make war, the House being in the control of the Whigs. Thus, he had to contrive circumstances in such a way as to give the country no choice but to go to war. And that is precisely what he proceeded to do.

As a first step he sent a small regiment under the command of Zachary Taylor deep into territory south of Texas that had long been claimed by Mexico. Then he waited for the force to be attacked so that he could say to the public that Mexico had been the aggressor. But the Mexican army never attacked. After waiting as long as he could bear Polk finally convened a cabinet meeting on May 9, 1846, and announced that he was prepared to go to the country with a demand for war based on two grievances: that Mexico had snubbed our emissary and that it had refused to settle three million dollars in claims filed by Americans for the loss of life and property resulting from skirmishes along the border. But then, coincidentally, just after the meeting with the cabinet, Polk got the news for which he had been waiting: Two weeks earlier Taylor's troops finally *had* been attacked, some had been wounded and killed.

Polk turned out to be a very, very good liar. At first he even had the Whigs fooled. The first lie concerned where the attack took place. This was critical. If it took place on Mexican soil, the Americans would appear to be in the wrong. If on American soil, then the Mexicans would. Polk correctly identified the spot where the incident occurred as near Matamoras, on the Rio Grande. But he mischaracterized the place. He said it was plainly on American soil. This was a flat-out lie. The American border was more than a hundred miles to the north on the Nueces River. While some Texans did indeed claim that the border properly extended to the Rio Grande, there was little evidence in support of their assertion except the fact that Americans had begun settling there. But that didn't make the land American. In any case, the most Polk could accurately maintain was that the land was claimed by both the United States and Mexico. But he wasn't about to sully his war message with inconvenient details that might raise questions about his intentions. The Whigs would immediately seize on such details. So he ignored them.

The second lie, growing out of the first, was his insistence that he had not wanted war. War had come, he said, "notwithstanding all our efforts to avoid it." Of course, this was utterly untrue. After Mexico refused to sell us its land, the administration had settled on a policy of war and pursued it relentlessly. And war had come.

The third lie he told was about our war aims. What he said he wanted was justice. What in fact he wanted, of course, was their land: the Southwest and California—about half of Mexico. But Polk didn't want to admit this. Not to the public, which would have to pay for the war. Not to the soldiers, who would have to die in it. Not to the Congress, which would have to declare it. For that, too, would undermine his case, give the Whigs something to argue about. And Polk didn't want to argue. He wanted to fight. The war would be nearly a year old before Polk finally alluded, and then obliquely, to the government's goal of acquiring land.

What Polk had going for him for awhile was, simply, tradition. Because no president had ever lied to the people, not brazenly like this anyway, the people and the Congress believed they were being told the truth. For all the changes that had taken place in politics—the tricks, the compromises, the image making—presidents were still believed, still trusted. It was the essence of the republican compact. Trust. A simple, childlike thing, but without trust, self-government seemed impossible. In foreign affairs especially trust was key. For the president was the only one in the government with access to all of the available information. Eventually information would become more widely available, as Congress reviewed the documents and as news from around the country slowly filtered in to Washington. But in those critical first days and weeks the president was in complete control of information. And as such was in complete control of the situation. Not for months would the Whigs finally begin to question the war, question what Polk had told them. By then they could hardly stop it. Once the boys were in the field the politicians back home had to support them fully. An important distinction could be made between support for the men in arms and support for the war and the Whigs tried very much to make it. But they could not order the army to be pulled back. That power was Polk's. And they certainly could not cut off funds to the war. That would leave the boys at risk. Which meant, drearily, that the Whigs could really do nothing at all. The lying had worked.

Though his enemies began calling him Polk the Mendacious, Polk wasn't by nature a liar, wasn't personally dishonest. It was just that in politics honesty was becoming less important. When politics was about individuals, honesty counted for a great deal. But the individual was becoming less and less significant. And as a result, values like honesty, values that human beings cherished in their relations with one another, values that were associated above all with the individual, became less and less significant. Much as presidents didn't want to admit it, they were fast becoming part of a vast machine. And a machine, an organization, didn't have values as such. It had goals. As the head of a party and the head of a government Polk was responsible for seeing to it that those goals were attained. If personal values like honesty and decency could be maintained during the pursuit of those goals, fine. If not, they would have to be dispensed with. For what was honesty next to the acquisition of California and the Southwest?

Under the circumstances a Polk could persuade himself that he had a duty to lie, a duty to see to it that the country expanded westward, and that next to that duty a few lies were as nothing. In this case, however, it wasn't in fact true that it *was* necessary to lie. The land was becoming Americanized anyway. After the gold rush of '49 California certainly would have become American. But Polk couldn't know that for sure, no one could. No one can ever be sure how history will come out in the end. Sometimes you have to give it a nudge. If, like Polk, you are ambitious, you may be anxious to give it a giant shove. Just to be sure things happen the way you want, just to be sure you remain in control of events.

Presidents do not like to lie willy-nilly. As much as anybody they are aware of the high cost of lying in a republic. But when the reward for lying was something as grand as the acquisition of nearly one and a half million square miles of territory, lying seemed of small consequence.

Once a president could believe that it was his duty to lie, he could believe a lot of things. Moral lines that once had appeared clear now could appear fuzzy. Formerly, for instance, it was pretty clear that a president's first responsibility was to the country and only afterward to his party. But as the party became an intrinsic part of government, this line blurred. A president could no longer pretend that he was above party, could no longer think, as George Washington had, that parties were bad. Parties were essential to democracy. As such, a president could convince himself that by helping his party he was helping democracy (and thereby helping the country). It wouldn't really be true. But a president could think it was, especially in wartime, if the party out of power happened to be opposed to the war that the president and his party were behind.

Polk's every inclination was to be a fierce partisan, so fierce that he even let friends use government funds to help establish a new party newspaper, an outrageous abuse. He owed everything to the party and to party leaders like Jackson. But for Jackson, his career would have been over after the Whigs finished him off in his two failed races for governor. Thus, it was no surprise really that when the war began he turned to his own party ranks to fill out the leadership positions that became available in the army. Generalships, brigadier generalships, and such. Congress created some thirty positions of high rank. Polk, to the horror of the Whigs, put Democrats in nearly every one (among them Franklin Pierce, the Democratic Party boss of New Hampshire), politicizing, as no one before him ever had, the nation's army.

He had just one frustration. And it was like a burr under his skin. While he could name Democrats to the positions opened up by Congress, the army remained under the command of two regular army officers: Winfield Scott and Zachary Taylor. And neither of them were Democrats. Worse, Scott was an avid Whig. That meant, irony of ironies, that the Whigs would benefit from Polk's war. For when the war was won (as surely it would be; there was never any doubt that the United States would triumph), Scott would be in a natural position to run for president. Americans liked to run war heroes as presidents. First there had been Washington. Then, Jackson. Then, Harrison. Scott would be next.

Polk was very angry about the possibility of his creating a hero for the Whigs. It was as if he would be betraying his own party. Determined to avoid this, he early on decided to give Taylor, whose party sympathies were unknown, a larger role in the war than Scott. Then, to his consternation, Taylor was revealed to be a Whig, too, and just as he was beginning to win on the battlefield. More fearful now of Taylor than Scott, Polk reversed course and turned over the main body of the army to Scott, leaving Taylor with just five thousand men. Inadvertently this *made* Taylor. For soon after Taylor, though outnumbered four to one, managed to stage a smashing victory over the Mexican army at Buena Vista. Overnight he became a national sensation and the obvious Whig choice for president at the next election. Polk, disconsolate, tried to outmaneuver the Whigs, asking Congress to create a new rank, that of lieutenant general, which he planned to fill with Democrat Thomas Hart Benton. The position would make Benton the highest ranking officer in the army, giving him a shot at hero status. But the Congress refused. Benton remained in the Senate. Taylor went on to win the Whig nomination for president. On inauguration day in 1849, Polk stood and watched as Taylor was sworn in as his successor.

The politicization of foreign policy was an old story in American politics, going back as far as the John Adams administration. And once the secretary of state's job began to be seen as a stepping stone to the presidency—Jefferson, Madison, Monroe, J. Q. Adams, and Van Buren had all served in the position—foreign policy had become intimately tied up with presidential politics. Of all the offices in the cabinet it was, although it had the least to do with daily politics, the most political.

Polk, ironically, had tried to take the cabinet out of politics, including the State Department, so that the administration wouldn't be sidetracked by political backbiting and so that he would be the only one with a political agenda. To that end he had demanded that each cabinet officer promise in writing not to make a run for the presidency in 1848. But James Buchanan, his choice for secretary of state, refused to sign. "I do not know that I shall ever desire to be a candidate for the Presidency," Buchanan insincerely claimed, but "I cannot proclaim to the world that in no contingency shall I be a candidate for the Presidency in 1848." Polk let him into the cabinet anyway. Buchanan was needed to secure Pennsylvania's support in Congress (Buchanan was from Pennsylvania).

Polk's lies had been new, as had been his politicization of the army. New and disturbing. But also inevitable. The more complicated the country became, the higher the stakes became. A small country could afford to have an honest president. A large and growing country often couldn't. Government was too important now to be hampered by a faithful adherence to the genteel standards of the eighteenth century. Sometimes now presidents would feel compelled to lie. Lie, and they could keep control of events. Tell the truth, and . . . well, sometimes things would work out, but sometimes they wouldn't.

Polk was followed by Zachary Taylor and Millard Fillmore, neither of whom was very memorable as president. Taylor died in office before he had a chance to leave a mark. Fillmore, his successor, was an inconsequential New York politician who lacked national clout; his chief contribution was to agree to the Compromise of 1850, which had been worked out by others, notably Stephen Douglas and Clay, and which Taylor had blocked.

Then came Franklin Pierce. Pierce was remarkable not for who he was or what he accomplished, for he achieved very little. But he was interesting as a representative of a new type, the politician for whom politics was everything.

Politics in the past, while demanding, had not been all-consuming. Often a politician for state office, for instance, could secure victory simply by providing free booze on election day to a couple of hundred voters. But as people power had expanded, the demands on politicians had become greater and greater. Constituents now had to be attended to. Political parades had to be organized. Even a politician's evenings were now taken up with politics. For it was often only in the evenings that voters could be reached at home or at their fraternal lodges. The party, with its thousands of worker bees, was helpful in keeping voters apprised of political developments, easing the politician's need to meet with each and every constituent on a regular basis. But keeping the party going—keeping the worker bees happy—was a monumental task in itself, exacting awful personal sacrifices.

It was not, ironically, the politicians themselves who suffered the most from the new obsessiveness required for success in politics. It was their families. The families of politicians had always suffered. Now they suffered more.

Franklin Pierce's suffered most.

5. The Story of Franklin Pierce

How a minor politician managed, through hard work, to make himself president at his family's expense

---

Franklin Pierce's presidency is utterly inconsequential and barely worth noticing. But as a type Pierce the man is both representative and interesting, for he is among the first politicians to have to choose between his family and politics.

He accomplished very little in his lifetime. Even his own (sympathetic) biographer admits that not a single achievement can be credited to his administration. At the end of his single term in the White House hardly anyone in the country desired him to run for a second. During the Civil War he let it be known that he objected to the course of Abraham Lincoln and as a result became a pariah. For nearly a hundred years after his death not even his home state of New Hampshire cared to erect a monument to his memory, even though he was the only president the state ever produced. Today many Americans have never even heard of him.[1]

But Franklin Pierce *was* likable—and always had been.

Shortly after Pierce was nominated for president in 1852 he received a touching letter from a barely literate woman with whom he had played as a child. One day, it seems, when he was a boy he had "meet with a little misfortune by falling into the River." Franklin, hoping he could keep the incident secret from his parents, had apparently then gotten the woman's mother to dry out his clothes and iron them. It seems her mother would have agreed to do almost anything for the young man:

"My mother said you was so agreeable in conversation that day she almost fell in love with you."

Among Pierce's qualities—his intelligence, his eloquence, his quickness—the most impressive was his ability to make people fall in love with him. By itself, of course, this ability might not have led to great prominence. But it was married to a burning political ambition he inherited from his father, Benjamin Pierce, a Revolutionary War hero who rose to become governor of New Hampshire. Building on his father's connections, Franklin enjoyed political success at an astonishingly early age. At twenty-four, just a few years out of college and training to be a lawyer, he was elected chairman of the town meeting in Hillsborough, New Hampshire, his hometown. At twenty-five, at a younger age than almost anybody else in state history, he was elected to a seat in the Great and General Court (the state legislature). At twenty-six, at a younger age than *anyone* else in state history, he was elected speaker of the court. At twenty-eight, at a younger age than almost anybody in American history, he was elected to Congress. "Frank Pierce," the local newspaper declared, "is the most popular man of his age." And so he seemed.

Popular but not entirely happy. The man whom everybody loved couldn't find a woman whom *he* loved romantically—or who loved *him* romantically. Following each election victory, after celebrating with friends and family, he would go home at night to an empty room. As the years went by, he grew more and more pessimistic that he would ever go home to a wife.

He might actually not have but for Jane Appleton. The year he ran for Congress she fell in love with him and he with her.

Both came of prominent families. But they were prominent in different fields. The Pierces were leaders in politics, the Appletons in the intellectual world and in business; Jane's father had been the president of Bowdoin College, where Pierce had earned his degree. Other Appletons had helped found the state's textile industry. Being prominent in different fields didn't matter nearly as much as the fact that they were prominent at all; in a small state like New Hampshire the leaders all knew one another. But Franklin and Jane were different not just in background, but as types—while he was outgoing and socially adept, she was shy and retiring; while he was largely indifferent to religion, she was openly devout; while he liked to drink, she was a teetotaler— and *these* differences mattered.

One difference mattered more than all the rest: He loved politics. She loathed politics.

In 1834, both twenty-eight, they married. They remained married some thirty years.

It was hell from the beginning. Jane hadn't simply married a man. She had married a politician. What that meant was to become clear to her from the very first day of their marriage. Any ordinary couple after their wedding would have taken a honeymoon. The day after *their* wedding, on Wednesday November 19, 1834, Jane and Franklin boarded a train to head for Washington, D.C., so that Franklin could begin his work as a U.S. representative. Three days later, on Saturday, they arrived in the capital. That left them one day to themselves, Sunday. Come Monday, Franklin had to go to work.

Their first home was a Washington boardinghouse filled with strangers. Jane, uneasy in a new city, remained cooped up in their small room most days. A woman of her time, she had wanted to set up house. But as Franklin Pierce's wife, a politician's wife, she could not. At night instead of serving her husband a hot meal at home before a cozy fire, she would descend with him to the main floor to eat dinner at a big wooden table along with the other guests of the boardinghouse, whom Jane, out of necessity, began calling "her family." Sometimes the Pierces would be asked to have dinner with government leaders. Jane usually objected. "We have an invitation to dinner to Gov [Lewis] Cass'* on Wednesday," she noted in a letter home, "which is accepted notwithstanding my predilections." Her main pleasure, she confessed, was a daily trip to church, and some days she couldn't even enjoy that; it would be too cold or windy to venture out. "I am disappointed," she wrote on one such occasion, "in my wish of going to church which is a deprivation to which I hope I shall not often be subjected." But, "these high winds are common here, and exceedingly disagreeable."

Somehow they muddled through, Jane coping as best as she could. But the following year, when Pierce came back to the capital, he came back alone.

They both thought it would be better if Jane stayed in Hillsborough while he was in Washington. Not only because she didn't like Washington, but because she was pregnant. But difficult as their life in the capital had been when they were together, it was worse being sepa-

---

*Cass, a native of New Hampshire, had been a hero in the War of 1812. Later he was appointed governor of the Territory of Michigan. He was by this time serving as secretary of war under Andrew Jackson.

rated. In January 1836, while Pierce was in Washington helping a constituent win a pension, his wife was at home giving birth to their first child. A few days later, the infant died. Pierce was still in Washington. Jane was alone. In her grief she began to hate politics more than ever.

Both of them felt the loss of their infant greatly, and felt the gulf that was widening between them greatly, too. Franklin began to drink more than he should, a lot more, so much that behind his back some people were beginning to whisper, *Frank Pierce is a drunk*. In a society of heavy drinkers—Americans then drank more heavily than almost anybody else—Pierce was one of the heaviest drinkers of all.

Jane, for her part, began to show signs of physical frailty and emotional turmoil. The summer after the loss of their baby she came down with an unspecified illness—probably tuberculosis—and sank into a deep depression. A doctor, adopting a conventional remedy, applied leeches to her veins to suck out the "bad blood." (Afterward he claimed it had worked.)

While the doctor was nursing Jane back to health, her husband, again in Washington, was nursing his political career. Her treatments lasted all summer. All summer Pierce remained in Washington.

Pierce wanted his marriage to work, wanted Jane to be happy. But he wanted a successful political career more. Blinded by his ambition, he focused on his career to the neglect of his family. His wife, ill and melancholy, simply went along, suffering largely in silence. Their marriage, in consequence, began to grow brittle.

As their troubles worsened, so did Jane's health. The periods when she was healthy always seemed shorter than those when she was ill. As time passed, the short periods when she was in good health grew even shorter.

The following winter, when Pierce again had to return to Washington, Jane went with him. Almost immediately she came down with a bad cold—a sure indication, it was now becoming obvious, of discontent. She spent most of the winter in bed. Friends commented that she was "wanting in cheerfulness."

Jane's pain was the greatest, but she was not alone in having to suffer for the sake of Pierce's ambition. So did other members of his family. His sister Harriet became gravely ill that winter; Pierce, engrossed in his political duties, declined to leave Washington to go to her aid. Nor did he leave even when word came that his father, suffering a devastating stroke, had fallen ill. Pierce was too busy building his political career.

It was a career that happened to be soaring. As he had once dazzled New Hampshire, he now dazzled Washington. In the winter of 1837, the New Hampshire legislature, impressed by his increasing reputation, elected him to the U.S. Senate. On March 4, just before Martin Van Buren gave his inaugural address, Pierce was sworn into office. At thirty-two he was the youngest member of the Senate.

The Pierces spent their third wedding anniversary at home in New Hampshire. But in December, Franklin returned to Washington to attend his first session as a senator. His wife went with him. Once again she fell ill.

Until now she had suffered quietly, but she was to do so no longer. Three years of life with Franklin, three years of almost unrelenting illness and melancholy, had taken their toll. Although stoical by nature, she found she had to tell someone that she didn't much care for her husband's astonishing political career, and in the spring of 1838 she finally did. "Oh," she confided to a friend, "how I wish he was out of political life!" She did not say that his leaving would be better for *her*, though obviously it would be. She insisted it would be better for *him*. She was alarmed, she wrote, that he had become involved, if only tangentially, in a duel involving a congressman.* But she may also have been worried about his drinking. Ever more loudly now could be heard the gossip, *Frank Pierce is a drunk.*

Their marriage, already cracking, now apparently began to come apart. Although they had two children in quick succession—Frank Robert in 1839 and Bennie in 1841—Mrs. Pierce went beyond complaining to her friends about his political career. She now complained directly to him. Finally, in the spring of 1841, around the time Bennie was born, she demanded that he resign from the Senate. He agreed, promising to give up a career in politics forever. But he refused to go quickly. For nearly a year he held off making the announcement. Then, in February 1842, at thirty-seven, he finally left, one year before the end of his six-year term.

Jane Pierce was hopeful now that she finally would have a normal life. And for five months she did. For five months Pierce devoted himself to the family and the law. For five months he did not make a sin-

---

*Pierce had helped prepare the instructions that would guide the duelists; the duel ended in the death of his friend and fellow Bowdoin classmate, Maine representative Jonathan Cilley, who had made a speech accusing a Whig editor of accepting bribes from the Bank of the United States.

gle speech. For five months he barely took notice of politics. And for five months, at her insistence, he even stopped drinking. Never had she been happier. For once they had a real home.

Those five months were, however, to be the only five months during their entire marriage when Pierce would be out of the public spotlight. For no sooner did summer arrive than he jumped back in front of it. While he had promised to give up a political career, he had not promised to give up politics altogether. That June he was elected head of the New Hampshire State Democratic Party. This post at least kept him home. But as a party boss Pierce was to become involved daily in back-room machinations to an extent he had never had to before. Evenings, during the months when the annual town meetings were held or the legislature was in session, Pierce would be completely consumed by political work. Involved as he had been before in politics, there now came long stretches during which he seemed even more so. In early 1843 he spent two weeks traveling around the state giving speeches. The following year, after again being elected party chairman, he made even more speeches. A year later he was elected chairman yet again. Once again he devoted himself to speechmaking and politicking.

In the summer of 1845 Pierce, after three terms as party chairman, tired of the work and breaking under the strain, declined to run for a fourth. No sooner had he quit, however, than another opportunity arose. In early fall the governor offered him the chance to return to the Senate to fill out the unexpired term of Levi Woodbury, who'd resigned to take a seat on the U.S. Supreme Court. Pierce, however, refused. A short while later President Polk offered to make him attorney general. Pierce again refused. Because of the promise he had made to his wife, he could not make politics his career.

Fortunately, he had a new ambition—to join the war against Mexico. And his wife hadn't made him promise to forgo military service.

Pierces had a history of military prowess and an attraction to war. When the British fired on the colonial patriots on Lexington Green in 1775, marking the outbreak of the American Revolution, eighteen-year-old Benjamin Pierce—then busy plowing a field—had instantly "stepped between the cattle, dropped the chains from the plough, and without any further ceremony, shouldered my uncle's fowling piece, swung the bullet-pouch and powder-horn and hastened to the place where the first blood had been spilled." A short time later Benjamin fought in the Battle of Bunker Hill.

Pierces loved military glory. Jane, an Appleton, did not. Appletons devoted themselves fervently to education, business, and religion—especially religion. Her own father was said to have died of starvation after undertaking a religious fast. But Jane was not the one deciding whether to go to war or not. Franklin was. Franklin, who grew up listening to his father tell war stories and yearned to have stories of his own to tell.

If Jane had been asked, it's plain she would have objected to Franklin's joining the fight—he was forty-two years old, after all, and had a family to support, a family that needed him now more than ever. Just three years earlier little Frank Robert had died after a bout of typhus fever, leaving Jane with just one last child, Bennie—but she *wasn't* asked. For a full year Pierce angled to get himself into the war as an officer. And for a full year he kept his wife in the dark about his efforts. He continued to keep her in the dark even after he was offered a commission as an infantry colonel.

In February 1847 he was told that he would probably be given command of a brigade with the rank of brigadier general—and then was given it. Even *then* he did not tell his wife what he was up to.

She alone, it would appear, was the only one he didn't tell. His correspondence at this time was full of his plans. People all over the country knew that he expected to get into the war at any moment: friends, congressmen, War Department bureaucrats. They all knew; only Jane didn't. "My purpose is fixed," he wrote Rep. Edmund Burke in February 1847 of his decision to join the war, "although I have not yet broached the subject with my wife."

In May, after accepting a commission as a brigadier general, he left for the front, following a two-month delay owing to the difficulty of raising a brigade in New England, where the war was unpopular.* Jane, devastated by his decision to join the war, slipped into a profound depression and had to be put in the care of a friend.

Pierce still loved his wife; in a touching letter he expressed what undoubtedly were his sincere feelings: "I can not trust myself to talk of this departure," he wrote. "My heart is with my own dear wife and boy, and will be wherever duty may lead my steps."

---

*From the start the war had been most popular in the South and the West. New Englanders were suspicious that it had been foisted on the country by slaveholders eager to expand the "peculiar institution" westward.

But he loved glory more.

Because the short war was nearly over by the time Pierce reached the front, he had little time in which to achieve glory. On the eve of what appeared to be the last major engagement of the war, his contingent briefly came under enemy fire, causing his horse to rear. Pierce was thrown and knocked unconscious. When he awoke he discovered that his knee had been badly damaged. The next day Gen. Winfield Scott relieved Pierce of his command. Pierce immediately appealed the order. "For God's sake, General," he pleaded. "This is the last great battle and I must lead the brigade." Scott relented.

The next day the Battle of Churubusco commenced. Pierce at last had a crack at gaining the military glory he craved. Unfortunately, just as the battle got under way, he injured his knee again and fell off his horse. He was left on the field (at his own insistence), and the battle raged around him. At battle's end Pierce was alive. His men forever after remembered him for his bravery. Shortly afterward the war ended.

Pierce returned to New Hampshire in time for the beginning of the presidential election of 1848. But it was not Brig. Gen. Franklin Pierce who emerged as a contender that year, but another general from the war, Zachary Taylor.

At this point the odds that Franklin Pierce would ever be considered a serious contender for the presidency appeared remote—as even Pierce himself apparently realized. When a friend tried to advance his name in the presidential sweepstakes, Pierce made no effort to help. The boomlet died.

But even had he had a chance at winning the nomination, Pierce could not go after it. He could not even run for governor of his home state, although at this point, as a war veteran, he certainly could have been elected. For he was still bound by *that* promise—the promise he had made to his wife six years earlier—the promise that had forced him to resign from the Senate in 1842 and that had subsequently forced him to decline appointment to the Senate in 1845, the same promise that had stopped him from accepting President Polk's invitation to become attorney general: the promise to forgo a career in politics.

The story of Franklin Pierce in these years is the story of a man at war with himself: a born politician who could not be a politician. Every day, it sometimes seemed, he was being importuned to run for office. Pierce himself yearned to run. And yet he could not—because of the promise.

He was not just a born politician but the son of one. Like many sons, he had wanted to follow in his father's footsteps. Benjamin Pierce had become one of the most beloved figures in New Hampshire history, serving as governor not once but twice, in a state that usually preferred to rotate its governors in and out of office after serving a single term. But Franklin Pierce would never have an opportunity to match his father's popularity in New Hampshire—because of the promise.

If only he could take back the promise—but he could not. Jane wouldn't let him. She had made him make it, and she was determined to make him keep it.

He had already kept it for seven years. For three more years he continued to keep it.

Devoting most of his time to the law, Pierce for the next three years operated largely behind the scenes in New Hampshire politics. He didn't even run for party chairman.

Although he held no formal position in the party, he remained, of course, a powerful figure. His biographer believes that Pierce may have been more powerful at this period of his life, because of the prestige earned during his service in the war, than at any time previously. Evidence of his power is abundant. At one point Pierce actually wrested the Democratic nomination for governor away from one candidate, whom he believed was flirting with the Free-Soilers, and gave it to another.*

But if Pierce was powerful enough to make someone else governor, he could not make himself governor. And all because of that promise.

For ten years he kept that promise despite having all manner of reasons for breaking it. And then the eleventh year came. And that year he found it more difficult than ever to keep it.

Levi Woodbury had died.

The eleventh year, 1852, was a presidential election year. No New Englander had had a chance of becoming president for years. The last one who had, John Quincy Adams, had been elected in 1825, and he had won only because the election had gone to the House of Representatives; Andrew Jackson had won more popular and electoral votes. Thus, the last New Englander to win the presidency through the

---

*The Free-Soil Party, organized in 1848, opposed the extension of slavery. Its platform declared bluntly, "[N]o more slave states and no more slave territories."

regular election process was actually *John* Adams, who had won in 1796—and he was the only one who ever had.

But in 1852 a New Englander seemed almost destined to win the Democratic Party's nomination. The most serious threat to the party that year was presented by the Free-Soilers, who were attracting huge numbers of Northern Democrats. To win, the Democrats had to find a way to keep these Northerners from defecting. That eliminated the nomination of a Southerner, for any Southerner, whether proslavery or not, would almost certainly drive Northern antislavery supporters away. A Westerner might be able to retain the Northern support, but the party had nominated a Westerner at the last election—Lewis Cass—and he had lost. The Middle States might plausibly provide a winning candidate, but since Free-Soilers were strongest in New England, it made the most sense to nominate a New Englander.

On the eve of the election one New Englander above all others seemed to have a lock on the nomination: Levi Woodbury, a former New Hampshire senator who had been appointed to the U.S. Supreme Court. But then Woodbury died.

Rarely before had there been such a golden opportunity for a New Englander to become president. Never again, perhaps, in Pierce's lifetime, would there be another such opportunity. If he was to become president of the United States it would probably be in 1852—or never.

If there were any doubts that this was his year, the state Democratic convention held in January was to erase them. Just a week into the new year, the party endorsed Pierce as its favorite son for the national convention to be held in Baltimore six months later. That action immediately put Pierce's name into play. From then on he would be one of the politicians who would always be named when lists—the long lists—of possible presidential candidates were drawn up. But his was not an impossible dream. Of the five presidents elected in the past twenty-five years, three had been war heroes: Andrew Jackson, William Henry Harrison, and Zachary Taylor. Almost certainly the next president would be a war hero, too. And Pierce, in his own limited way, qualified.

For a time Pierce put off making a decision. After the state convention he actually issued a statement that seemed to indicate he did not want to be president. The use of his name "before the Democratic National Convention at Baltimore," he wrote, "would be utterly repugnant to [my] tastes and wishes." But nobody took his protestation too seriously. All candidates for the presidency claimed that they didn't want to be president—that, it was widely recognized, was the only way to *become* president.

Still Pierce had to decide—and he could not. Each choice had its merits. Stay out of the race, and he could remain true to his wife and to his promise. Stay out, and he could avoid the horrific strains that inevitably would accompany a race for the presidency. New Hampshire politics had proved too great a strain when he was party chairman, and he had quit; national politics undoubtedly would pose even a greater strain. Stay out, and his wife would probably retain the fragile grasp on good health she had been able to maintain during the last couple of years.

All these factors added up to a powerful argument against running. But compelling as the argument against running was, the argument for running seemed even more so. Run, and Pierce could at last return to the career he loved best—and was best at. Run, and he could show the world—and himself—what Pierces were made of. Run, and he could satisfy the deep longing he felt for political recognition and success.

As winter turned to spring Pierce finally made up his mind. He *would* run. His ten-year absence from elective politics was at an end. Early in April, using the coded language presidential candidates then employed, he told his friends, who had been pressuring him to run, that *they* should decide "what is my duty and what may be the best interest of the party." As *they* had already declared that he should run, Pierce's statement was tantamount to an open declaration.

The promise that had sabotaged his Senate career and stopped him from running for governor and accepting the offer to become attorney general—the promise that had haunted him for ten long years—would not now stop him from running for president. The promise he had kept all those years he now was prepared to break.

What of Jane, to whom he had made the promise? What did she think of his decision?

Jane didn't express an opinion because Pierce didn't tell her about it. As earlier he had concealed from her his decision to fight in the Mexican War, he now concealed his decision to run for president of the United States. Once again he told friends and politicians what he was up to. Once again he declined to tell his own wife.

Over the next two months Pierce was busy almost daily promoting his candidacy. Although he left the actual maneuvering required to advance his cause up to his friends—it was their job to confer with key constituencies in other parts of the country, to make political overtures to disaffected groups, to puff up Pierce in the press—to keep the effort going Pierce himself had to stay in constant touch, writing letters almost daily and occasionally even arranging to meet supporters. In April there was one such meeting, in May three; on one of these occasions he traveled to Boston to visit with delegates from Massachusetts and Maine.

It was, of political necessity, a stealth campaign. As a minor politician with little national recognition, Pierce had no hope of competing with the major contenders and didn't want to try. It was in his interest not to appear to be a serious candidate at this early date. By remaining in the background he could escape the brutal punches the main candidates—James Buchanan, Stephen Douglas, and Lewis Cass—were throwing at one another. With a little luck, they might bloody one another so badly the Democrats would have to turn to somebody like Pierce as a compromise choice. (And Pierce was the only New Englander of the bunch.)

By mid-May, his managers reported, the strategy seemed to be working. "His name is familiarly talked of by the members as the man upon whom all could unite," one of them wrote. "[And] he has not been thrust forward in a manner to obtain the enmity of the friends of other candidates or any clique of the party." Some of the managers had worried that the campaign was being conducted so quietly that the convention delegates wouldn't consider Pierce visible enough to warrant the nomination. But while that might have been the case two months earlier, when the campaign got under way, it no longer was. "I am now satisfied," the manager wrote, that "he must be in the minds of all."

His candidacy *was* in the minds of all—all, that is, except Jane, his wife. She still knew nothing.

Jane knew, of course, that people were talking him up for president. It was in all the local papers. One day a friend, running into her, even jokingly addressed her as the "future Presidentess." But Jane believed it was just talk. Her husband had assured her that he didn't want the nomination—and told her he was making no efforts to secure it.

That, of course, was a lie. But Jane didn't know it.

Jane and Franklin were at home in New Hampshire when the Democratic National Convention was gaveled to order on Tuesday, June 1, to nominate a presidential candidate for the election of 1852. Tuesday and Wednesday the delegates fussed over the rules, approving one that required the winning candidate to have a two-thirds majority—a rule that this year almost guaranteed a deadlocked convention. On Thursday the balloting began. On the first ballot Cass was first, then Buchanan, then Douglas. On succeeding votes Cass initially maintained his edge, but as the day proceeded he slipped back and Buchanan pulled ahead. But Buchanan, like Cass, couldn't seem to win a two-thirds majority. After seventeen ballots the frustrated delegates finally went home for the day.

On Friday there were sixteen more ballots. The day began with Buchanan ahead, Douglas gaining, and Cass nearly extinct. Then Douglas took the lead. But he, too, was unable to command a two-thirds majority. As the day wore on, Cass's candidacy suddenly revived. At the end of Friday he remained in the lead. Perhaps he would win the Democratic nomination in 1852 as he had in 1848.

Pierce, still in New Hampshire, was following the news at the local telegraph office in Concord. His name hadn't even come up in the balloting, but he was optimistic. Just as his managers had predicted, the convention was deadlocking. The longer the deadlock lasted, the greater his chance of winning. If the delegates didn't reach a decision soon, they would grow weary of voting and turn to a fresh face—his, he hoped.

On Friday afternoon, as Cass's forces were staging a comeback, Franklin and Jane traveled to Boston for the weekend. On Saturday morning, the convention still deadlocked, they went for a drive in Cambridge. Jane was relaxed; Franklin pretended to be.

Shortly after they left, things at the convention suddenly took a dramatic new turn. On the thirty-fifth ballot, Virginia, egged on by Pierce's managers, who had persuaded the state's delegates that Pierce was a friend of the South—as good a friend as they were likely to find in the North—nominated Pierce for president. That didn't immediately start a stampede in his direction. But over the next few hours Pierce came to seem more and more inevitable: As a Northerner who had supported the Wilmot Proviso, which would outlaw the introduction of slavery into the territories newly acquired from Mexico, he could help keep the antislavery Democrats loyal to the party; as a war hero—of sorts—he could attract general support; as a firm believer in the Compromise of 1850 he could hold the South.* Supporters in the other camps moaned that

---

*The Compromise of 1850 provided for the admission of California as a free state and abolished the slave trade in Washington, D.C., but Southerners enthusiastically backed the deal because it required Northerners to return fugitive slaves.

Pierce wasn't sufficiently well known, but ballot after ballot he slowly gained strength. Just before the forty-ninth ballot took place, a delegate from North Carolina and a Pierce supporter stood up to give the kind of stem-winding speech for which the convention had been yearning. Minutes later the delegates, bone weary after five days of talking and voting, nominated Pierce as their candidate to be the fourteenth president of the United States.

Just what Pierce had hoped might happen had happened. But he didn't yet know: He and Jane were still out on their morning drive.

And then at noon, as they were leisurely taking their carriage through the woods outside Cambridge, a rider on a horse suddenly appeared, making straight for them at a furious pace. It was a Colonel Barnes. *Sir, there's news,* said Barnes. *The Democrats have nominated you for president!*

Pierce smiled broadly and shouted with joy in delight. Jane nearly fainted.

His dream was her nightmare. That weekend, her nerves fraying, she plunged into depression. A few days later she received a note from son Bennie, who echoed her own concerns. "I hope he won't be elected," the eleven-year-old wrote, "for I should not like to be at Washington." "And," he added, "I know you would not be either."

Andrew Jackson's wife, Rachel, had prayed that her husband would not be elected president. Zachary Taylor's wife, Peggy, had prayed that hers wouldn't. Now Pierce's wife prayed that *hers* wouldn't.

There was never really any chance that Pierce would lose. He had strength in every section of the country; the Whig candidate, Mexican War hero Gen. Winfield Scott, because of the growing weakness of the Whig Party, had little strength anywhere. Within four years the Whig Party would disappear, to be replaced by the Republican Party. A third candidate, nominated by the Free-Soilers, barely attracted any votes. In November, Pierce triumphed handily, winning 254 electoral votes to Scott's 42.

Nor was there ever really any chance that Jane would suddenly approve of her husband's political career. Her hatred of politics ran bone deep. No matter what he said or did now, he and she were destined to be at odds—and she was destined to turn melancholy. Nothing that had happened since his nomination had changed her feeling of gloom. While she tried hard to be cheerful, she wasn't and couldn't pretend to be. Life was all politics now, politics, politics, nothing but politics, damn politics. And she hated politics.

Little in Jane's life had gone as she had wished. Her first child had died in infancy. Her second, Frank Robert, had died at age four. All she had now was Bennie, dear little Bennie, her one source of joy.

And then one day in early January, as she and Franklin and Bennie were riding a train out of Boston, their car suddenly derailed, tumbling down a hill. She and the president-elect survived without injury. But Bennie, right before her eyes, flew through the air and became mangled in the train's wreckage. He died instantly.

There had never really been any chance that Jane would be happy as First Lady—but there had been a chance that she would at least be able to endure as she had endured her many other troubles. But after Bennie's death there was no chance of this at all. Merely retaining her sanity would be difficult.

Deeply religious, she soon found a way to take some comfort from Bennie's death. God, she came to believe, had taken Bennie away so that her husband could focus on his duties as president without distraction.

On this thin thread—this fragile thread—hung her hopes of remaining sane. And then this thread, already pulled taut, was to be stretched to the breaking point—and then beyond.

On March 2, just two days before Pierce was to be sworn in as president, Jane had a chance meeting in Baltimore with a cousin, Charles Atherton, the senator-elect from New Hampshire. At this meeting Jane discovered the secret her husband had kept from her for the past year. Not only had he wanted to be president, he had actively worked to become president.

Pierce had tried to keep the truth from his wife for one full year. And for one full year he had. Now, almost literally on the eve of his inauguration, she had found out the truth: The past year had been a lie. The office hadn't sought Franklin, he had sought the office. When he had said he was surprised to be nominated, he wasn't surprised at all. Behind her back he had been scheming to be elected president.

Most important, perhaps, he had broken his promise.

As she began to absorb this vital news, Jane, whose grip on reality had grown weak since the death of her last son, began to slip into a world of her own. First she had lost Bennie. Now she had lost her trust in Franklin. It was all too much. Never again would she be able to function as she had before. The truth was, she would barely be able to function at all.

On March 4 Franklin Pierce was inaugurated as the fourteenth president of the United States. His wife, ill, angry, and incoherent, refused to attend. Not until the end of the month did she finally agree to move into the White House.

By custom she was now First Lady, although the title would not come into use for another four months. But Jane was always to remain indifferent to the duties of a president's wife. A pitiably shy woman by nature, she had never had much of an interest in socializing. Now, her shyness accentuated by the tragedies that had befallen her, she walled herself off from the outside world, choosing to remain secluded during the four years of her husband's presidency in the family living quarters on the second floor of the White House. During the first two years she joined her husband in public on just one occasion: New Year's Day, 1855. Thereafter she appeared only occasionally. In her absence an aunt took over most of the traditional duties of the First Lady, aided on occasion by Varina Davis, wife of Secretary of War Jefferson Davis.

Since Frank Robert had died Jane had lived her life for Bennie. With Bennie dead, she seems to have had no interest in the living at all. Bennie and only Bennie seemed of interest to her. As the years went by she became obsessed with his memory, idly passing her days in loving communion with his dead spirit, repeatedly twisting a locket of his hair in her fingers or writing him long, loving letters—long, pathetic letters. Several times she even held seances in the White House in an attempt to get in touch with him.* Bitter and disillusioned, she came to blame Bennie's death on her husband, apparently concluding that God had taken away Bennie to punish Franklin for his lies.

What of Franklin, the man she no longer trusted?

He no longer trusted himself. Guilty about the lies he had told, devastated by Bennie's death, and unnerved by Jane's retreat into unreality, he became a shriveled presence in Washington, unable to exercise leadership. To an extent never seen in Washington before or since, his

---

*In mid-nineteenth-century America, seances were becoming popular. Mary Todd Lincoln was to hold at least one, maybe more, in the White House when she was First Lady. Like Mrs. Pierce, Mrs. Lincoln lost several children.

cabinet ran the government; Pierce came to be seen as largely inconsequential.

The presidency has been known to break several men. Pierce arrived in office already broken. Not long after he began his term there once again could be heard—whether with good reason or not, it is difficult to say—the old whisper, *Franklin Pierce is a drunk.*

A likable man from youth, Pierce remained likable throughout his presidency. Likable, he aroused little hatred. If anything, because of the tragic death of little Bennie, Americans seemed to feel real warmth for him. Few presidents had ever seemed so human, so vulnerable.

But as the country's troubles over slavery began to accumulate and grow more intense—these were the years, after all, when Kansas began to bleed, as pro- and antislavery factions came to blows—Pierce came to be regarded as a man who was ill suited for the high office to which he had been elected. As the slavery crisis grew worse, Americans looked forward to the day when Pierce would leave.

So, too, did Jane.

Pierce—over Jane's objections—made a pathetic attempt to win a second term, but was, of course, unsuccessful.

No one in the United States was happier at that than his wife. At last they would be through with politics. Not because he wanted to be through with it, but because the politicians and the people were through with him.

6. The Slavery Crisis

How the fight over slavery divided the country and vastly complicated politics, forcing a desperate James Buchanan to countenance extraordinary measures to retain control, including meddling in the affairs of the Supreme Court and allegedly using bribery to affect the course of legislation in Congress

Nobody had expected much of Pierce. But James Buchanan was different. Buchanan had credentials, classic presidential credentials. He'd been a congressman, senator, minister to England, and secretary of state. He was the kind of man Americans used to elect president before people power, before they began turning to splashy war heroes or mere vote getters. His very bearing was presidential. A man obviously accustomed to responsibility and power.

True, he was old by the time of his election in 1856, sixty-five, but that was held to be an asset. The United States needed somebody old at the helm, somebody who knew their way around Washington and politics. For the country was unraveling. The fight over slavery, which Jefferson had referred to in 1820 at the time of the Missouri Compromise as the fire bell in the night, had finally begun to ring.

To hold the country together in the face of such complex forces was, to be sure, a vastly ambitious undertaking. But then, James Buchanan had always been vastly ambitious.

He had been born dirt poor, in a log cabin in rural Pennsylvania, his mother an impoverished American country girl, his father an orphaned Irish immigrant sent by his relatives to America to live with an uncle. But Buchanan was the kind of young man with a knack for getting ahead and he quickly pulled himself out of poverty, even managing to get himself into college. Arriving in Lancaster, Pennsylvania, as a young man determined to become a lawyer, he apprenticed himself to one of the leading attorneys in town and in no time at all, through hard work, became one. "I determined that if severe application would make me a good lawyer," he was to recall, "I should not fail in this particular; and I can say, with truth, that I have never known a harder student than I was at that period of my life. I studied law and nothing but law." In the evenings—"almost every evening"—he took lonely walks along the edge of town so that he could practice speaking, since a lawyer needed to know how to speak extemporaneously.

Buchanan loved Pennsylvania and wanted to remain there; his father had insisted he should remain there. But so eager was he for advancement that one day he decided to move to Kentucky in the hope that opportunities there would be bigger. He quickly returned, however; expecting to find little competition in Kentucky, he discovered to his astonishment that the competition there was even greater. "Why, sir, they were giants," he remarked after recalling his encounters with the likes of Henry Clay and others, "and I was only a pigmy. Next day I packed my trunk and came back to Lancaster—that was big enough for me. Kentucky was too big."

At age twenty-one he was admitted to the bar. By age twenty-two he became a partner of Lancaster's chief clerk and was appointed a prosecutor in nearby Lebanon County. And then, at age twenty-three, Buchanan ran for a seat in the state house of representatives.

Shortly after he announced for the legislature, the British burned down the Capitol. Buchanan immediately volunteered to join a citizens' militia to try to force the British out. It appears he volunteered merely as a means to win election. Within two weeks he was back in Lancaster. The war went on without him.

He insisted in his campaign speeches that he was running because he believed in Federalist principles and wanted to see them enacted into law. But his real reason for running, he confided to his father, was that he wanted to improve his legal practice; as a legislator he could attract better clients.*

Buchanan ran a hard race and in the end, despite his youth, won. The following year, still in need of clients, he ran again—and won again. But that was his last race for the state legislature, for by then he

---

*His father, initially opposed to his running, came to agree he should after learning of Buchanan's real purpose.

had become an established presence in the legal community. While he continued to show an interest in politics, he devoted most of his time to building his practice. Then, a few years later, he finally began to look around for a wife.

Whom a person marries says a lot about a person. Whom James Buchanan chose to marry said a lot about him. But then Buchanan always chose well.

In all of Lancaster when Buchanan arrived there was just one Yale graduate his age, Amos Ellmaker. Buchanan had managed to join his circle, eventually becoming his best friend.

In all of Lancaster there was just one son of a rich governor Buchanan's age, a young man by the name of Molton Rogers. Buchanan had succeeded in becoming his friend, too.

And in all of Lancaster there were just a handful of rich families with rich daughters available for marrying. Molton Rogers was to marry one of them. And in 1819, at age twenty-eight, Buchanan announced that he was to marry another, her cousin Ann Coleman.

Rogers's girlfriend was the daughter of Cyrus Jacobs, a Pennsylvania ironmaster, one of the innovative breed of Americans responsible for the birth of the Industrial Revolution. Buchanan's girlfriend was the daughter of Robert Coleman, an Irish immigrant, who was also in the iron business. Cyrus Jacobs was rich, reputedly one of the richest men in Pennsylvania. Robert Coleman was even richer—one of the richest men in the United States. In reference books he's identified as one of the country's first millionaires. When Buchanan would be invited to the Coleman mansion, he always had to ask which one. There were half a dozen, named after the ironworks on which the family fortune was based: Elizabeth Furnace Mansion, Speedwell Forge Mansion, Hopewell Furnace Mansion, among others.

Old man Coleman himself had become rich by marrying into the family of one of the richest ironmasters in his day. Now Buchanan was to become rich by marrying into the family of one of the richest ironmasters in *his* day.

History, it seemed, was repeating itself—seemed, but perhaps wasn't. For no sooner had Buchanan announced his engagement to Ann Coleman than it suddenly appeared to be in jeopardy, undermined by a devastating whispering campaign. Behind his back people began to say that James Buchanan was marrying Ann solely for her money.

Almost anybody who pursued Ann would have been followed by

the same rumor. Other young men had chased after her, and *they* had been followed by it. So it was not damning that the rumor followed Buchanan. But what was damning was that in Buchanan's case people in society circles believed that it was true.

Buchanan simply struck many as a man who was almost too eager to make it; they got the sense he would do *anything* to make it. As the news of his engagement spread they became sure of it. And as they did, their comments turned nasty. There was, to begin with, the matter of his birth in a log cabin. *A log cabin!* Of course, others had been born in log cabins, too, but they had the sense not to go after the richest girl in town. *Who does James Buchanan think he is, anyway?* And what about those parents of his? *His mother was so pretentious she named her second son George Washington, for heaven's sake!* And his father—*Why, he's a sharp trader, honest but conniving.* Buchanan himself said that his "father was a man of practical judgment"; wasn't Buchanan "practical," too? *Bet James Buchanan thinks marrying that Coleman girl is "practical."*

But if Buchanan's intention in going after Ann was to get his hands on her fortune, he was to learn it would not be easy.

Old man Coleman, then seventy-one years old, was protective of his daughter Ann and skeptical of the motives of all the young men who courted her. Soon it wasn't just the neighbors who were saying that James Buchanan was after her for her money, Coleman himself was. And once he became suspicious, it was just a matter of time before Ann became suspicious as well. And soon she did.

Buchanan himself seemed to give her a reason. In the months leading up to their engagement in June, he had tended to her every need, visiting her often. But almost immediately after the engagement he found he didn't have time for her. In July, it seemed, he had to travel home to see his family in Mercersburg. In August he suddenly announced that he would be going to Bedford Springs for several weeks to vacation. Beginning in September and extending through October, November, and into December, he was deeply involved in a complicated legal case that required him to make frequent trips to Philadelphia.

Ann naturally became distraught. Perhaps he *was* just after her for her money. Buchanan kept explaining that he simply didn't have spare time because of the pressing nature of his big legal case, in which many of Lancaster's most prominent citizens had an interest. But Ann wasn't so sure. After all, he continued to be active in politics; in November he had even attended several meetings called to debate the Missouri Compromise. To be sure, the Missouri Compromise was important, but so, in Ann's view at least, was his impending marriage to her.

Buchanan in his own self-interest should have resumed courting Ann in the manner she desired. It is one of the big mysteries of his life that he did not.

In early December, no longer able to stand his neglect, Ann sent Buchanan a note to complain. A friend, recounting Ann's reproach, wrote that she felt "Mr. Buchanan did not treat her with that affection that she expected from the man she would marry." Consequently she had begun to conclude that "it was not regard for her that was his object, but her riches."

Faced with this insult to his honor Buchanan reacted passively—and entirely uncharacteristically. On the many previous occasions in his life when he had felt mistreated or misunderstood, Buchanan typically had fought passionately to defend himself. It was one of his most noteworthy traits. As a seventeen-year-old, for instance, he had been kicked out of college for disruptive behavior. But, convinced that he had been wronged—his grades after all were better than anybody else's, and his behavior had been no worse, he believed—he had appealed on his own to the president of the board of trustees to be reinstated (and was). Subsequently, on account of his record of disruptiveness, school officials had denied him the right to deliver an honors address at graduation even though the leading honor society had unanimously chosen him to give it. ("I have scarcely ever been so much mortified at any occurrence of my life as at this disappointment," he was to recall years later.) Once again he appealed their decision and once again he got it reversed. And passionate as he had been as a teenager, he was even more passionate as a young man. In the legislature he gave a speech one day that inadvertently alienated many of his warmest supporters, raising concerns that he was a wild-eyed radical. Buchanan, in response, protested loudly that he'd been misunderstood. Then he gave another, deeply conservative speech, just to make *sure* he wasn't misunderstood.

But on this, the occasion of his latest contretemps—and undoubtedly the most important of his life to date—he mysteriously declined to assert himself. Instead of showering Ann with affection and love, he merely wrote her a short, cold note insisting that he wanted her, not her money.

Their relationship continued as it had before despite Ann's reservations, but given her feelings almost any incident might lead her to end it. Shortly there was one. Returning from Philadelphia on yet another of his business trips, Buchanan briefly stopped to visit a friend, William Jenkins. Jenkins wasn't home but his wife was, accompanied by a "pretty and charming young lady." Nothing of consequence happened, but when Ann heard that Buchanan, who claimed not to have time for *her*, had spent an afternoon in the company of this "pretty and charming young lady," she became disconsolate and then enraged. Shortly thereafter she wrote him a hasty note breaking off their engagement. He received it at the courthouse. Witnesses said that his face turned pale as he read it.

Buchanan may have wanted to try to change her mind, but he was never to get the chance.

After breaking off the engagement Ann became wretched, utterly miserable. Alarmed, her mother persuaded her to go to Philadelphia to visit her sister, hopeful that the change of scenery would get her mind off her failed romance. But moody as she had been in Lancaster, in Philadelphia she grew moodier still, quickly slipping into a profound depression. Doctors and friends didn't know what to do with her. Her behavior baffled them. One minute she might be chatting happily as if nothing were wrong, appearing to be, in the words of a judge who happened to visit, "in the vigor of health." The next she could be ranting incoherently, "attacked with strong hysterical convulsions."

Shortly after midnight on Thursday, December 9, just six days after arriving in Philadelphia, Ann Coleman suddenly died. The doctor called to the house said it was "the first instance he ever knew of hysteria producing death." To this day no one knows what the cause of death was. Possibly she killed herself.

Buchanan heard the news early Thursday, morning—and immediately sank into a morbid depression of his own. In a note to Ann's father he confided, "I may sustain the shock of her death, but I feel that happiness has fled me forever." In the coming days he was so overcome by grief that he could barely communicate. Trying to write an obituary for Ann, he found he could not and turned the task over to a friend.

In his note to her father Buchanan indicated that "she, as well as I, have been much abused." And he added: "God forgive the authors of it." But if Buchanan felt abused, the Colemans—and the Colemans' many friends—felt that it was they and Ann who had been abused, not he. Little as they had cared for him before, then he had merely been a social-climbing parvenu after Ann's money. Now he came to seem something more sinister—in the words of one harsh observer, "her Murderer."

Her father returned Buchanan's note unread. When the funeral was held Buchanan did not attend. He was not invited.

Broken in spirit, bereaved by Ann's death, Buchanan returned home for Christmas to his family in Mercersburg. The weeks he spent at home were among the worst in his life, so terrible he refused ever to speak about them.

In January he was back at his law office in Lancaster. At the beginning of February, when the Lancaster court reopened after the holiday recess, Buchanan could be found roaming the hallways, buttonholing clients and hobnobbing with community leaders just as he had before. He was obviously hurting, but not hurting so much as to neglect business.

Buchanan, who had put so much of himself into his business, could not afford to neglect it. What else did he have? At age twenty-eight he was alone, and it appeared now that he was to remain alone.

As he contemplated his prospects, however, he could see that they were not what they once had been. The legal practice he had been building up for years depended on his friendly relations with the Lancaster establishment. And it appeared that he had seriously damaged his standing with the establishment—or at least with that segment of the establishment associated with the Colemans. As the days went by he discovered that he was seeing fewer and fewer of the clients he had signed up through the Colemans' influence. Now instead of knocking on his door for help, they knocked on someone else's. All he had worked for seemed in jeopardy. "It is now thought," a friend wrote, "that this affair will lessen his consequence in Lancaster as he is the whole conversation in town."

For many years he had known nothing but success, the steady onward and upward success of a man on the make. Now he was slipping backward, a man on the make who was not making it. His close friends—his close friends outside the Coleman circle—stood by him during these days of anguish. But Buchanan realized having these friends was not enough. He needed new friends, new influential friends to replace the ones he had lost, and he needed them quickly if he was to survive the Coleman tragedy and scandal.

Slowly he began finding them. The Colemans, while powerful, had enemies. And as it sank in that Buchanan was now an enemy of the Colemans, *their* enemies became his friends. But no matter how many new friends he seemed to pick up, he never seemed able to pick up enough. As February turned into March and March into April, Buchanan became desperate to find new friends.

It was at this point that Buchanan showed just how ambitious a man he truly was. In early April 1820, just four months after Ann's death, he announced that he was running for a seat in Congress. By running for Congress he could find new friends. "I saw," he admitted years later, "that through a political following I could secure the friends I then needed." As once he had run for a seat in the state legislature to improve his legal career, now he would run for the national legislature to do the very same thing.

A Federalist at the time, Buchanan had the great fortune to be running in a district favorable to Federalists. After locking up the nomination, he sailed to victory.*

Buchanan became a popular representative and remained in Congress ten years. During all that time he occasionally dated, but he never married.[1]

Why did he never marry?

Two explanations have been offered over the years to account for his behavior. One—*his* explanation—was that he simply was never able to recover from Ann's death ("I feel that happiness has fled me forever"). The other, advanced recently by some writers, was that he was homosexually inclined and uninterested in marriage—which, if true, would account for much that is mysterious about his engagement: his uncharacteristic passivity, the empty excuses he made for not seeing Ann, his inability to offer evidence of his love when she demanded it and needed it. Though it was obviously in his financial interest to make every effort possible to save the relationship no matter how he felt about his own sexuality, he may have been so confused about his sexual identity that he was unable to press ahead with his plans to marry. In effect, he may have been at war with himself, one very powerful force pushing him toward Ann, another pushing him away, leaving him, pathetically, in a desperate state of ambivalence.

The evidence for his homosexuality is weak but intriguing. His closest relationship in life seems to have been with William King, the Alabama senator with whom he roomed in the capital, whom some cruelly referred to on occasion as "Miss Nancy."

It was not unusual for members of Congress to room together, but Buchanan's arrangement with King apparently was. President Polk's law partner, Tennessee representative Aaron Brown, referred to Buchanan

---

*Ironically, the Federalists also ran Ann's brother, Edward Coleman, for office, as a candidate for the state senate.

and King as a couple, "Mr. Buchanan & *his wife*." At the time Brown made this observation, however, Polk and King were competing for the vice presidential nomination of the Democratic Party in 1844; there had even been talk that King and Buchanan might run as a team. Thus, it is clearly possible that Brown, as a friend of Polk's, may simply have been indulging in political rhetoric. Politics seems to have been very much on Brown's mind. In the same letter Brown speculated that King might be better off running for vice president on his own, unshackled to Buchanan; a "newspaper puff which you doubtless noticed, excited hopes that by getting a *divorce* she [King] might set up again in the world to some tolerable advantage. . . . *Aunt Fancy* [King] may now be seen every day, triged [*sic*] out in her best clothes and smirking about in hopes of securing better terms than with her former companion."[2]

Buchanan repeatedly over the years pushed the Democratic Party to select King as its vice presidential nominee. But whether anything should be read into this is doubtful. On each of the three occasions when Buchanan behind the scenes fanned the candidacy of his erstwhile roommate—in the elections of 1840, 1844, and 1852—it was in Buchanan's political interest that King be chosen. Since King had no presidential aspirations of his own, his selection as vice president would help deprive other candidates who did of a national forum. And, of course, it was Buchanan's hope to have as few rivals as possible to compete with in the presidential sweepstakes.

Nothing in Buchanan's correspondence indicates he was a homosexual. But in King's papers there exists a tantalizing hint that *he* was. In a letter to Buchanan, written after King was appointed minister to France, King confessed: "I am selfish enough to hope you will not be able to procure an associate who will cause you to feel no regret at our separation." And then, he added: "For myself, I shall feel lonely in the midst of Paris, for here I shall have no Friend with whom I can commune as with my own thoughts." On another occasion King reflected that the United States might be better represented in Paris "by someone who has more the spirit of a man."

In 1852 William King was elected vice president of the United States—the only bachelor vice president there ever was.

In 1856 James Buchanan was elected president of the United States—the only bachelor president there ever was.

James Buchanan made winning the presidency his life's work. He ran for president four times, putting himself forward as a candidate in every election from 1844 to 1856. Each time he lost he couldn't believe it. He was sure that he was better than the fellows who had won. And it ate at him, turning him bitter and hard. By the time he finally assumed the office he was very hard, ruthless almost. And, of course, as always, very, very practical. Given what was happening to the country these were not inconsiderable qualities. They seemed to be just the qualities a president needed to deal with the burgeoning slavery crisis. But Buchanan was to discover that toughness wasn't enough. To do what had to be done to save the Union he needed raw power. And he lacked power.

Slavery had been bedeviling American politicians from the beginning. But always before the politicians had been able to compromise their way out of danger and division. In 1776 Jefferson had agreed to drop from the Declaration of Independence language suggesting that the slave trade was immoral. In 1820 Henry Clay had successfully brokered the Missouri Compromise, the deal that prohibited slavery from the northern part of the Louisiana Territory. In 1850 Clay and Douglas working together had successfully arranged for the admission of California as a free state, even though its admission upset the traditional balance of power in the Senate between the free and the slave states.* Then in 1854 there had been the crisis over Kansas and Nebraska. And that had changed everything. For there didn't seem to be a way out of the crisis that left anybody happy. Not even Stephen Douglas was happy. And it had been Douglas, in a way, who was responsible for the mess.

Douglas was a very proud, very determined man, the senior senator from Illinois, and in the 1850s the greatest orator in the Democratic Party. By everybody's estimate an able and wily politician, he was almost certainly a presidential prospect. Douglas hadn't wanted to step on the hornet's nest of sectional politics. All he had wanted was to get a railroad built westward through Illinois and the Nebraska Territory. But nothing was simple in the supercharged atmosphere of sectional politics in the 1850s. Douglas couldn't get his railroad unless he first got a government established in Nebraska. And he couldn't get a government there unless he got Congress's approval. And he couldn't get Congress's approval unless he got the support of the South. And he

---

*Since 1796 the Senate had been evenly divided. For each slave state that was admitted, a free one was. As of 1850 there were fifteen free and fifteen slave states. California was, thus, the sixteenth free state.

couldn't get the support of the South unless he agreed to the repeal of the Missouri Compromise (which prohibited slavery in the Nebraska Territory), and which had become for the South one of the many symbols of federal oppression.

In 1854 Douglas finally succeeded in obtaining a bill that created a government in the Nebraska Territory (actually two territories, Kansas and Nebraska). But the passage of the Kansas-Nebraska Act alarmed the North. Instead of slavery being outlawed under the Missouri Compromise, it would be left to a vote of the people. For Northerners the Missouri Compromise was sacred; repealing it was unthinkable, and they considered the Kansas-Nebraska Act a repeal. And now *they* felt oppressed. Douglas tried to reassure Northerners that because the people of the two territories would have the final say in deciding whether to have slavery or not, slavery would be banned because they would surely vote not to have it. As he also pointed out, he was doing nothing more in Kansas and Nebraska than he and Clay had done in California, where the question of slavery had also been left up to a vote of the people, a practice known as popular sovereignty. But Northerners remained unconvinced. More to the point, as far as Douglas was concerned, so did many of his own constituents.

The year of the presidential election, in 1856, the conflict turned bloody in Kansas as pro- and antislavery forces battled for control of the territory. First proslavery agitators sacked the town of Lawrence, a haven of antislavery Northerners. Then, three days later, in retaliation, fanatical abolitionist John Brown swooped down on a community of Southerners in Potawatomie and slaughtered five farmers.

The very same week, a thousand miles away, in an uncanny parallel development, the fight in the U.S. Senate over Kansas and slavery also turned violent. Massachusetts senator Charles Sumner, an abolitionist, was seated at his desk in the Senate when he was suddenly attacked by a proslavery congressman. Using a gutta percha cane, the congressman repeatedly struck Sumner about the head and shoulders so many times and with such force that the senator was unable to push himself away from his desk to defend himself, leading to his incapacitation for years. (The attack, of course, was indefensible. It was not, however, without provocation. Sumner had recently given a fiery speech in which he denounced "the barbarism of slavery," which aroused Southerners everywhere.)

By this time the entire country was aroused—and dividing into two camps, splitting families, churches, parties. The Methodists split into

Northern and Southern factions. The Whigs destroyed themselves, not even fielding a candidate for president in the election of 1856. Only one major institution was left by then which crossed the sectional divide, drawing support from both Northerners and Southerners, the Democratic Party, and it was so riven by strife that some feared it would go the way of the Whigs. (The Republican Party, which replaced the Whigs, was wholly based in the North.)

The main question preoccupying people was not whether slavery should or should not exist. Most Americans were willing to let the thing remain where it was; over time, it was felt, slavery would slowly peter out. The issue was whether it should be extended into the new territories the country had recently acquired. Northerners, by and large, felt it shouldn't; Southerners, by and large, felt it should.

Complicating matters was the fact that the whole dispute had become symbolical, imbuing every little controversy with apocalyptic significance. In the South a victory by the North was considered evidence that the federal government wanted to abolish slavery. In the North a victory by the South was considered proof of the existence of a slave-state conspiracy. Southerners especially felt put upon, and in their defensiveness began tightly clutching the "peculiar institution" as they had never clutched it before. Where once they had regarded it simply as a necessary evil, they now began calling it a positive good. The more people argued, the harder it became for the politicians to practice the art of compromise.

Things were now so out of control that hotheads in the South were demanding that Southerners secede from the Union and hotheads in the North were responding that they should.

From the beginning Buchanan knew he was in a new world, this world in which Americans were settling their differences violently, in which the old parties were collapsing, in which compromise was proving feeble. And he decided that with a new world came new rules. To save the Union, to keep things together, he would have to do things no president before ever had. If Northerners and Southerners couldn't work out their differences peaceably and rationally then he, as president, would have to find a way to work things out for them. To take the contentious issues out of politics altogether.

It was an extreme idea. And it was antithetical to the principles of self-government. Americans were supposed to possess enough good sense to be able to reach settlements themselves. But they hadn't been able to with regard to slavery. And Buchanan didn't want to wait to see if they could. Wait, and things might spin completely out of control. So, as he was putting together his government in the months before his inauguration, he hatched a secret plan to end the crisis over slavery by getting the Supreme Court to intervene. If the Court came out with a ruling saying that the Missouri Compromise was unconstitutional, then all the squabbling about the compromise would suddenly become moot. Those Northerners who had been using the repeal to keep the pot of sectional politics boiling would discover that the fire under the pot had gone out.

To be sure, there was more to the conflict than just the controversy over the Missouri Compromise, but it was Douglas's bill to repeal the compromise that had turned the fire up under the controversy, giving the fanatics on both sides an issue susceptible to extreme emotionalism. It so happened that the Court had a case at hand in which such a ruling could plausibly be made, as Buchanan well knew. It was the case of Dred Scott, a slave who had used the federal courts to sue for his freedom on the ground that he had lived for a time in the area covered by the Missouri Compromise where slavery was excluded.

Presidents aren't supposed to meddle in the internal proceedings of any court, and especially not in those of the Supreme Court. If found out, a president who did very possibly could face impeachment charges. But Buchanan didn't care. *Something* had to be done. So in February, just a few weeks prior to his inauguration, he wrote a couple of his friends on the Court, explaining his views and asking for action. By luck the Court was already planning to do the very thing Buchanan wanted. While the majority initially had decided to make a narrow ruling denying Dred Scott the right to sue in federal court on the grounds that he was a Negro and as such wasn't a full citizen of the United States—Negroes, Chief Justice Roger Taney would write, "had no rights which the white man was bound to respect"—they had learned that two justices on the minority side planned to include a defense of the Missouri Compromise. Angry, the majority had decided to declare the compromise unconstitutional on the grounds that it infringed the property rights of slaveholders.

Buchanan was very smart and very shrewd and thought that he was being smart and shrewd in asking the Court to intervene in the Missouri Compromise controversy. But it was a fool's dream to think that a Supreme Court opinion could defuse the crisis over slavery. Things had *already* gone too far for that. When, two days after Buchanan's inauguration, the Court issued its opinion in Dred Scott, there was a firestorm of criticism. In this new world not even the Supreme Court, powerful as it was, was powerful enough to restore calm. Buchanan had miscalculated. The crisis would not simply go away. Kansas would remain to fester. Politics, instead of growing more civilized, would grow less and less so.[3]

Even if slavery had not been an issue in the 1850s, the country would have been difficult to govern. For it was changing rapidly, at a ferocious, befuddling pace. As a people Americans were used to change. The population had been doubling every twenty-three years on average since the Revolution. But change as it was now happening was of a scale not even Americans could wholly grasp. Most fundamentally of all, perhaps, the transportation system was in the process of being revolutionized. In 1840 there had been fewer than three thousand miles of railroad track in the entire country. By 1850 there were more than twelve thousand, making it easier and easier for farmers to get their products to market. Farmers more than anybody experienced change the most. Inside two decades the wooden plow had been replaced by the iron plow, and the iron plow in turn had been replaced by the steel plow. These improvements along with the invention of the McCormick reaper had radically altered the lives of those who worked the fields.

Because of these changes and because of the great spurts in population growth, farm production rose dramatically. In the space of barely two generations (1823 to 1861) sugarcane production rose from 20,000 to 270,000 tons. Between 1850 and 1860 cotton production more than doubled. And as the railroad pushed west so, too, did farm production. In 1840 it was South Carolina that made cotton king, growing more cotton than any other state. In 1860 Texas had overtaken South Carolina in cotton production. Wheat and corn farming also moved west. In 1840 most wheat was grown in New York, Pennsylvania, and Ohio, by 1860 in Illinois, Indiana, and Wisconsin. In 1840 most corn came from Kentucky, Tennessee, and Virginia, by 1860 from Illinois, Ohio, Missouri, and Indiana.

Just behind these economic changes in importance were the changes in American demography. In 1820, 72 percent of the American people made their living from agriculture. By 1860 just 40 percent did. As Richard Hofstadter observed, the United States was born in the country and moved to the city. In 1820 there was just one city with more than a hundred thousand people. By 1860 there were half a dozen. Much of the increase was due to the great unprecedented growth in immigration. The magic year was 1842, the first year immigration broke a hundred thousand. Just five years later, in the wake of the Irish potato famine, it broke two hundred thousand, in 1850 three hundred thousand, in 1854 four hundred thousand.[4]

The United States was still an overwhelmingly white, overwhelmingly English-rooted country. Not until the end of the century would immigrants arrive in such numbers as to shift the balance substantially. But already immigrants were having a profound effect on politics. Almost everywhere they settled politics became rougher, tougher, and dirtier. Before the arrival of the first great wave of immigrants, vote stealing had been a rare occurrence in American politics. Beginning in the 1850s, as the political parties realized the ease with which they could manipulate immigrants, it became almost common in places.

The election of 1856 was important, apart from the issue of slavery, because it was the first in which there was massive vote fraud. Virtually all the fraud took place on the Democrats' side, Buchanan's side (the immigrants mostly voted Democratic). As it happened, Buchanan didn't need stolen votes; he won by nearly half a million. But neither he nor his managers knew going into the campaign that victory would come so easily. While the opposition was divided between Republican John C. Frémont and Know-Nothing* Millard Fillmore, nobody knew how well they would run. Neither party had ever run a national race before, and neither of their candidates had ever run for the presidency. It was all new territory—confusing territory.

In the circumstances Buchanan wouldn't have been human if he hadn't been tempted to tolerate vote stealing on his behalf. It was, after all, his fourth try at the office and certainly his last. Lose this chance and there'd never be another. And vote stealing was a distinct possibility. Because immigrants wanted to be naturalized, and because the Democratic machine controlled the naturalization bureaucracy in some of the biggest cities, Democrats figured out that they could control thousands of votes. Naturalize the immigrants, throw in some cash as a bonus, and they'd vote as they were told.

Buchanan, devious by nature and desperate to win, desperate to con-

---

*The Know-Nothings believed in stopping immigration, to keep the United States for native Americans (that is, for white Anglo-Saxon Protestants). Committed to secrecy, members were instructed—when asked questions—to say that they knew nothing.

trol politics in a situation that was otherwise out of control, gave in to temptation. The chief agent behind the vote fraud was John Forney, an old Buchanan friend, who was headquartered in Philadelphia, and who was to help win Pennsylvania. Forney was later to have an unseemly falling out with Buchanan, but at this time the two were very close, keeping constantly in touch through the mail. At one point Forney all but admitted in one of his letters to Buchanan what he was up to, boasting, "We have naturalized a vast mass of men. . . . The Opposition are appalled. They cry fraud. Our most experienced men say *all is well*." Buchanan, no naïf, certainly understood what was happening but did nothing to stop it.

After the election, prosecutors investigating the Philadelphia naturalization bureau uncovered widespread evidence of abuse. At trial a clerk admitted he helped print up 2,700 phony forms for the Democrats to use to naturalize immigrants illegally. Others testified that immigrant voters were given bribes to vote Democratic.

All of this cost lots of money, more than several hundred thousand dollars, it's estimated. Where did the money come from? Quite a bit apparently came from Wall Street, which was just emerging as a financial powerhouse. Some came from federal officeholders, who were required to kick back about 3 percent of their salaries to the Democratic Party. And some came from military suppliers for the navy.

Several years later it was revealed that Buchanan himself had authorized George Plitt, a friend, to lean on suppliers in exchange for campaign donations. In one case a supplier of live oak (a particularly durable hardwood), W. C. N. Swift of New Bedford, Massachusetts, was pressured into giving $16,000. After Buchanan was elected, navy officials rigged a bid on live oak in such a way that Swift and only Swift was guaranteed to win it. Other money may have come from Democrat Isaac Fowler, the head of the New York City post office. As was subsequently revealed, over a period of years he stole more than $160,000 from the post office. He himself was broke. Most of the money, it was surmised, went into the coffers of the Democratic Party.[5]

Buchanan, always astute, made sure to conceal his involvement in the various schemes concocted to put him in the White House. During the campaign the cover-up held, encouraging Buchanan to think that he had successfully controlled the process, successfully manipulated the electoral system as no candidate before ever had. It was almost as if the great changes sweeping the country had been an advantage. But for the breakup of the Whigs, the rise in immigration, and the sudden availability of abundant new sources of campaign cash, Buchanan might never have been able to make it to the White House. But the promise of control was to prove illusory. The forces at work in the country were simply too great, too unwieldy, to be reined in. In the end they would bring down Buchanan, and shortly afterward, nearly succeed in breaking apart the Union.

The year 1856 had been a good one for James Buchanan. The next, 1857, would prove to be one of the worst years any president ever had. It was slavery that made it a terrible year, and it was slavery in Kansas in particular that made it so. What was at issue was, simply, whether Kansas would come into the Union as a free or a slave state. Buchanan early on privately told the governor of the territory that the choice should be left up to the people of Kansas. On this, he said, "I am willing to stand or fall. It is the principle . . . at the foundation of all popular government." But during the summer Buchanan, facing great pressure, changed his mind. The pressure came from the South and in particular from four states in the South: South Carolina, Alabama, Georgia, and Mississippi. Let Kansas come in free, these four states hinted, and they would vote to secede from the Union. At this point it was just a threat, but it was a threat any president had to take seriously.

Given the political realities, Buchanan had to take it *very* seriously. For his political base was in the South. It was the wily Louisianan John Slidell who had manipulated the Democratic convention into nominating Buchanan. And it was the South that gave him his margin of victory. Buchanan had carried every Southern state except Maryland, which had gone for the Know-Nothings. The three most powerful figures in his cabinet were Southerners: Secretary of the Treasury Howell Cobb, Secretary of the Interior Jacob Thompson, and Attorney General Jeremiah Black. Alienate the South, and Buchanan might as well resign.

An incident at the dawn of his administration had indicated just how dependent he felt himself to be on Southern power. He had wanted to install his good friend John Forney—who had helped deliver Pennsylvania and who had orchestrated the wonderfully efficient vote fraud in Philadelphia—as the editor of the administration paper, the *Union*. But the Southerners had objected and Buchanan had caved. "Sir," Forney had complained, "you have surrendered me." "I have been compelled to do so," Buchanan had responded.[6]

There was a lot of talk that Buchanan had caved simply because he was weak. Not weak politically, but morally, spinelessly weak. And there seemed to be something to this contention. A lot of people thought so, even his Southern friends on whom he so depended. On one occasion Howell Cobb was supposed to have remarked contemptuously, when asked about the opposition to a particular piece of legislation he favored, "Oh, it's nothing much; only Buck is opposing the administration," as if Cobb were the administration and Buchanan merely a figurehead.

Actually Buchanan wasn't morally weak, as many thought, but politically weak. Like other presidents he was discovering that events can push a leader to do what he doesn't want to do, to do what he knows he should not do, but cannot in the end help doing. A really politically strong president, a Jackson, could confront events and try to mold them, but Buchanan was not a Jackson. He lacked Jackson's power. Dealt a weak hand, Buchanan, time and again, would have to cave in to Southern pressure.

There is no doubt, if a fair vote had been held, that Kansas would have opted to become a free state. It's estimated that fewer than two thousand voters favored slavery and that ten thousand or more opposed it. But a fair vote could not be held in Kansas in 1857. Things had gotten that far out of whack. Several votes were held to decide the question. But on each occasion the process was utterly corrupt. At the October election, proslavery forces engaged in massive fraud, pumping up their numbers dramatically through wholesale manipulation of the ballots. In one county alone, McGee, which is located on the southern border of the state, more than twelve hundred voters were reported to have cast proslavery ballots even though there weren't twenty bona fide voters in the entire district. It was subsequently revealed that the names of many of the voters on the rolls had been signed in the same handwriting.

In December the voters were told that they were being given the chance to vote on a proslavery constitution drafted by a proslavery convention held in the fall in Lecompton. But the vote was rigged. The way things were arranged, Kansans would have slavery whether they voted yes or no. Vote yes, and slavery would be permitted as it was throughout the South. Vote no, and no new slaves would be admitted to the state, but the masters with slaves already there would be allowed to keep them. There were only about two hundred slaves in all of Kansas at the time, but antislavery residents believed, quite reasonably, that they should have the choice not to have any. And besides, they didn't like the fact that the constitution had been written by proslavery forces. Thus, when the vote was held the antislavery majority boycotted the election. In early January, in response to a public outcry, yet another election was held, and this time it offered a fair choice. A simple up or down vote on the Lecompton constitution. But even this election was marred, proslavery forces refusing to participate, which discredited the results, a lopsided rejection of Lecompton, by a vote of 10,226 to 162.

The January vote, flawed as it was, demonstrated amply that Kansans did not want slavery. But in February Buchanan, pressed by his Southern supporters, recommended to Congress that Kansas be admitted as a slave state under the rejected Lecompton constitution. It was a galling decision, the kind a king in medieval England might have lost his head over. But Buchanan thought he could get Congress, where he had Democratic majorities in both houses, to go along.

It was a serious miscalculation, the worst in his long career. And it cost him dearly, damaging his reputation and losing him old friends. The remarkable thing is Buchanan didn't seem to care what happened to him. Fast running out of options, he felt that the course he took was the only one he could. He simply had to accommodate the Southerners again. And he had to do so no matter what the cost. Fail and the Union would break up. So Buchanan wouldn't let himself allow for failure. He would fight as he had never fought before, fighting as if to the death, fighting as no president ever had.

Buchanan had already proved on several occasions that he felt entitled to break rules when they got in his way, as he had to engineer his election. But now he acted as if there were no rules at all, or none anyway that he was bound to respect. It was as if his credo had become *Anything to win, but above all, win.*

It was at the very start that Buchanan signaled his willingness to do anything to win. Douglas had come to the White House in December to register his opposition to Lecompton. The two had once been friends but now, it was to become apparent, they would be enemies. Douglas told the president that Lecompton was wrong, morally wrong, and that he, the senator who had championed popular sovereignty, would have to make a fight of it, even if that risked splitting the Democratic Party apart, even if it meant breaking with the administration. Buchanan listened quietly, then announced, calmly but determinedly, that if Douglas insisted on attacking Lecompton he would be crushed, ruined, destroyed.

Douglas wasn't used to being talked to this way, wasn't used to being threatened, and he instantly turned red-hot angry, telling Buchanan that he would oppose Lecompton. Then he stormed out of the White House. Thus began a fight to the finish between the two titans of the Democratic Party, one a president of the United States, the other a man who very much wanted to be, and who, in 1860, would indeed run for president.

That the fight was going to turn ugly was inevitable, not only because both sides were determined to win, but because the fight was in the family. It wasn't a battle between Republicans and Democrats; the Republicans would oppose Lecompton and nothing could be done about that. It was a battle between Democrats. And that was to make it especially bloody. For there was plenty that a Democratic president could do to a Democratic member of the legislature who opposed him.

As a baby step he could stop inviting the legislator to social functions held at the White House. And that was promptly done, which was not insignificant. Legislators like to be able to boast about their intimacy with the president. Less so in those days than now, because presidents in those days lacked the power and majesty they have today. (Pierce had been distraught to learn on the night of his inauguration that he had to make his own bed when he went to sleep; nobody had bothered to do it for him.) But the loss of White House social privileges was something.

Of far more importance to a legislator was patronage, the lifeblood of a party politician. And federal patronage then was almost totally controlled by the president. Of the 26,274 civilian jobs then in existence in the federal government, 25,713 were directly controlled by the chief executive. And he could fill them with almost anybody he wanted. Almost anybody—for there was as yet no civil service. That gave him enormous leverage with politicians. Presidents were to complain it was a terrible ordeal being in charge of so many jobs; Polk said he worked on almost nothing but the patronage his first nine months in office. And they frequently moaned that for every person they hired they alienated ten whom they didn't. But none volunteered to give up the responsibility for it was the source of great, overwhelming power; the one in ten who *was* hired usually did everything he could to keep the administration in power. Jackson had used the patronage to build up the Democratic Party. Tyler had used it to try to take over the Whig Party. And now Buchanan would try to use it to get legislation passed.

Use of the patronage in this way was not revolutionary. Jackson had used the patronage in a limited way to win the bank war. But Buchanan was to go much farther than Jackson ever had, much farther than any other president had ever dreamed of going. As the fight wore on, he dismissed dozens and dozens of federal employees to punish the legislators who had sponsored them. Some employees had been friends for years. But Buchanan didn't stop to worry overmuch about this. They would go because they had to go. And that was that ("I have been compelled to do so").

It was Douglas he was after most, of course. And it was Douglas he decided to make an example of. Douglas was especially vulnerable at the time because he was up for reelection in 1858 and facing a tough opponent, Abraham Lincoln. To keep the Democratic Party unified in Illinois in the face of Lincoln's strong challenge, he needed to control the federal patronage. But he could not. Buchanan controlled it.

Buchanan exercised his control like a tyrant, taking cruel pleasure in his humiliation of Douglas. Buchanan's most egregious act was to replace the Douglas man in charge of the Chicago post office with Isaac Cook, an enemy of Douglas's who had held the job before. Cook was a known swindler who had boldly attempted to pass off a phony land title on the federal government to pay off an old debt. The attorney general under Pierce had caught Cook red-handed and insisted on cash payment. But Buchanan's attorney general accepted the title! Chicagoans were appalled and predicted that Cook would again try to swindle the government. Unsurprisingly, he did, stealing thousands of dollars from registered mail. Still, Buchanan refused to remove him.[7]

It was all unprecedented and it was terribly destructive, self-destructive, for it did terrible damage to the Democratic Party, weakening its value as a national institution, and leading ultimately, in 1860, to its division into Northern and Southern factions, one to be headed by Douglas, the other by a Southerner, John Breckenridge, Buchanan's vice president. It was a tragedy for the country, a tragedy for Buchanan.

In the end Buchanan was unable to win the approval of Lecompton. The Senate approved, the House refused. A compromise measure was then quickly adopted to, in effect, start all over, requiring Kansans to vote yet again on Lecompton. To encourage them to vote yes the Congress offered the people a bribe. Approve Lecompton, and the new state would be given five million acres of public land for free. In August the people of Kansas went to the polls. By an overwhelming majority they voted against Lecompton.

But at least the Southerners had not seceded. As long as Kansas wasn't admitted to the Union as a free state there was still hope they wouldn't.

Terrible as his first two years as president had been, Buchanan's last two were worse. In the fall elections the one Democrat he wanted defeated, Stephen Douglas, won. And a raft of Democrats he wanted to win, lost, costing the party control of the House. The Republican victory in the House gave Buchanan's critics the opportunity to use the full power of Congress against him and they proceeded to do so almost immediately. Already the Republicans had uncovered evidence of corruption in the administration that had proved embarrassing. Representative John Sherman, intelligent and honest, had uncovered the cozy deals struck with the military suppliers who'd contributed to the Buchanan campaign in 1856. But what was to come out now was shocking and outrageous and it was forever to stain the Buchanan name.

Republican representative John Covode was not, in 1860, a well-known figure, either nationally or even in his own state of Pennsylvania. But he was disturbed by what he had been hearing about the Buchanan administration, particularly a claim that Buchanan had attempted to bribe two members of Congress in the vote on Lecompton, and when he could bear sitting on the sidelines no longer he decided to do something about it: to open a formal, detailed, investigation of the president of the United States.

Covode had not planned on a career in politics. A poor boy who had made good despite a lack of formal schooling, he had accumulated a fortune as a mill owner and railroad investor. But he had done such a good job as a justice of the peace, in which he had earned the nickname "Honest John," that he had been propelled forward politically as many successful businessmen have in American history, serving first in the Pennsylvania state legislature and then in the U.S. House of Representatives. He was not lacking in partisan fervor. He deeply believed in the Republican Party and was a member of the Republican Executive Congressional Committee, which was responsible for electing Republicans to Congress. But he was driven in 1860 by what appears to have been a simple honest zeal for the truth. He was, in short, that most dangerous kind of person to people in power, a man on a mission, and a man Buchanan was soon to despise.[8]

The Covode Investigation Committee held its hearings in secret. But as with all congressional hearings there were leaks, and these quickly became the subject of one sensational news story after another. The lead-off witness was Cornelius Wendell, an erstwhile friend of Buchanan's and for years the editor of the administration's party newspaper, who had made hundreds of thousands of dollars off the government as the printer of official documents. He revealed that over the past four years he had diverted more than a hundred thousand dollars into the coffers of the Democratic Party. Some of the money, he said, had gone to the Know-Nothings in 1856 in Pennsylvania and New Jersey to draw votes away from the Republicans. Other substantial amounts, he testified, had been used to help prop up the party's newspapers. Buchanan himself, he said, had personally picked out the papers that were to receive subsidies.

The involvement of the government's printer in political activity was an old story, going back as far as the Washington administration, when Hamilton had awarded government printing contracts to his pet newspaper. Somebody, after all, had to print government documents. Might as well be a political friend. Polk had even arranged to have a friend buy a newspaper for the Democratic Party with funds advanced by the government itself, which were to be paid back out of the money earned on the contract. But nobody outside the government had quite realized until Wendell came along how profitable the contracts had become and that they had become a rich source of campaign boodle.[9]

That was new and shocking. But what really alarmed Americans was another of Wendell's allegations, that at Buchanan's instigation he had spent between thirty and forty thousand dollars to bribe members of Congress to vote in favor of Lecompton. Cynical as Americans were becoming about politics—and they were becoming more and more cynical the more they learned—they had discounted the accusations of bribery that had prompted Covode to convene his hearings. Bribery seemed totally outside the American political experience—and was. Never before had there ever been any evidence of a president resorting to bribery to influence the outcome of a congressional debate; only Fillmore had ever been accused of bribery and he had subsequently been cleared.

Buchanan adamantly denied the charge, and for a time Americans thought it might not be true. After all the hearings had been held in secret. Nothing seemed official. What they knew they knew from leaks published in the partisan press. But then the committee published the results of its investigation in a fat book, more than eight hundred pages long, and that seemed to settle the matter. For the book was filled with so much detail, so many facts, as to be very convincing, despite a lame attempt by Democrats to discredit it.

According to the report: some congressmen had accepted payments directly; some had made Wendell deposit the money in fictitious accounts; one had demanded—and gotten—a gigantic sum, some fifteen thousand dollars. Democrats contended that the money had been used to pass a private patent bill, not Lecompton, but the one person who could have proved that refused to testify. Wendell himself subsequently denied that Buchanan had anything to do with the effort, contradicting his own statements that Buchanan had, but that was after Wendell had been subjected to extreme pressure.

Anyway, it wasn't just Wendell who was saying that money had been used for bribes. John Forney, Buchanan's old friend from Philadelphia, who had been denied the editorship of the administration paper and gone on to take over another, testified that he had been offered an eighty-thousand-dollar printing contract if he came out in favor of Lecompton. The bribe had been offered by Jeremiah Black, Buchanan's attorney general. "You can get all that post-office printing," Forney said he was told, "if you will write an editorial as long as your hand."

Of course, both Forney and Wendell might be lying. Both had had a falling out with the administration and neither of their accusations was verifiable. But the Covode report included one other sensational finding that *was* provably trustworthy and because *it* was, everything else seemed trustworthy, too.

Publicly Buchanan had always maintained that the December 1857 vote on Lecompton, in which the people had had no real choice on slavery (no choice because they'd have slavery whether they voted yes or no), had been legitimate. It was, after all, the vote on which he had hung his entire proposal to admit Kansas as a slave state. But the committee in April produced a letter in Buchanan's own handwriting that proved he privately believed, contrary to what he had publicly been saying, that Kansans should have the right to vote slavery up or down, as they did in that second January vote, a vote he deliberately ignored because of the South's objections. In effect, the committee had caught Buchanan red-handed. And people reasoned that if Buchanan, under pressure from the South, could defend a vote he really didn't believe in then he was totally untrustworthy, the kind of man who probably could countenance the bribery of legislators and newspaper editors. In June the House of Representatives voted to censure Buchanan.

Buchanan had gone into the fight over Lecompton promising to ruin anybody who crossed him. But he himself had been ruined. He left the presidency a broken man. By then the secession movement had become unstoppable. What Buchanan had feared would happen, was happening, the South breaking away from the Union. Buchanan believed secession was illegal, unconstitutional. But he also believed that under the Constitution he lacked the power to do anything about it. After Lincoln's election, the states prepared to leave. Buchanan watched and did nothing.

Certainly his constitutional scruples hobbled him. But even if he had decided he legally could take military action to keep the states from seceding he probably could not have stopped them. With the Republicans hounding him in Congress and the country sharply divided, he could not have called out the army on his own authority. His political base had been in the South; to call out the army he would have needed a base in the North. All he really possessed in the end was his willingness to connive to achieve his ends, and conniving wasn't nearly enough.

Because Buchanan had been a bachelor, he had installed his niece Harriet as his White House social director. Harriet was a vivacious young woman who was an immediate hit in Washington, but she wasn't always popular with Buchanan himself. One of their big disputes was over dancing. Buchanan had banned dancing from the White House; Harriet wanted the ban lifted. Another dispute had to do with the mail, *her* mail. An inveterate busybody, the president enjoyed reading his niece's mail, even private letters she exchanged with her best friend. Harriet naturally had taken exception to Buchanan's interference. Eventually she found a way to write her best friend in secret: She sent her mail out via the kitchen steward's empty tub of butter.

Events had gotten beyond Buchanan's control. So in the end had even his young niece Harriet.

* * *

Buchanan, one of the weakest presidents in American history, was followed by Lincoln, one of the strongest. But they were not as different as they appeared. Though Lincoln was far more honest, he, too, was a scrappy, conniving, wire-pulling politician. And like Buchanan, he was also very, very ambitious.

7. The Story of Abraham Lincoln

How a boy who started out with nothing became the shrewdest politician of his generation, successfully restoring the Union by assuming powers no president before had ever dreamed of using

In the early 1830s New Salem, Illinois, where Lincoln lived as a young man, had but one chance of becoming something. One very small chance. So small a chance that the community's great optimism about the future was almost laughable. For its one chance rested on the hope that the village could become a port of call for steamships traveling on the narrow Sangamon River. And as yet only one steamship had ever passed by New Salem on the Sangamon, let alone the dozens of ships that would have to pass by for the village to benefit. And all indications were that dozens of ships would not be passing by. For there were only three months—March, April, and May—when the river was passable at all. Before March the river was too full of ice to permit safe ship travel; after May it was too low. Of course, theoretically enough ships might pass by during those three short months to make New Salem prosperous, but after that first ship came through, it became increasingly apparent that no others were going to follow. And no other ships did. For the river as it then existed was difficult to navigate. The one steamship that had gone through had gotten stuck on a dam built at a New Salem mill; Lincoln, a hired hand on the ship, had had to work for hours to get it unstuck. Moreover, improving the river for ship travel would take money, and no one had any.[1]

As the limited future of New Salem became more and more evident, its population, which had been climbing quickly, soon stopped climbing. Then it began to decline. And shortly this village, which had been founded in 1829 with fervent hope, was out of hope. In 1839, it died. Everyone moved out.

New Salem had not become anything, but at least it had had a slim chance. Abraham Lincoln in the early 1830s seemed to have hardly any chance at all. He had nothing. He was nothing. And it seemed likely he always would have nothing and be nothing.

Like New Salem, though, he too was full of optimism. His, too, was also almost laughable. Neither his father, his mother, nor his stepmother had ever amounted to anything. Nor had his grandparents. Nor had anybody in the Lincoln line except for an uncle—his father's oldest brother—who, because he was the oldest, had inherited just enough money to get a satisfactory start in life, a satisfactory enough start that he was able to break out of the poverty that was so common for Lincolns. But he was the only one. And he wasn't around when Abraham was growing up, so Abraham did not benefit from his uncle's good fortune.

Lincoln's father, orphaned at age six, and unable under the law to inherit anything, had had to make his own way in the world without help. Like many in frontier America, he tried hard, but never seemed able to succeed. Several times as Lincoln was growing up his father uprooted the family to make a fresh start, moving the clan from Kentucky to Indiana to Illinois. Each time the family's troubles, already terrible, grew worse. Shortly after they moved to Indiana Lincoln's mother died, plunging the family into chaos. Then they moved to Illinois, and every member of the family came down with "the ague"—malaria. Their first winter there was one of the worst in the history of Illinois, ever remembered as the winter of the "deep snow." First the blizzards came, then the freezing rain, then the howling winds. The temperature remained so cold so long—twelve below zero for nine weeks—that the farm animals, the horses and the cows, froze where they stood—froze and died. Even the deer died.

Many people on the frontier lived with hardship as Lincoln had, and took pride in surviving. Lincoln did not. In his frontier roots he felt not pride, but shame. Shame because his people were poor and undistinguished. Shame because his people were ill-educated—his mother couldn't even write her own name—and shame, above all, because his people had not only a history of poverty and ignorance but also a history of immorality. His mother's sister, his aunt, had had an illegitimate child, Dennis Hanks, who lived for a time with the Lincolns. And Lincoln's own mother, Nancy Hanks, had been an illegitimate child. His whole life Lincoln refused to talk about her. It would be too simple to say he despised her. He despised no one. And anyway, he seemed to think that his own gifts, which seemed absent in any other Lincolns, could be traced to the man who had fathered her out of wedlock: a well-to-do Virginia planter.

Lincoln thus was eager to escape his family, but for a long time he could not because they needed his help. As was the custom of the time, his father even hired him out to earn cash. As was also the custom, when the young man returned from work, his father pocketed every cent Lincoln had earned.

And then finally the day came when Lincoln was able to light out on his own, ending up in New Salem. Twenty-two years old, he was free, hopeful—and terribly disadvantaged. For not only was his family poor, he was poor. And not only was his family ignorant, he was ignorant. He had spent only a few winters in school. His biographers estimate that in all he had no more than a year's formal education. Lincoln himself recalled that he had learned only "to read, write and cipher."

Lincoln, of course, hoped, as many young men did, to be able to improve himself. And like others, he hoped to improve himself quickly. But unlike others, he had, at age twenty-two, a plan.

New Salem was not a place where the average young man could hope to improve his position very quickly, even if he had a plan. If he was lucky, he might after a few years as a farmhand accumulate enough money to buy a little property and build a cabin. If he was really ambitious, he might dream of starting his own farm. But the movement up and ahead, though usually steady, was usually slow. To become a success a person had to wait, and then, often, wait some more.

And Lincoln was not the waiting kind. Waiting was what his father had done, and it had gotten him nowhere. That was the trouble with waiting: There was no guarantee that it would bring success.

Lincoln at twenty-two was skilled in only one area, unfortunately, farm work. And the waiting was longest for farmers. Only after years of toil could a farmer become successful. And sometimes, after years of work, success didn't come. A man might work and work and work, work and wait until his hands became gnarled and old, and in the end still have nothing. One bad winter could bring ruin. And, likely as not, there would be bad winters, winters like the one the Lincolns faced when they first arrived in Illinois. And bad summers, too, summers when the rain never came, summers when the crops died and shriveled.

Besides Lincoln hated farm work. Hated it even though he was good at it, so good that he became famous for his farm work. His wood chopping especially was legendary. "My how he could chop," a friend remembered. "His ax would flash and bite into a sugar tree or sycamore, down it would come. If you heard him felling trees in a clearing, you would say there were three men at work, the way the trees fell." And it wasn't just chopping wood that he was good at. He was also good at cornhusking. Recalled another friend: "He was the best hand at husking corn on the stalk I ever saw." But Lincoln wasn't interested in chopping wood or husking corn. Wasn't interested even though that was what a young man with his background ordinarily would be interested in. And not only would be interested in but would *have* to be interested in, because there was no other thing for which he was suited. While there was plenty of opportunity on the frontier, it was difficult to take advantage of that opportunity if you didn't have an education, and Abraham Lincoln did not have an education. Nor did he have any of the other advantages that might make up for a lack of education: connections, money, or power. He truly was a person without advantages. As he himself later said, he was at the time he entered New Salem a "friendless, uneducated, penniless boy."

America was the land of opportunity, but for a boy like Lincoln the opportunities were thin.

Unless, that is, the boy had brains—brains and an ambitious plan. Lincoln had both. With luck, the plan would help him leap to success, help him go from being a nobody to being a somebody.

Just how ambitious he was no one guessed when he first arrived in New Salem at age twenty-two. To those who first met him he seemed to have no ambition at all. In those days, as in ours, a man was often judged by how he dressed. Lincoln dressed badly. A fifteen-year-old at the time noted that he wore "a ragged coat, an old drooping hat, and a pair of tattered jean pants, the half of one leg which was then off . . . [and] a coarse pair of gaping shoes."

But if Lincoln at first did not strike people as ambitious, he soon enough showed them he was. On March 15, 1832, little more than six months after arriving in New Salem, he made a startling announcement in the pages of the local newspaper, the *Sangamo Journal*. He was running for a seat in the state legislature.

He was, by then, all of twenty-three years old.

Six months before, when he had moved to New Salem, no one had known who Abraham Lincoln was. He had been so unknown that he had had to fight Jack Armstrong, the local bully and leader of the

Clary Grove Boys, to prove himself. (He had trounced Armstrong in a wrestling match. They quickly had become friends.)

And now he was running for office.

What was remarkable about this, his first campaign for office, was not how he ran. He ran the way candidates always ran. He issued a platform, gave speeches, and traveled around the district to meet the voters. No, there was nothing extraordinary about the way he ran, but that he ran at all.

And that he ran for this particular office. Other young men interested in politics are also known to have run for office. A few even ran at his young age. But when they ran at that age they ran for local office, to fill an opening as an election clerk or as a constable. Lincoln in his very first drive ran for a seat in the state legislature.

To run for any lesser office would have meant waiting longer for success. And Lincoln did not want to wait.

Lincoln knew that his decision to run would no doubt shock people. He tried to explain it: "Every man is said to have his peculiar ambition," he said in his announcement:

> Whether it be true or not, I can say for me that I have no other so great as that of being truly esteemed by my fellow men, by rendering myself worthy of their esteem. How far I shall succeed in gratifying this ambition, is yet to be developed. I am young and unknown to many of you. I was born and have ever remained in the most humble walks of life. I have no wealthy or popular relations to recommend me. My case is thrown exclusively upon the independent voters of this county, and if elected they will have conferred a favor upon me, for which I shall be unremitting in my labors to compensate.

Years later, after he had become a success, some young men in his party complained that "older men" were keeping them down. How, they wanted to know, could they make their mark as he had made his? "You must not wait to be brought forward by the older men," he told them. They had to rely on no one but themselves. "Do you suppose that I should ever have come to notice if I had waited to be hunted up and pushed forward by the older men," he asked. The answer was obvious. He had done it all himself. No one else was pushing Abraham Lincoln at age twenty-three to run for the state legislature but Abraham Lincoln.

The odds, unfortunately, were against him. Not only was he uneducated. Not only was he poor. Not only was he young. He was also new to the community. Those voters he had been able to meet in the six months since he had moved to New Salem he had befriended. But he had not been able to meet many voters. There were fewer than twenty-five families in New Salem. To win one of the four seats from his district, he would have to get to know hundreds and hundreds of families. And he would have to get to know them quickly: There were just five short months until the election.

And then, as brief a time as he had to get to know all those people, he suddenly had even a briefer time. In April, just a few weeks after he entered the race, war broke out with the Indians, and Lincoln signed up for duty. It was a minor war—the Black Hawk War—and Lincoln never saw any action; afterward he would joke that his greatest fight had been with the mosquitoes. But by the time he got back from the war in July there were, instead of just a few months to campaign, just a few weeks.

A few weeks to put the past behind him, a few weeks to make his mark in the community and become somebody, a few weeks to begin putting real distance between himself and his hated frontier roots, a few weeks to end the waiting. A few short, frustrating weeks.

And he didn't even have anyone to help him. No wife to aid in spreading the word, no party leaders to give him advice, no friends to pass out handbills (his friends were busy working). Lincoln didn't even have a horse to ride. His horse had been stolen in the final days of the war.

All he had was himself.

In future campaigns he would be his own greatest weapon, for he was a genius at politics. But the Lincoln who would one day bedazzle the voters, winning them over with his logic and anecdotes and jokes, was not the Lincoln who faced them in the election of 1832. The Lincoln the voters saw that year was, one of them was to remember, "a very tall and gawky and rough looking fellow." His pantaloons didn't even "meet his shoes by six inches." Lincoln himself was well aware of his inadequacies. In his March announcement he had referred to himself as a "youth." "Considering the great degree of modesty which should always attend youth," he had admitted after a lengthy dissertation on the necessity of internal improvements, education, and lower interest rates, "it is probable I have already been more presuming than becomes me."

And yet even then, even at age twenty-three, he showed promise. For even then he possessed his ability to communicate. Not the ability he was to show later, of course, but ability nonetheless. The very same person who noted that Lincoln's pantaloons were six inches too short also noted that "he made a sensible speech." And even though his speech-making was to improve, "the manner of Mr. Lincoln's speeches then was very much the same as his speeches in after life. . . . In the election of 1832 he made a very considerable impression upon me as well as upon other people."

He had eighteen days to make "sensible" speeches—the eighteen days between his return from the war and the date of the election. And they proved to be not enough days. When the election finally came on August 6, Abraham Lincoln lost, ending for now his use of politics as a way out of poverty and obscurity.

But he had not lost badly. There had been thirteen candidates in the race. The top four had won election to the legislature. Lincoln had come in eighth.

And in the New Salem precinct—*his* precinct, where people knew him best—he had won in a landslide. Of the 300 votes that were cast, he had received 277, feeding his appetite to be "truly esteemed of my fellow men."

But it was just a taste. And Lincoln, his appetite whetted, wanted more. Some poor boys on the make spend their lives hungering after money. Others after power. Still others after position. Lincoln shared with these boys, poor boys like himself, the desire for money, power, and position. But above all at this time he craved the approval of the crowd. The election showed that he had the approval of his New Salem neighbors, but their approval was not enough. He needed more, needed to get the approval of his neighbors' neighbors, needed to get the approval of the people who had voted against him. To get their approval he would have to run again.

So he did.

But this time he was to see to it that he had a real chance of winning. The last time he had let the Indian war get in his way, which kept him away from the voters until the final weeks of the campaign. This time he would do everything to win.

Everything meant everything: studying the law, becoming a presence on the local political scene, even, oddly, becoming a postmaster.

He had considered becoming a blacksmith and actually became a surveyor. But his great professional ambition was to become a lawyer; lawyers seemed to run things. Over the next two years, even though he had no legal training, Lincoln began to argue cases before the local justice of the peace. He even drew up legal documents: a mortgage for William Green, a deed for Jesse and Christiana Baker, a bond for David Rutledge (brother of Ann, with whom Lincoln was supposed to be in love).

To increase his visibility in the community he began to attend political functions, taking on burdensome and unrewarding political tasks. In early 1834 he even agreed to be the secretary at a political meeting called to nominate a candidate for governor.

To further increase his visibility he became the New Salem postmaster. What Lincoln realized was that the postmaster was in a particularly advantageous position for a politician. The job would give him access to the local newspapers, allow him to become friends with a wide variety of voters, and give him the power to help people; if they could not afford to receive their mail (at the time the recipient of mail instead of the sender paid its cost), he could use his official frank to let it through for free. Using the frank in this way was not legal, but Lincoln seems not to have worried about that. He frequently used his frank to help people, even though each time he did so he risked a ten-dollar fine. (He never got caught.)

Lincoln later described the postmaster job as insignificant, but getting it was a coup. Jobs in the post office then—and for many decades after—were handed out to party loyalists as a reward for service. Lincoln as a loyal party worker could expect to receive a place in the post office—but only if his party was in power. And his party was not in power. There were two political parties then in existence: the party of Henry Clay, which would evolve into the Whig Party, and the party of Andrew Jackson, which became the Democratic Party. And Lincoln was firmly aligned with Clay, and Clay had not won the presidential election in 1832. Jackson had. And Jackson men believed in using the power of patronage to fill federal offices with members from their own ranks. Fortunately for Lincoln, the Democrats were willing to allow him to have the job of postmaster in New Salem, probably because it was not a terribly lucrative job and no Democrat seems to have vied for it. But obviously an important consideration was that Lincoln, despite his affiliation with the Whigs, was acceptable to the Democrats.

This was not because Lincoln was a likable fellow, though he was. It was because he had shrewdly decided it was in his interest at this time to be acceptable to everybody. Reviewing the election results from his first race, he had concluded that what was important was not engaging in partisan wrangling, tempting as that was, but becoming better and better known. Where he was known best—in New Salem—he had run best.

Avoiding the wrangling had been difficult as Lincoln had become more and more political. But throughout 1833 he had avoided it. And the next year he continued to avoid it even as his political work, already heavy, grew heavier still. And then came the next campaign, in the spring of 1834. And even then he avoided it. The last time out he had issued a long three-point platform; almost laughably long, it took up several pages and included more than seventeen hundred words—longer than President Jackson's recent inaugural address. This time there would be no platform at all, nothing to indicate where he stood on the issues, nothing to indicate how his approach to politics differed from the Democrats'. The last time, even though he had had little opportunity to make speeches because of the short period during which he was able to campaign, he had given a lot of speeches. This time, even though he had months and months and months to give speeches, he gave hardly any at all. Instead of giving speeches he went out and shook hands. Once, he came upon a large group of field workers taking in wheat. Lincoln, rather than give a speech, pitched in to help. He got their support.

He got nearly everybody's support. Even the Democrats agreed to support him. He did not give up his Whig ideas. But he kept them to himself.

And this time he won.

Once again thirteen candidates ran, but Lincoln, who had come in eighth before, now came in second, giving him one of the four seats to which Sangamon County was entitled.

He was twenty-five years old.

Twenty-five years old and broke, more broke now than when he had first moved to New Salem. He had been debt-free then. Now he was burdened with debts. On impulse he had bought a half interest in a New Salem country store, paying for the investment with a promissory note. At the time it had seemed like a smart move; owning a store would give him respect and a steady income. But it would prove to give him neither. He and his partner, an alcoholic who regularly got drunk, were terrible businessmen. In the hole to begin with, they soon got in deeper. The longer they stayed in business the more they lost. And then his partner died, leaving Lincoln to pay not only his own half of the debt load, but his partner's as well. The sum came to more than one thousand dollars. Lincoln referred to it as his "national debt."

To get out of debt Lincoln had taken up surveying, teaching himself the rudiments of the profession the same way he had begun teaching himself the law: by borrowing books and studying them until he knew them by heart. But not even becoming a surveyor could help make much of a dent in his debts. Lincoln, poor when he got to New Salem, was remaining poor, so poor he could not even afford a permanent home. He would live a few weeks with one family, then a few weeks with another. Sometimes he would stay a few months. There were about twenty-five families living in the village then. Lincoln was to live with many of them, renting a back room or just a bed.

The legislature was scheduled to meet in December. Nine days before the session began, the Sangamon County sheriff paid Lincoln a visit. A man who was owed $211 wanted to be paid. *Could Lincoln pay? No.* The sheriff would then have to confiscate his property: his saddle, bridle, and surveying instruments—even his horse.

Lincoln would show up at the first session of the legislature broke, but he would not look broke. Before he arrived he borrowed some money from a friend to buy a new suit—his first suit. It cost a whopping sixty dollars.

What Lincoln did after he arrived is revealing. He drove himself relentlessly. Just four days into the session, he introduced his first bill. Within a week he wrote another bill for a fellow legislator. Soon this self-taught man, who had never taken a single law class or even worked a day in a law office, this man who had not yet received a license to practice law, was writing laws for other legislators who either couldn't write as well as he or who lacked his discipline. Like other members Lincoln was appointed to a permanent committee. Like others, too, he was appointed to special committees. But he was appointed to twelve special committees, more than almost anybody else. And when a report had to be written about the state's contingent fund, a complicated report requiring a deep knowledge of the state's finances, Lincoln, whose own finances were shockingly pitiful, was picked to write it. At the end of the session he was paid $258 for his services. It was more money than he had ever earned.

He was still just twenty-five years old, but it was difficult for people to remember that. He seemed much older. The following December, at twenty-six, he was appointed chairman of a committee.

And then it was time for the next election.

In the last election it had been to Lincoln's advantage not to be too political; despite temptation, he had avoided being political. In this one, just two years later in 1836, it was to his advantage to be highly political. The political parties, formerly inchoate, now suddenly loomed powerful, forcing politicians with an eye on the future to curry the favor of party leaders. And because this was a presidential election, national issues surfaced that added an element of passion to the local political debate.

And now he who had formerly abjured rank partisanship gave himself over to it. He blamed the Democrats for all that was wrong in the country, denounced a critic as "a liar and a scoundrel" (promising "to give his proboscis a good wringing"), and engaged in a raw and ugly debate with a prominent lawyer who had switched his political allegiance to the Democrats after receiving a profitable federal job. "The gentleman commenced . . . by saying that this young man will have to be taken down," Lincoln remarked in rebuttal. "I am not so young in years as I am in the tricks and trades of a politician." Then, referring to his adversary's recent purchase of a lightning rod, Lincoln sneered that he "would rather die now, than, like the gentleman, change my politics and simultaneous with the change receive an office worth $3,000 per year, and then have to erect a lightning-rod over my house to protect a guilty conscience from an offended God."

Of the four candidates to win election from Sangamon County the last time, Lincoln had come in second. This time he came in first. People liked him.

And he was going places. Just a few months before, he had finally received a license to practice law. A few months later he was registered as an attorney with the Illinois Supreme Court. A few weeks after that he moved from New Salem, which by then was well on its path of decline, to Springfield, where he joined the law office of John Todd Stuart, one of the leaders of the Whig Party. (Stuart had recently run for Congress but lost; he would run again in 1838, against Stephen Douglas, and win.)

Both inside and outside the legislature Lincoln was becoming a powerful figure, but especially inside. Although he still struck many as odd looking, and in repose appeared listless, once he began speaking his face became animated and interesting. And at this session, as his confidence grew, he began speaking more and more often. The newspapers even began printing his statements from the floor.

But it was not for his speaking that he was best known. It was for his work behind the scenes. Two proposals dominated the legislature that session: first, the plan to use state funds to finance a colossal ten-million-dollar internal improvements project, including the construction of railroads, canals, and roadways; and second, a scheme to move the state capital from Vandalia, which was located in the southern part of the state, where the population had remained stagnant, to Springfield, which was located in the middle, near where the population was booming. Lincoln was a leader in both causes. Both won the support of the legislature.

In the next session Vandalia's representatives made a desperate attempt to sabotage the Springfield move. Lincoln singlehandedly derailed the effort.

As Robert Caro has pointed out, practical politicians, when they talk with one another, commonly refer to "easy votes" and "hard votes." Lincoln's votes to support internal improvements and the move to Springfield were "easy votes," easy because, while they were controversial in some parts of the state, they were not controversial in his: The internal improvements would directly benefit his county, as would the move to Springfield, which was located in his county. Voting in their favor thus seemingly posed no threat to Lincoln's career.

Almost all the votes Lincoln cast in the state legislature fell into this same category. One, however, clearly did not: a vote on a measure to approve of slavery in the Southern states. For Lincoln it was a "hard vote," hard because his constituents were in favor of the measure and he was against it. If he voted his conscience, as he wanted to, he could wreck the career he had been working so hard to build up. Years might go by before he was forgiven—if he was ever forgiven.

The people of Illinois were not themselves in favor of slavery. Their own constitution outlawed it. But they were content to let it remain where it already existed. Content not because they were indifferent to the immorality of slavery—though some were; at the time most of the people who lived in Illinois came from the South and lived in the southern half of the state—but because they could not envision how the country could handle the freeing of millions of black slaves. Where would they go? While many white Americans had long hoped that freed blacks could be shipped back to Africa—a hope Lincoln himself was to share —there was no expectation that they would be. That left open the possibility that emancipated blacks would roam freely over the country, settling wherever they chose—perhaps even in Illinois.

And that prospect frightened and appalled the great majority of the white people of Illinois.

The prospect that blacks might freely roam the country also frightened and appalled white Southern slaveholders, of course. But they had a more immediate fear than that their slaves might someday be given their freedom. They lived in fear that their slaves might try to make themselves free *now*. In 1831 Nat Turner had tried to lead the slaves of Virginia in a violent and bloody rebellion. Might not others try to do the same? Soon there were rumors that others were. For two years, from 1835 to 1837, white Louisianans read in their newspapers that their slaves were secretly plotting to rebel and kill their masters. The reports were not true, but whites in Louisiana and all over the South believed that they were. Frightened as whites had been after Nat Turner's revolt, they now grew even more so.

There were many reasons for the slaves to revolt, the most obvious being, of course, that they were not free. But because slaveholders had told themselves for years that their slaves were happy, they felt compelled to find another cause for their slaves' discontent and they did: the birth of abolitionism, which conveniently coincided with the reports and rumors of mass slave uprisings. The more white Southerners thought about abolitionism, the more they became convinced that it was the root cause of the perils facing the "peculiar institution." Had not William Lloyd Garrison, the abolitionist editor of the *Liberator* (he who had defiantly declared: "I am in earnest—I will not equivocate—I will not excuse—I will not retreat a single inch—and I will be heard") begun publishing his newspaper the very same year as Turner's revolt? He had.

As the cries of the abolitionists began to be heard in the 1830s, whites in the South tried to smother the sounds before they grew louder. They tried to persuade Congress automatically to table petitions seeking the abolition of slavery. They enacted laws to restrict the shipping of abolitionist literature through the mails. They drove home-grown abolitionists out of the region. When those measures proved inadequate, they went further. In an effort to choke off abolitionism at its source—in the North—they began enlisting the state legislatures in the North in their cause. Many Northern legislatures—including that of Illinois—quickly joined, passing resolutions in defense of slavery.

Illinois's resolutions were among the most extreme. They declared that under the Constitution, Southerners had a "sacred" right to own slaves, a right the federal government was powerless to abridge. The resolutions went on to declare that slavery could not legally be abolished in Washington, D.C., as the abolitionists were demanding, and charged that abolitionism was a menace to the Union.

The Founding Fathers—though embarrassed by slavery—had considered the institution a necessary evil. These resolutions made slavery seem almost like a positive good (an argument Southerners themselves were to advance in the 1850s as the institution came under renewed attack).

Seventy-seven legislators voted in favor of the resolutions, six against. Lincoln was one of the six.

He would later say that he hated slavery, hated it as much as the abolitionists did, and that he had always hated it. As a young man traveling down the Mississippi he had been horrified at the sight of slaves chained together like animals. His conscience, therefore, required him to vote against the resolutions, even if doing so put his career at risk.

But he did not intend that it should. There had to be a way to keep both his conscience and his seat. Blessed with a powerful mind, he figured out how: It was pure political genius.

In a statement he and a fellow legislator issued at the end of the session, he agreed with the legislature that slavery was legally protected in the states where it then existed. He also agreed that abolitionism was destructive, which protected him from the ruinous charge that *he* was an abolitionist, that *he* wanted the federal government to abolish slavery immediately. But unlike the legislature, he held that "the institution of slavery is founded on both injustice and bad policy," and argued that the federal government could abolish it in the city of Washington, D.C., with the consent of the citizens there.

There was at the time a feeling that one could take just two positions on the issue of slavery: either for it or against it. Lincoln showed that there was a third. His was not a novel position, however. The Founding Fathers had shared it, believing that slavery was evil and that it should be allowed to remain in existence for the time being in the expectation that, like a noxious weed, it could be left to die on its own instead of being pulled out. But in the heated politics of the 1830s most Illinois politicians, frightened by the rhetoric of the abolitionists and unwilling to risk being identified with them, rushed unthinkingly to the proslavery position.

Lincoln undoubtedly ached to strike a blow against slavery, ached to join the abolitionists in fighting it. If he had not been a politician, he very well may have. Slavery deeply offended him. If slavery is not a sin, he would subsequently intone, nothing is. But he was a politician. So when he tackled the issue of slavery, he tackled it carefully.

It was wise that he did so. The politics of slavery, already explosive, were to become in 1837, the year of Lincoln's statement, more explosive still. There were many explosions around the country in 1837. Illinois was where the worst one took place.

It happened in Alton, a small town on the Illinois side of the Mississippi River, deep in the southern half of the state—the southern and conservative half. There in July, Elijah Lovejoy, an antislavery publisher, began producing articles in his newspaper in support of abolitionism. It was just words he wrote. But given the combustibility of those words, he may as well have been using dynamite. In August a mob of Southerners destroyed his printing press, dumping it in the Mississippi. Subsequently friends helped him obtain another press. In September the new printing press was destroyed and dumped in the river. Again he obtained another press. And in November that one, too, was destroyed. In the course of this attack the mob fired on Lovejoy and several of his supporters. Lovejoy was killed.

There were many ways a politician could exercise leadership in this crisis—if he had power. Lincoln did not yet have power. He could not command the army to put down the mobs that were upsetting the peace. He could not convene a conference where proslavery and antislavery adversaries could try to work out their differences. And not for a long, long time would he have power. So he did what politicians without power and without resources have done throughout history when they were eager to try to affect things and make a name for themselves: He gave a speech. It was in January 1838, just a little more than two months after Lovejoy's death.

The speech was delivered before Springfield's Young Men's Lyceum, a local cultural institution, and because it was delivered there Lincoln made sure he sounded statesmanlike. Following in the tradition of the great statesmen of the era—Webster, Clay, and Calhoun—he adopted a noble stance. The speech was filled with references to liberty and justice and character. Eager to strike a patriotic note, he frequently cited the Founding Fathers and even managed to work in references to both the Declaration of Independence and the Constitution. He did all this so successfully that his speech could almost have been mistaken for one by Webster, Clay, or Calhoun—at least the main part could have.

The main part of the address was concerned with the outbreak of violence. Alluding to the Lovejoy incident among others, Lincoln denounced the "vicious portion" of the population which "gathered in bands of hundreds and thousands" to "burn churches, ravage and rob provision stores, throw printing presses into rivers, shoot editors, and hang and burn obnoxious persons at pleasure, and with impunity," warning that if they were not stopped "this Government cannot last." He then argued that the only way to make sure the government did last would be to instill in "every American, every lover of liberty" a "reverence for the laws": "Let every man remember that to violate the law, is to trample on the blood of his father."

The second part of the address, however, was pure Lincoln.

Compelling as the recent events of violence were, in this second part Lincoln could not resist addressing another subject, one more immediate to him even than the fatal riot that recently had rocked nearby Alton: ambition. His stated reason for bringing up the topic was his belief that the country faced a grave threat from the rise of "men of ambition and talents" whose goal in life would be to aspire for something "beyond a seat in Congress, a gubernatorial or a presidential chair." As he put it somewhat hysterically:

> What! Think you these places would satisfy an Alexander, a Caesar, or a Napoleon? Never! Towering genius disdains a beaten path. It seeks regions hitherto unexplored. It sees no distinction in adding story to story, upon the monuments of fame, erected to the memory of others. It denies that it is glory enough to serve under any chief. It scorns to tread in the footsteps of any predecessor, however illustrious. It thirsts and burns for distinction; and, if possible, it will have it, whether at the expense of emancipating slaves, or enslaving freemen.

Ambition was not at the time a subject on most people's minds. Violence was. But it was evidently a subject on Lincoln's mind, and his real reason for bringing it up would appear to be that he was obsessed by it. He was still very young, just twenty-eight: a young man full of ambition—and evidently uneasy with it. In the speech he asked: "Is it unreasonable . . . to expect, that some man possessed of the loftiest genius, coupled with ambition sufficient to push it to its utmost stretch, will at some time, spring up among us?" And that this man would, like a Caesar, subvert our liberties? It was not unreasonable at all. "Distinction will be his paramount object; and although he would as willingly, perhaps more so, acquire it by doing good as harm; yet, that opportunity being past, and nothing left to be done in the way of building it up, he would set boldly to the task of pulling down."

Was Lincoln thinking of himself? There is no way of knowing if he was. But what is known is that he was very much concerned at this time with the theme of ambitiousness. To his one close friend, his only close friend in life, Joshua Speed, he had confessed on a recent occasion that he was ferociously ambitious, driven by the desire to link his "name with something that would redound to the interest" of his fellow men. Thus far, he had conceded, he had not fulfilled his ambition, having "done nothing to make any human being remember" that he had lived. But he remained hopeful.

Understanding Lincoln's ambition is essential to understanding Lincoln. But it was not a subject on which he himself was open. Although he was a prodigious writer—his collected works run to ten stout volumes numbering thousands of pages—only a few of those pages include the admission that he was ambitious, so very few that the same passages get quoted over and over and over again by historians eager to come to grips with his relentless struggle to improve himself and his position. His law partner for two decades, William Herndon, complained that Lincoln was the most "shut-mouthed" man who ever lived. This is also quoted over and over again by historians in explanation of the dearth of revealing quotes from Lincoln. And Herndon was not the only person to believe that Lincoln was "shut-mouthed." So did many of the others who came into contact with him. Leonard Swett, who rode with Lincoln on the circuit for eleven years, remarked that Lincoln "always told enough of his plans and purposes to induce the belief that he had communicated all; yet he reserved enough to have communicated nothing." David Davis, a political confidant for years, related that Lincoln "was the most reticent, secretive man I ever saw or expect to see." Lincoln himself admitted: "I am rather inclined to silence."

His ambitiousness then is not to be found in his words for he spoke few words on the subject. It is to be found instead in his deeds. And they speak loudly. Barely out of his teens, and with no education, he ran for the state legislature. Then he became a surveyor. After that he taught himself the law and became a lawyer. And, of course, he continued running for the state legislature.

In 1838 he ran for a fourth time—ran and won, of course. As he had in the last election, he won more votes than any of the other candidates in his district. What he had once wanted most—to be "truly esteemed by my fellow men"—he was finally getting.

And still it was not enough.

Lincoln, who had succeeded in winning the approval of his county's voters, now was to try to win the approval of fellow legislators. When the legislature met he immediately announced his intention to run for speaker of the house.

He could not win. The Democrats outnumbered the Whigs. And he did not win. But if he could not be speaker, he could still try to become a party leader. And he became one: the Whig floor leader.

He was twenty-nine years old: still broke, still deficient in his education—but profoundly ambitious, and slowly but surely, becoming more and more successful. As Herndon was to say years later, "His ambition was a little engine that knew no rest."

And like the good solid engine it was, it kept pumping and pumping and pumping.

And then Lincoln, who for years had been more and more successful, suddenly sensed success eluding him.

It was 1840, another election year. Again Lincoln wanted to run for the state legislature. Again he announced that he was a candidate. But this time there were chilling indications that the party he wanted to lead did not want *him*. He told friends he thought he might not even be nominated. As he gave the matter greater thought, he finally began to conclude it was likely he would not be nominated. He did not think his "prospects individually are very flattering, for I think it is probable I shall not be permitted to be a candidate."

The reason he felt this way is unclear. But it was probably because of his strong identification with the colossal internal improvements project he had helped get passed. In the years since it had proved a disastrous mistake. Hugely expensive, it had almost forced the state into bankruptcy. (A year later the state, dead broke, found it could not pay off its bondholders. After great delays, the bondholders finally received fifteen cents on the dollar.) And even though the people had been united behind the program, and both parties had supported it, and his party had made the passage of the project its chief legislative goal, Lincoln as the house Whig party leader was blamed when things went badly.

Despite his fear that he would not receive a nomination, he did. And he was grateful. But it was for a seat in the lower house, the same seat he had held since he was twenty-five, six years earlier. There had been an opening for a seat in the higher house. Lincoln had been passed over for it. A friend with no legislative experience at all had been selected to run for that post.

What had once seemed such a promising political career now suddenly seemed a little less promising. And there didn't seem an obvious way to turn things around. He had no hope of a state appointment, no hope of a federal appointment. Although he was an executive member of the Whig central committee, the state committee that ran the party, it was not his ambition to be a party boss. Perhaps, it began to occur to him, it was entirely possible that he would never amount to much in politics, that all he would be he already was: a member of the minority party of the lower house of the state legislature of Illinois.

His one hope that he would amount to something more rested on his renown as an above-average speaker. His speaking ability, already impressive at age twenty-three when he first ran, had steadily become more and more impressive. Even voters who disliked him for his stand on internal improvements liked to come hear him speak. He was fun to listen to. Unlike many other stump speakers he did not put on airs, and he knew how to connect to folks. He used jokes and anecdotes to make his points as well as logic and facts.

Lincoln was sure it was his speaking ability alone that had persuaded the Whigs to renominate him to the state legislature. "The fact is," he wrote his law partner John Stuart, by now a congressman, "the [rural] country delegates made the nominations as they pleased; and they pleased to make them all from the country, except [Edward] Baker & me, whom they supposed necessary to make stump speeches."

Winning over audiences had always been important to Lincoln. But it had never been as important as it was now. His career seemingly depended on his ability to connect with his audiences. It would do Lincoln little good, however, to keep connecting with audiences in Sangamon County. He had gone as far as a politician could go in Sangamon without being named to either Congress or the state senate, and he had not been named to either of those posts and likely would not be, not at the present anyway. But it could do him a great deal of good to begin giving speeches in the rest of the state. Fortunately, this was a presidential election year. And the Whigs in the rest of the state needed strong speakers. When they began asking him to come and speak, he readily accepted.

The Whigs had not had any luck electing a president. Andrew Jackson had beaten them twice (before they had adopted the name Whigs and were still known as National Republicans) and then his hand-picked candidate, Martin Van Buren, had beat them, too. But in 1840 the Whigs, borrowing from the Democrats' victory book, had found their own Andrew Jackson to rally around, the aging William Henry Harrison.

Harrison was an appalling presidential prospect. Lincoln, ever the loyal party politician, supported him enthusiastically. In the summer, when the crusade to elect Harrison began in earnest, Lincoln, shrewdly exploiting the opportunity to make a statewide reputation, traveled as he had never traveled before. "He stumped all the middle and lower part of the state," a friend remarked in awe, "traveling from the Wabash to the Mississippi in the hot month of August, shaking with the ague one day, and addressing the people the next." And he continued to campaign into the fall, even though doing so took him away from his law practice.

He not only traveled more this election, he campaigned more, giving more speeches than he ever had before. And as he campaigned he began giving different speeches than he ever had. Previously in campaigns he had refrained from demagoguery. In his speech to the Young Men's Lyceum in Springfield he had insisted that emotion had no place in politics. "Reason," he had declared, "cold, calculating, unimpassioned reason, must furnish all the materials for our future defense and support." But now, impelled by ambition to make a bigger name for himself—and scared that it was only his ability to connect with audiences that could help him win a bigger name—he embraced emotion.

Of all the emotional appeals a politician can make in the United States, one always has been guaranteed to provoke a powerful audience reaction: the invocation of race prejudice. Such an appeal was especially powerful at this time, given the raw tensions aroused by the debate over abolitionism. And this was just the appeal Lincoln now made. In debates with Stephen A. Douglas, then a rising star of the Democratic Party, Lincoln denounced Van Buren for having once voted to give blacks the right to vote in New York State. In an early debate Douglas retorted that Van Buren had not in fact done so. But at a dramatic moment Lincoln subsequently produced a letter from Van Buren admitting that he had. A furious Douglas spluttered at him in rage.

There were some parts of the state, the northern parts, where the exploitation of race prejudice, though possibly effective, would not be advisable; the north was being settled by New Englanders who generally shared a more benevolent attitude toward blacks. But Lincoln was not speaking to audiences in the northern parts. He was speaking in the southern parts to audiences largely made up of Americans who (like him) had migrated from the South. And there racism played well.

Lincoln did not dwell on the subject of race. And no one at the time seems to have felt he was exploiting it. Had Douglas not reacted so strongly to Lincoln's introduction of the Van Buren letter, creating a dramatic incident for people to talk about, it is possible that no one would have taken any notice of Lincoln's remarks on race at all. In most of his speeches he concerned himself simply with standard economic Whig doctrines: the importance of state banks and internal improvements. But Lincoln rarely did anything in politics unthinkingly. Evidently he included the racist appeals because he felt he had to. He had to connect, had to win over his audiences as he never had to before. If he did not, he might not have the future he hoped to have.

Eight years earlier he had taken a courageous stand on slavery. And in the future he would continue to do so whenever he could. But on the issue of racism he was unwilling to take a courageous stand. Racism, Lincoln believed, was well-founded as slavery was not. Like most Americans at the time he shared the belief that blacks were inferior. So now, when opportunities arose to curry the favor of audiences by appealing to peoples' racism, Lincoln exploited them for political gain.

And then the returns came in.

And instead of coming in first in Sangamon County as he had the last two times, he came in fifth, barely holding on to his seat. It is difficult to say why he had lost ground. It may have been because of his association with the internal improvements boondoggle. Or it may have simply been that he had spent so much time in the southern part of the state campaigning for Harrison that he had not done enough campaigning for himself. But he *had* lost ground.

And his efforts on behalf of Harrison had not helped him gain ground. For his efforts, while prodigious, had not been sufficient. Harrison lost Illinois. (Of course he won the presidency.) And his loss in Illinois reflected on those who had tried to help him win—tried and not succeeded, although it was also well known that Illinois leaned heavily Democratic; no Whig candidate for president would ever win Illinois.

By any standard Lincoln's political future was in some jeopardy, even though he had done well as a speaker. Feeling insecure, he made a damaging political mistake, the worst so far in his career. It came in November, just after the election, when he attended a special session called by the governor to save the state from bankruptcy.

Under the law, state-chartered banks, which were run by Whigs, had been allowed to use paper money instead of silver or gold to pay off their debts because of the current weak economic climate; the internal improvements projects had overburdened the state with debt just when a panic had struck, resulting in a devastating depression in Illinois. The law was scheduled to expire at the end of the current legislative session. But the Whigs had come up with an ingenious plan to extend the law: They would simply not allow the legislative session to end. They would keep the legislature in special session until the regular session began in December.

That was the plan, anyway—the Whig plan. The Democrats had another. They would adjourn the house before the beginning of the regular session. The Whigs, aware of this, stayed away from the legislature to prevent the quorum needed for adjourning. But to make sure a quorum wasn't simply declared in their absence without the taking of the ayes and nays—as it could be under the rules—they kept two Whigs on the floor at all times to demand a vote on any quorum call. (The Democrats lacked enough votes to make a quorum on their own.)

Lincoln was one of several Whigs on the floor one day when the Democrats declared a quorum was present. According to the Whig plan he promptly called for a vote, but then realized that enough Whigs were present to make a quorum. He and another Whig tried to make it out the door, but could not; it was blocked by a burly sergeant at arms.

At this point Lincoln might simply have remained and accepted defeat. But he panicked. He and a fellow Whig jumped out the second-story window.

For this act—a useless act since he had already been counted as voting, giving the Democrats the quorum they wanted—Lincoln became a laughingstock. "By his extraordinary performance on this occasion," one of the state's leading newspapers commented, "Mr. Lincoln will doubtless become . . . famous. We learn that a resolution will probably be introduced into the House this week to inquire into the expedience of raising the State House one story higher, in order to set in the third story so as to prevent members from jumping out of windows!"

These were troubled times for Lincoln. And they were about to become more troubled. He had been scheduled after the election to marry Mary Todd, a Kentucky belle who came from an aristocratic family. But on the eve of the ceremony he broke off the engagement and slipped into a profound depression, brought on by questions of self-worth and his readiness for marriage. Always insecure about his background, he was now more insecure than ever. Friends worried that he might be suicidal. There were rumors that he'd gone crazy. He stayed in bed for an entire week. "I am now the most miserable man living," he lamented. "If what I feel were equally distributed to the whole human family, there would not be one cheerful face on the earth. Whether I shall ever be better I can not tell; I awfully forebode I shall not. To remain as I am is impossible; I must die or be better."

He was to get better, of course. Two years later, having slowly regained his self-confidence, he finally married Mary Todd. But getting better had cost him time, precious time for a man in a hurry as Lincoln was. And now he had not only himself to care for, but a family. As ambitious as he had been before, he now became even more so.

He did not again run for the state legislature. Instead in 1843 he set his sights higher, on a seat in Congress.* "If you should hear anyone say that Lincoln don't want to go to Congress," Lincoln wrote a friend, "I wish you . . . would tell him . . . he is mistaken. The truth is, I would like to go very much."

As a Whig, Lincoln had, fortunately, an advantage. Whigs dominated the seventh district, where he lived. Winning the party nomination would be tantamount to winning the general election. But first, of course, he would have to win the nomination, and winning wouldn't be easy. Two other deserving Whigs also wanted the nomination: Edward Baker and

---

*The election for Congress was held in Illinois in 1843 instead of 1842, as would normally have been the case, because of a delay in reporting the census. The next election, reverting to tradition, was held in 1844.

John Hardin. And both of them seemed to be in a better position to secure it. Baker, a state senator, was a practiced orator known for a crowd-pleasing gimmick. At stump speeches he would bring along a trained eagle. When he came to the part in his speech where he denounced the Democrats, the eagle would suddenly lower its head and drop its wings. When he began speaking about the glorious future Americans could have under the Whigs, the eagle would magically raise its head and majestically spread its wings, letting out a dramatic scream. Hardin was less impressive as an orator, but perhaps an even tougher opponent. A college graduate, he was extremely well connected; his father had been a U.S. senator from Kentucky.

Lincoln, too, was a tough opponent, despite his obvious failings as a state legislative leader. A self-taught lawyer, he spoke well and he had lots of friends. But enough friends?

There were two stages in the nominating process. First there were individual county conventions, then a district-wide convention. At the convention held by Sangamon County—Lincoln's county—the contest came down to Lincoln and Baker. Here, if anywhere, Lincoln should have been able to win since he had many friends here. It was here in Sangamon County that he had first begun running for office at age twenty-three, here that he had run up his impressive first-place finishes in two state legislative races.

But he did not win. Baker did, Baker who had campaigned with an eagle. To compound Lincoln's troubles, the Sangamon County Whigs then picked him to serve as a delegate for Baker to the district convention. Lincoln, the dutiful Whig, complied, but he complained to a friend that he was " 'fixed' a good deal like a fellow who is made a groomsman to the man what has cut him out, and is marrying his own dear 'gal.' "

Unable to win his own county's support Lincoln immediately began trying to win the support of nearby Menard County. At the least, with its support he could be a player at the district convention. At best, he might even emerge with the nomination—if the Baker and Hardin forces deadlocked.

There was just a small chance that Lincoln would walk away with the nomination, but it was a chance he was willing to take. For it might be his only chance to win for years and years. The seventh district was the most solidly Whig district in the state. Whoever won the nomination that year would probably be elected to Congress not just that term but the term afterward and the term after that and the term after that. The current holder of the office, Lincoln's onetime law partner John Stuart, had won two terms in a row and could easily have won a third; only his desire to retire had stopped him from running and winning a third time. Whoever took Stuart's seat could, therefore, expect to remain in Congress for years. If Lincoln did not win this time, he could be out of the running for many, many years. Lincoln, who hated to wait, might have to wait and wait and then, like his father, wait some more—wait and perhaps never savor the political success he so very much wanted.

Lincoln was still quite young: thirty-four. After waiting five or even ten years he would still be relatively young. But he would no longer be the wunderkind of Whig politics in Illinois that he had been. Already he was being outdistanced by men who were younger. (Hardin was thirty-three, and Baker thirty-four.) And as the years went by there would be more and more young men yearning to outdistance him.

As he expected, Lincoln won the support of Menard County, putting him in the unique and uncomfortable position of being a delegate for Baker from Sangamon County and a candidate himself from Menard. But for him to win the nomination Baker and Hardin would have to deadlock, and it quickly became apparent that they would not. Hardin clearly had the edge. And he would probably win. And even if he did not win for some reason, then Baker would. The one thing that was clear was that Lincoln would not.

Lincoln had planned. His plan had gone awry. But now as he contemplated the bleak situation in which he found himself, he came up with a new plan, a brilliant plan that addressed the chief danger to his career: the real possibility that whoever won the election this year would keep winning elections for years and years, blocking his own advancement for years and years.

Previously it had been the custom among the Whigs in Illinois (as in most other parts of the country) to let a candidate who was a proven winner keep on winning, election after election, even at the risk that he might become so entrenched that he grew deaf to the people's concerns. Lincoln now was to challenge that custom. If, he reasoned, he could persuade the Whigs in Illinois to reject on philosophical grounds the automatic reelection of incumbents—as some Democrats had in a few other states—then it might not matter who was elected this time. For at the next election the Whigs would insist on someone else.

The district convention in 1843 was held in Pekin, Illinois. It was there that Lincoln put his new plan into effect. Taking advantage of the split in the convention between the Hardin and Baker forces, Lincoln proposed that Hardin, Baker—and he himself—take turns as the party's nominee. Hardin could run this time around, then Baker in 1844, then Lincoln in 1846. (Lincoln could plausibly put himself in the line of succession because of his victory in Menard County, but of course he was farther back in the line than either Hardin or Baker.) After some debate the convention accepted the Pekin Agreement, as it came to be called, on the grounds that "turn about is fair play."

Lincoln, who hated waiting, would now have to wait some more. But unlike the farmer who has to wait and wait and then, perhaps, discover that his waiting was in vain, Lincoln's wait would not be in vain. Through a deft political maneuver he had succeeded in winning a mortal lock on the nomination in 1846, guaranteeing him relief from uncertainty, a kind of relief seldom known by an American politician. All he had to do was wait.

Hardin, as expected, went on to win the election in 1843. Baker, in turn, went on to win in 1844. Then came Lincoln's turn.

If ever there was a sure thing in Illinois politics, it would seem to have been Lincoln's nomination to run for Congress in 1846. The two candidates who would have had the best chance of defeating Lincoln had both agreed at Pekin that they would not try. And nobody else seemed to want to either. The Pekin Agreement, of course, did not legally bind any future convention; the Whigs could decide on impulse to nominate somebody else. Politics often is driven by impulse. But there was no indication the Whigs wanted to nominate somebody else. Lincoln could relax.

But, being Lincoln, he could not relax—and he did not. Beginning in the fall of 1845, nearly a full year before the election, he carefully began lining up support for his own nomination. There were five Whig newspapers in the district; he won the support of four of them. As he rode the legal circuit he stopped in at the homes of important Whig leaders, gathering their endorsements. And then, even though both Baker and Hardin had approved the Pekin Agreement, Lincoln went to each of them directly to find out if they intended to honor it. Baker, the incumbent, expressed the hope that he could remain in Congress. (Lincoln felt that Baker was popular enough so that he probably could remain if he wished.) But Baker agreed that he had an obligation to step aside and promised he would. Then there was Hardin.

Hardin did not say that he would be a candidate for Congress. But neither did he say he would not be. Lincoln drew the conclusion that Hardin would be. And soon there was an indication Lincoln was right. A paper under Hardin's influence suddenly began touting Lincoln as the next Whig candidate for governor. This was a devious ploy. Lincoln did not want to run for governor, and there was no chance in any case that he could be elected. The purpose of floating his name for the office was simply to try to derail his congressional candidacy. If enough people talked him up for governor, he would have trouble pressing his campaign for Congress.

In January, Hardin openly announced his candidacy for Congress. But by then it was too late. While Hardin had dithered, Lincoln had accumulated wide support—accumulated it and then kept quiet about it. Keeping quiet was essential. His silence meant that Hardin was in the dark about the extent of his efforts while there was still time to undermine them. In February Hardin withdrew. In May, Lincoln was nominated. Finally he would have his seat in Congress, the seat he had worked to get for four long years.

Then, just ten days after he won the nomination, the United States went to war with Mexico.

And that seat he had wanted so much, the seat that seemed so valuable as a symbol of glory and success, suddenly seemed less so. Now it began to appear that the way to make a name for oneself was not by winning a seat in Congress but by getting a place in the war effort. Soon the most ambitious men in Whig Illinois politics, the very men Lincoln had been competing with for years—Hardin and Baker— joined the army to fight in the war. Baker's decision to join was especially noteworthy. To join he had to resign the very seat in Congress that Lincoln was now nominated to fill.

Lincoln could have joined Hardin and Baker on the battlefield, but he seems never to have given it any thought. His place was in politics. Nothing would get in the way of his dream of political glory, not even war.

Especially not war. Lincoln despised war, despised killing. And he especially seems to have despised this war, believing it to have been provoked by President Polk's decision to send Zachary Taylor's army into territory long claimed by Mexico.

When exactly Lincoln reached this bold conclusion is not known.

But he clearly entertained reservations about the war from the outset—entertained them and kept silent about them.* During the election campaign, which coincided with the first four months of the conflict—the four months when battle fever was at its peak—Lincoln declined to say anything against the war: not a single word.

As careful as he was to conceal his real views in public, he was even more so in private. And he was more careful because he had to be. In public settings nobody questioned Lincoln closely about his views, never put him in the dock and cross-examined him. After fifteen years in public life he was considered to be so well known that he did not need to be cross-examined. Voters felt that they knew who he was and what he stood for. But in private there was one person who routinely *did* put him in the dock and cross-examine him: his own law partner, William Herndon. And Herndon was an enthusiastic backer of the war. Lincoln had to be extra careful when talking with him about his views if he was to keep his private reservations truly private. ("He was the most reticent, secretive man I ever saw or expect to see.")

Lincoln had evidently made two decisions during the campaign, both of them highly significant for what they say about him and his ambitions. His first decision was that he would stay in the race rather than volunteer to fight in the war. His second was that he would decline to express any doubts about the war. Neither the war nor his incipient opposition to it would get in the way of his drive for a seat in Congress.

Once in the distant past he had let a war—the Black Hawk War—

---

*In a speech he was to give as a congressman on January 12, 1848, Lincoln alluded to the existence of early reservations about the war. Accused by Democrats of having concealed his reservations, he defensively remarked: "When the war began, it was my opinion that all those who, because of knowing too *little*, or because of knowing too *much*, could not conscientiously approve the conduct of the President, in the beginning of it, should, nevertheless, as good citizens and patriots, remain silent on that point, at least till the war should be ended. Some leading Democrats, including Ex-President Van Buren, have taken this same view, as I understand them; and I adhered to it, and acted upon it, until since I took my seat here; and I think I should still adhere to it, were it not that the President and his friends will not allow it to be so." To deflect criticism of his patriotism Lincoln, like other National Whigs, continued to vote supplies for the troops in the field while continuing to deny they should be in the field. See Roy Basler, ed., *The Collected Works of Abraham Lincoln* (1953–55), vol. 1, p. 432.

get in the way of his plans. This time he would not.

Lincoln, all agreed, was an extraordinarily honest man, unusually honest for a politician. In his years in the state legislature he had built up such an unassailable reputation for honesty that even Democrats trusted him. "Honest Abe," he was called—by both Whigs and Democrats alike. Never once in his career, even at the height of the controversy over his role in the passage of those budget-breaking internal improvements projects, had his critics doubted his honesty. But on the occasion of his election to Congress, the most important election in his career thus far, and one he had looked forward to for four long years, he seems to have been less than honest with the voters about his views on the most important issue facing them, a matter literally of life and death: the Mexican War.

His opponent, Democrat Peter Cartwright, an itinerant preacher, never suspected that Lincoln might have been a closet opponent of the war. Nor did anyone else. During the four months of the campaign Lincoln had to defend himself against many charges, some extremely damaging and all of them untrue—that he was an infidel, that he approved of drunkenness, that he socialized exclusively with the rich—but never once did he have to defend himself against the charge that he opposed the war, which would have been the most damaging of all and the one charge that was true.

In August, Abraham Lincoln, who had settled in New Salem nearly fifteen years earlier, a "friendless, uneducated, penniless boy," became Illinois congressman Abraham Lincoln. He won by a huge majority, 6,340 to 4,829, a bigger majority than either Hardin or Baker had received.

Being a congressman was what Abraham Lincoln had always wanted to be. And for a year and a half it appeared that he was happy being one, although he did admit to his friend Joshua Speed that his election had "not pleased me as much I expected." But then the thirtieth Congress, to which he had been elected, finally met at the end of 1847. And as soon as it did, Lincoln began using his position as a U.S. representative to help him go on to something else.

As a first step he decided to try to raise his political profile.

On December 22, just two weeks after being seated, he issued an audacious and cavalier challenge to President Polk, demanding to know exactly on what spot of American soil American blood had been shed in the battle that began the Mexican War. (Polk had insisted that the Mexicans had started the war, "shedding the blood of our citizens on our own soil.") In a paper that came to be known as the "Spot Resolutions," Lincoln asked: Was American blood shed or not shed on territory wrested from Spain by Mexico during its revolution? Was it or was it not shed on soil inhabited by Mexicans in a settlement that "had existed ever since long before the Texas revolution"? And then in his final two queries, he asked:

> Seventh: Whether our citizens, whose blood was shed, as in his messages declared, were, or were not, at that time, armed officers, and soldiers, sent into that settlement, by the military order of the President through the Secretary of War—and
>
> Eighth: Whether the military force of the United States, including those citizens, was, or was not, so sent into that settlement, after Genl. Taylor had, more than once, intimated to the War Department that, in his opinion, no such movement was necessary to the defence or protection of Texas.

And then, two weeks later, he took the floor of the House of Representatives to follow up his "Spot Resolutions" with a dramatic speech in which he scathingly denounced the Polk administration for getting the country into the war, flatly declaring that the conflict was both unnecessary and unconstitutional. Going further than he had in his resolutions, in which he had suggested that American blood *might* not have been shed on American soil, he now pointedly charged that it *had* not been. In a devastating conclusion, he criticized President Polk for failing to say when the war might be over. (The war had by then gone on for twenty months; at the outset Polk had implied that it would not last even four.) "He is a bewildered, confounded, and miserably perplexed man," Lincoln declared. "God grant he may be able to show, there is not something about his conscience, more painful than all his mental perplexity!"

His friends were appalled. However the war had started, however long it had gone on, it was now nearly over. And however great the cost in blood and treasure, it seemed certain to extend the country's borders to the Pacific. And whatever one thought of President Polk's handling of the war, the people were for it.

To his political friends, that was the main thing Lincoln should have kept in mind and evidently had not: The people were for it—Illinois people—his people. And because they were for it, he should have been politically savvy enough to know not to oppose them—not on such a matter as this, a matter of war. If, however, he felt compelled by his conscience to oppose them, he should have done so quietly. Instead he had issued a series of searing resolutions challenging the president's right to go to war. And then (in his friends' eyes) he had compounded his error by giving that fiery speech in Congress in which he had charged that Polk had lied the country into war. In Morgan County, Illinois, a citizens' meeting openly rebelled at Lincoln's leadership, charging that he had "acted in direct contradiction to the wishes of his constituents of the whole state as well as of this district." Soon there were other meetings, and at these, too, Lincoln was denounced.

To no one was Lincoln's behavior more puzzling than to his own law partner back in Springfield, William Herndon, who had sat across a desk from him for years, and in particular, for the past year and half since the war broke out, and during that year and a half had learned nothing of Lincoln's true views. To Herndon, Lincoln's actions were incomprehensible. *What was he thinking? What of the reaction of his constituents? What of his future?* "If you had been in my place, you would have voted just as I did," Lincoln wrote in reply. "Would you have voted what you felt you knew to be a lie? I know you would not." (Of course, Lincoln had done more than vote; he had issued his "Spot Resolutions" and then had gone on to give a major speech.) "Would you have gone out of the House—skulked the vote? I expect not. If you had skulked one vote, you would have had to skulk many more, before the end of the session." (Herndon had not suggested Lincoln skulk the vote, but had implied Lincoln might have been wiser to vote and say nothing.) "[Illinois representative William] Richardson's resolutions [in favor of the war] . . . make the direct question of the justice of the war; so that no man can be silent if he would. You are compelled to speak; and your only alternative is to tell the truth or tell a lie." (Lincoln had not only voted against Richardson's resolutions, he had devised the "Spot Resolutions" as a bold counterpoint to them.)

To his friends the Lincoln now on view was a new Abraham Lincoln, a Lincoln who did not seem to care for the political consequences of his actions, a Lincoln who after just a few weeks in Washington seemed to have lost touch with his home constituency and his own political interests—a Lincoln they plainly did not understand.

Previously Lincoln had almost always done everything he could to stay close to public opinion. When the people were in favor of the internal improvements projects, he favored the projects. When the people turned against internal improvements because the state could not afford them, Lincoln, though he took his time, joined them in their opposition. (The delay was on account of a natural unwillingness to admit that his previous position had been a mistake.) When he campaigned in 1840 against the election of Martin Van Buren, he embraced a blatantly racist appeal to try to win over the support of his Southern-born audiences.

On one important issue and only on that issue—slavery—had he taken a position obviously at variance with that of his constituents. At the height of the hysteria over abolitionism, he had dared to vote against a proslavery resolution that denounced abolitionism. But even then he had been careful not to alienate his supporters. During the debate he had kept silent about his views, refusing to be drawn into a public defense of his stand, leaving his vote as the only record of his dissent. And then, at the end of the session, to inoculate himself against the charge that he was an abolitionist sympathizer, he had submitted a carefully worded denunciation of abolitionism.

As they reviewed the statements he had made about the war since entering Congress, his friends came to conclude that by his actions he had virtually forfeited a future in Illinois politics. And that seemed downright odd. For the Lincoln they knew—the Lincoln who had run for the state legislature at age twenty-three and then kept on running—had always done everything possible to ensure he had a future. ("His ambition was a little engine that knew no rest.")

Lincoln's friends were right, of course: He no longer did appear to have much of a future in Illinois politics. But that was not because he had sabotaged his future by his conduct in Congress. His future as an Illinois politician had been in doubt from the moment he was elected to the House. For at that precise moment he had achieved about all he seemingly ever could hope to achieve as a Whig in Illinois. Whigs were in the minority there and were seemingly destined to remain in the minority for many years to come. All he could probably ever be in Illinois, he already was: a congressman.

In the foreseeable future he could not hope to be elected governor. No Whig had ever been elected governor (and none ever would be).

Nor could he have much hope of being elected to the U.S. Senate. No Illinois Whig had ever been elected to the U.S. Senate either (and none ever would be).

In his fifteen years in politics he had held nearly every position a Whig in Illinois could hold—election clerk, party official, member of the state house of representatives, Whig leader in the house of representatives, and finally U.S. representative—and now he had no place else to go in politics, not if he meant to continue going up the political ladder. The only rung he had not grabbed that other Whigs had—state senator— he had succeeded in climbing past. Running for a seat in the state senate now would mean reversing course and taking a step down the ladder. And of course he could not go back to the state house of representatives; after Congress that would mean descending even lower.

Harshest of all realities, he could not even remain where he was, in Congress. The clever maneuvering that had put him in Congress—the arranging of the celebrated Pekin Agreement, by which the party had settled on a policy of rotation in office—now was to keep him from holding onto the office. The great stone wheel of rotation that had put Hardin in office and then turned him out, and had subsequently put Baker in office and then turned him out, had now put Lincoln in office and would turn him out, too. Both Baker and Hardin had felt the rough edges of that wheel pressed heavily against their backs—and had had second thoughts; Hardin had tried to run again, Baker had wanted to. Now Lincoln felt the weight of the wheel—and he had second thoughts, too. But it would roll over him as it had rolled over them, crushing his dreams as it kept on rolling. He had promised to serve just a single term; now he would have to keep his promise. ("Turn about is fair play.") The only way he could run for reelection was if nobody else volunteered to run against him. Then he could stay in Congress without anyone accusing him of inconsistency.

That would never happen, of course. There would always be someone who wanted to run. More than likely there would be many. But one day as he sat in Washington, D.C., contemplating his future, Lincoln, who long ago had committed himself to "cold, calculating, unimpassioned reason," surrendered for a moment to a dreamy thought. "If," he wrote his law partner, William Herndon, "it should so happen that nobody else wishes to be elected, I could not refuse the people the right of sending me again." But he could run only if no one else did. ("To enter myself as a competitor of others, or to authorize any one so to enter me, is what my word and honor forbid.") And of course others *would* run. And he knew it—even if on this one occasion he indulged the fantasy that maybe, just perhaps, others would not.

The principle of rotation did not, of course, preclude his running and winning again after a two-year hiatus. And after two years in, he could drop out again, then come back in again, then drop out again—in a never-ending cycle. But this way was risky. Even if by some political miracle he managed to be reelected repeatedly in this cyclical manner, he could never hope to become a powerful figure in Congress. To become a power in Congress one had to put in years and years of service. And no matter how many times he ran and won in alternate elections he could never serve enough time to achieve equality with those who ran and won every election.

His friends had been mistaken, then, in thinking that Lincoln had changed. He had not changed. He was the same ambitious man he had always been, ambitious and prudent. But his circumstances had changed—and as they changed, so had his behavior. Realizing that he had no place in Illinois politics, he instead had tried to use his genius at speechmaking to find a place in national politics by becoming a leader in Congress. He had not deliberately tried to write off his Whig friends back home; he was in fact disturbed when they felt he had. But he had been willing to risk alienating them in order to advance his reputation on the national stage.

His attempt to win a place in national politics by attacking the Polk administration did not, however, succeed. Polk, of course, paid no attention to the attack at all; the ranting of a first-term Whig congressman from the Democratic state of Illinois—and the *only* Whig congressman from Illinois—could safely be ignored. Polk may never even have heard of Lincoln's jabs: The newspapers barely covered them. Fellow Whigs in Congress heard, but they, too, ignored them. Lincoln's argument, while controversial in Illinois—the local papers had begun referring to him as "Spotty Lincoln"—was common in Washington. For months Whig leader Henry Clay had been taking the very same line in speeches around the country. (Lincoln himself had apparently heard one of these speeches in Lexington, Kentucky, in November during a visit to Mary Todd's relatives.) And Abraham Lincoln was no Henry Clay—not yet.

Lincoln may possibly have won more national attention giving a prowar speech than an antiwar speech. Not many national Whigs were giving prowar speeches then. But delivering such a speech would have helped him only in Illinois. It would not have helped him with the Whigs in Washington, where Whigs held a slim majority in the House of Representatives and were daily doing battle with the Democratic administration of James K. Polk. And it was the Washington Whigs whose help he badly needed. Besides, he sincerely opposed the war; as an honest politician he could not in good conscience vigorously support it even if doing so might be helpful to his career.

Speechmaking happened to be one of Lincoln's great strengths. But by February there was no longer an excuse for delivering speeches against the war: By then the war was over.

Once Lincoln realized that he could not use the war to advance his career, he gave up making controversial speeches and concentrated instead on politics—presidential politics. But small as his chance had been of making a name for himself as a national speaker, his chance of becoming a player in the presidential sweepstakes was smaller yet. At least as a speaker he could use his gift for language to attract attention— that awesome gift he had first used as a young man of twenty-three and that he had continued using throughout his career in politics, much to his vast benefit. As a budding presidential kingmaker he had almost nothing to offer. There were 115 Whigs in the House of Representatives in 1848. Of them all, Lincoln was in nearly the worst position to help the national Whigs win the presidency. Outside Illinois nobody had ever heard of him. Inside Illinois people had heard of him, but as a result of his opposition to the war they now disapproved of him. Earning their enmity had always been the risk Lincoln had run in taking an active antiwar line; now he was discovering just how substantial a risk it was.

Even if he had retained their support, his usefulness to the national Whigs would have been limited. A kingmaker has to be able to deliver votes—enough votes to help contribute to electoral victory—and Lincoln could not. Illinois had never yet voted for a Whig presidential candidate and probably never would (and never did).

Being a Whig from Illinois always seemed to be a hindrance to Lincoln's plans. Because he was an Illinois Whig he had had to give up any immediate hope of ever being elected governor or U.S. senator. Now because he was an Illinois Whig he had to give up hope of being a kingmaker.

But if he could not be a kingmaker, he could still help pick the next king. And if he did, then *that* might lead to something else for him.

Every week that went by made the necessity of finding something else more and more paramount. When he was elected he'd had thirty-one months—the time between his election to the thirtieth Congress in August 1846 and the end of that Congress in March 1849—to come up with something else. Now he had just twelve months to find another political post to satisfy his ever-growing ambitions. Twelve months to reach even higher than he already had for something—something not even he could as yet name, but something. If something did not come along, he would have to return to Illinois to resume his career as a lawyer. Lincoln told friends that he looked forward to practicing law again, and because he had by now become quite successful as a lawyer, there was much to look forward to. But still he was not quite ready to give up politics. It was in his blood—and had been nearly his whole adult life.

Several candidates vied for the Whig nomination for president in 1848, but two quickly became the front-runners: Henry Clay, the old party warhorse, and Zachary Taylor, the Mexican War hero. Clay by any standard would have made the better president (though Clay by now was old, over seventy). Taylor knew nothing about politics; he'd never even voted for president. But if Clay was the better man for the office, Taylor made the better candidate—and every Whig knew it. The Whigs had only won the presidency once, when they ran Gen. William Henry Harrison. Their only chance of winning again in 1848 was to run Gen. Zachary Taylor. Taylor would be the practical choice. With him at the helm the Whigs would be protected from the charge that their opposition to the war was unpatriotic.

In February, just as the war was ending, Lincoln, whose personal hero was Henry Clay, came out in favor of General Taylor—and immediately busied himself in the Taylor campaign.

With the passage of every week he became more and more involved in the campaign, so involved that he began to complain to his wife, Mary Todd, that he lacked the time to attend to her and do all that he had to. Eventually she went home to Springfield.

Lincoln's work as a congressman was work enough for two men. A member of two committees, he was continually attending hearings, studying documents, and making speeches about committee business. One day found him making a report about a Mr. H. M. Barney, whose post office had been destroyed by fire. (Lincoln was a member of the post office committee.) Another day found him giving a speech concerning "military bounty lands." (He was also a member of the Committee on Expenditures in the War Department.) On still another day he took the floor to make a speech concerning the admission of Wisconsin to the Union.

But busy as he was attending to official business, he was never too busy to pursue his political work. And the political work could be grueling, so difficult Lincoln was to refer to it as "sweating blood." It included making contacts with the Taylor campaign, corresponding with Illinois politicians about Taylor's prospects, and at least once writing the draft of a speech for Taylor.

In early June, Lincoln left the capital for the first time since his arrival six months before. It was to attend the Whig National Convention in Philadelphia. By then the race had broadened to four candidates: Zachary Taylor, Henry Clay, Daniel Webster, and Gen. Winfield Scott. Fortunately for Lincoln the convention settled on Taylor, the man he had backed.

Lincoln was by now devoting a great part of his day to Taylor's campaign. But no matter how much time he put in on his political work, he always felt he hadn't put in enough. And he was right. There usually wasn't enough time to do all that he wanted to. "Excuse this short letter," he wrote to Herndon. "I have so many to write, that I can not devote much time to any one." And even when he had the time, he felt he still wasn't doing enough—not if his work in the campaign was somehow to lead to something else. And he was right about that, too. And as he realized he was unable to do enough, he became frustrated—frustrated and desperate. As Taylor's chances of winning began to seem brighter and brighter, Lincoln became more frustrated with his inability to make a major contribution—and more desperate. In letters home to rally support for Taylor he began to sound almost hysterical. When the "sanguine men here, set down all the states as certain for Taylor" they always leave out Illinois. "Can not something be done, even in Illinois?" Ten days later he wrote of attending a Whig caucus: "The whole field of the Nation was scanned, and all is high hope and confidence." But then he had received Herndon's letter saying that the Whigs had made "no gains" in Illinois. It was, Lincoln wrote, "heart-sickening." He then recommended that Herndon "gather up all the shrewd wild boys around town, whether just of age, or little under age," and form a Rough and Ready Club. (Taylor was known as Old Rough and Ready.) "Let every one play the part he can play best—some speak, some sing, and all [holler]." And then: "Don't fail to do this."

By the late spring there were, instead of twelve months for Lincoln to find something to do after Congress, just nine. Just nine short months either to make a name for himself in the East or return home to Illinois, an ex-congressman who had gone as far in politics as he probably ever would.

He had tried speechmaking—and speechmaking hadn't worked to lift him from obscurity. And then he had tried playing at presidential politics. And that hadn't worked either. So now he tried doing both simultaneously. When the Whigs in Congress formed an executive committee to drum up support for Taylor around the country, Lincoln joined. When the Washington, D.C., Rough and Ready Club held a meeting to celebrate Taylor, Lincoln gave a speech. One day he took the time to write Horace Greeley, the editor of the leading Whig newspaper in the country. Another day he corresponded with Thaddeus Stevens, the political boss (and future congressman) of western Pennsylvania. In late July he gave a conspicuously partisan one-hour speech on the floor of the House in defense of General Taylor's selection as the Whig candidate for president.

And then, instead of there being nine months to prove himself, there were just six. Six months to do what he had not been able to do in the past six—or the six before that, or the six before that, or the six before *that*.

And now, in case there had been any doubt of the necessity of winning a national reputation if he were to remain in politics, there came news from home that was to erase those doubts completely. Stephen Logan, who had been nominated by the Whigs to fill the seat Lincoln would be vacating, lost the congressional election, which in Illinois was held in August. It remains unclear to this day why Logan lost. And it was unclear at the time. Some, with reason, laid the blame on Logan himself; arrogant, irascible, and inarticulate, he made a terrible candidate. But others, including Herndon, blamed Lincoln, "Spotty Lincoln," who, against the wishes of his constituents, had made one controversial statement after another against the Mexican War. Lincoln denied responsibility for Logan's loss, but he knew that there would be many people now who said he was responsible. So even had he wanted to continue as a local politician, he knew almost certainly that he could not.

Now he would *have* to do well nationally—or abandon politics.

Pressed, as perhaps he had not felt pressed since his calamitous jump out of the statehouse window, Lincoln adopted a strategy that was new, risky—and audacious. Once again it involved speechmaking. But this time instead of making one, two, or maybe three speeches in the course of a month as he had been doing the past six months, he would make ten of them and he would make them virtually back to back, ten speeches in eleven days. And not just little stump speeches, but lengthy, formal full-blown addresses, lasting upwards of an hour and a half.

Politicians today often deliver more speeches over the same period of time. But the speeches rarely last as long, and in any case they are usually written by someone else. Lincoln was to write all of his speeches himself. And although the speeches would sound the same themes, each one had to be tailored to fit a particular audience. Furthermore, unlike politicians today, he would not have the assistance of a public address system to help carry his voice to the furthermost reaches of the crowd. To make himself heard, especially at the bigger events, he would virtually have to shout his lines, shout them until he was nearly hoarse, and he would have to do this day in and day out, for eleven days in all (with one day off for the Sabbath).

Lincoln had delivered numerous long speeches on similarly tight campaign schedules before, most notably in the 1840 presidential race. But every one of these speeches he had delivered in Illinois, in front of crowds composed of people very much like him. This time he would not return to Illinois to speak, or even to the states around Illinois—Indiana or Kentucky—which he knew so well from having lived there as a youth and where his accent and country ways would seem familiar to his audiences. Instead he was to give his speeches in Massachusetts, where his accent and country ways almost certainly would put off audiences. (Biographer Stephen Oates reports that Lincoln his whole life pronounced "sot" for "sat," "thar" for "there," "kin" for "can," and "heerd" for "heard.")

Where precisely in Massachusetts he chose to give his speeches is even more suggestive of his audaciousness. He would go not only to the small towns of Massachusetts, backwoods towns like New Bedford, Taunton, and Worcester, where he could expect to find friendly, easygoing crowds, but also to the state's most urbane places, intellectual havens like Cambridge and Boston, where the crowds could be expected to be demanding, used as they were to great speakers like Daniel Webster, Charles Sumner, and William Seward.

The first of his ten speeches he delivered on Tuesday, September 12, in Worcester, on the eve of the meeting of the Whig State Convention. In substance it was decidedly unremarkable, and included an utterly conventional defense of General Taylor's fitness for the presidency. To the constantly reiterated charge that Taylor, for instance, lacked principles, Lincoln responded that "Gen. Taylor consents to be the candidate, and to assist the people to do what they think to be their duty, and think to be best in their natural affairs, but because he don't want to tell what we ought to do, he is accused of having no principles."

On every single issue Lincoln stuck to the Whig Party line. On this, his first national speaking tour, he had to. His political future now depended on the willingness of the national Whigs to reward him with something. He well knew they would do so only if he seemed safe. He had been willing to give offense to his Whig friends back home in Illinois because they no longer held his political future in their hands. But these national Whigs did in theirs.

Age thirty-nine now, Lincoln, to remain on the political fast-track, could not afford to alienate the national party as he had alienated many in the Illinois state party. But neither could he afford to deliver speeches that were humdrum. If he was to have any chance of attracting attention, he would somehow have to shine.

Now.

Desperate to shine, Lincoln gambled. Scared that he was fast running out of time to make his mark, and unable under the circumstances to say anything controversial to make an impression, he decided to resist the compelling temptation to be staid and proper on this, his first foray into the refined "part of the country where, in the opinion of the people of his section, everybody was supposed to be instructed and wise." Instead, he would just be himself—gangly, humble, unprepossessing—and utterly, recklessly, and primitively passionate. Even his critics had to agree he was passionate. "It was reviving to hear a man speak as if he believed what he was saying," the *Bristol County Democrat* reported, "and had a grain or two of feeling mixed up with it." The paper did not think that Lincoln made a logical argument, "but then he was more unscrupulous, more facetious" than some other speakers, "and with his sneers he mixed up a good deal of humor. His awkward gesticulations, the ludicrous management of his voice and the comical expression of his countenance, all conspired to make his hearers laugh at the mere anticipation of the joke before it appeared."

At each stop on his tour Lincoln put on display the same Lincoln the folks back home saw. When he reached Boston, perhaps the most literate city in the country, he found the courage to show even the people there his down-home side. The *Boston Atlas*, in a favorable review, noted that Lincoln had helped make his points through the use of "keen satire," a reference to his uncommon talent for mimicry, which he had developed to a high degree during his years as a frontier speaker trying to connect with unlettered audiences.

It was not just the *Boston Atlas* that was impressed, so were all of the papers that covered his speeches. The *Boston Daily Advertiser* called the speech in Worcester "masterly and convincing." The *Daily Journal* remarked that the speech in Lowell "was replete with good sense, sound reasoning, and irresistible argument, and spoken with that perfect command of manner and matter which so eminently distinguishes the Western orators."

Lincoln had hoped he would do well on this tour, and he did. And it was essential he did. Instead of there being six months left for him to find something, now there were just five.

In November, General Taylor became President-elect Taylor—and that was essential, too. Lincoln's political hopes had rested on Taylor's election and the expectation that *that* might lead to something.

But what? For the past twenty-seven months Lincoln had not been sure what that something might be. And he still wasn't. But if his political work was to lead to something, it would have to lead to something quickly. Now instead of there being five months left to his congressional term, there were little more than three.

Lincoln spent those last three months or so doing what he hated doing most—waiting. His only significant official act during all that time was the submission of a bill "to abolish slavery in the District of Columbia, by consent of the free white people of said District, and with compensation to owners." (The bill went nowhere.)

Lincoln was not the only Whig from Illinois who hoped for something from the incoming Taylor administration. One Walter Davis of Springfield wanted something, too—maybe a position as a postmaster?—and asked Lincoln, the sole Whig in Congress from Illinois, to help him get it. Edward Baker, Lincoln's old rival for Congress, wanted something, too—a cabinet post perhaps. So did many others. In reply Lincoln sounded hopeful. "If the distribution of the offices should fall into my hands," he wrote one of them, "you should have something, and I now say as much, but can say no more."

Lincoln was still hopeful in February. And he remained hopeful, even as the months left to his term as a congressman dwindled to weeks and then finally to days.

The Thirtieth Congress came to a close in the wee hours of Sunday, March 4. Lincoln stayed in his seat until the very end, which came at seven in the morning. The following day he joined thousands as Zachary Taylor was sworn in as the twelfth president of the United States. That evening he attended the inaugural ball, partying until three or four in the morning. Long after his friends finally went home, however, he continued wandering the streets of Washington—in search of his hat, his signature black top hat, which he'd lost. (He apparently never found it.)

Finally he was a congressman no more—and still he was hopeful. But over the next six months he was to become bitterly disappointed. After nearly "sweating blood" to help General Taylor get elected, he had to confess to supplicants that he lacked influence with the administration: "Not one man recommended by me has yet been appointed to any thing, little or big, except a few who had no opposition."

Not only could he not help others find a job, he could not find himself one—at least not one that was suitable.

The one post that the administration offered—federal land commissioner—Lincoln declined as it was wanted by several Whigs in Illinois whom Lincoln felt he had to accommodate. "There is nothing about me which would authorize me to think of a first class office," he wrote his old friend Joshua Speed, "and a second class one would not compensate me for being snarled at by others who would want it for themselves."

The land commissioner job was to remain open for months, however, as the result of endless feuding between the candidates. When it appeared likely it was finally to be given to a Chicago Whig whom he detested, Lincoln decided to go for it himself; competitive as always, the chance to defeat a rival gave direction to his ambitiousness. For weeks he did all he could to attract wide support, corresponding with Whigs across the country for backing. To clinch the job he even returned to Washington, traveling by train and stagecoach, in a mad-dash effort to beat his rival to the city.

But for Lincoln, having finally decided what he wanted to do, the opportunity to do it did not materialize. The land commissioner post he could have had on a platter months before went instead to his rival, who had received the endorsement of Daniel Webster. Upon hearing the news, according to one touching account, Lincoln stole off to his hotel, went up to his room, threw himself on his bed, and lay there for more than an hour.

In the fall, long after he had returned to Springfield, the administration finally offered him another position, but it was comically beneath him: the secretaryship of the governor of Oregon. Lincoln's friends howled in protest about the meagerness of the offer, prompting the administration to proffer Lincoln the post of governor. But by now Lincoln was demoralized, and he declined it; it was, anyway, not a position which would do him much good politically. Oregon then was the remotest place in the entire United States, the last place a man of ambition would want to go.

His political career seemingly over after seventeen years, Lincoln took up the full-time practice of the law.

For the next five years Abraham Lincoln could be found most days either in an Illinois courtroom or in his law office off Springfield's lively main square. As once he had worked at politics, he now worked at the law. As he was to put it later, "From 1849 to 1854, both inclusive, [I] practiced law more assiduously than ever before."

In elective politics he had gone about as far as a Whig could in Illinois. Now as a lawyer he was to go about as far as a lawyer could, becoming in the space of those five brief years one of the leading attorneys in the entire state. These were good years for Lincoln and he was happy. By 1854 he was earning thousands of dollars a year; once he pocketed a five thousand dollar fee on a single case. As his reputation spread he became more than the country lawyer he had started out as, eventually representing many of the most powerful corporations in the country. As his ambition grew he increasingly began arguing cases before the state supreme court. Soon he was arguing more cases there than anybody else.

During those five years Lincoln so devoted himself to the law that he almost forgot about politics. But he could not forget about it completely. And when the sectional conflict broke out anew in 1854 over Stephen Douglas's Kansas-Nebraska Act, Lincoln's passion for politics sprang back to life, arousing him, he was to recall, "as he had never been before."

Lincoln was still a Whig and being a Whig in Illinois was still a handicap. But not for long. Douglas's decision to back the repeal of the Missouri Compromise as part of a deal to build a railroad through the Nebraska Territory was so controversial that it began to undermine the hegemony of the Democratic Party. In twelve short months it began to cripple the party in Illinois. Lincoln, who had left Illinois politics after it became clear to him that as a Whig he had no future, made the decision to return, hopeful he finally did have a future again.

And then the repeal of the Missouri Compromise, which was crippling the Democratic Party, killed the Whig Party outright, hopelessly dividing it into pro- and antislavery factions. In its place there arose, briefly, a new antislavery Whig Party known as the Anti-Nebraska Whigs. It in turn was quickly replaced by an entirely new organization, the Republican Party, a loose coalition of businessmen, anti-slaveryites, Know-Nothings, and immigrants.

In this new political environment. Lincoln, first as an anti-Nebraska Whig, and then as a Republican, was able to thrive as he had never been able to before. For the Republicans' base was broader than the Whigs' had been, giving the Republicans a chance to become the majority party in Illinois.

Only one office interested him: a seat in the U.S. Senate, the next rung on the ladder he had begun climbing twenty-two years earlier. Twice he ran for the Senate and twice he lost, the first time to a fellow Anti-Nebraska Whig, and then to Democrat Stephen Douglas.* But with each loss his popularity, instead of waning, increased, as more and more people nationwide began to see in him the leader they were looking for.

Unable to grab the rung on the ladder he wanted—a Senate seat—Lincoln made an audacious grab for the rung beyond.

In 1860, two months before the Republican presidential nominating convention, Lincoln, who rarely admitted his ambition, confessed, "[T]he taste is in my mouth a little." At the convention he who had not been able to win a Senate seat succeeded in defeating well-known senators and governors to win the party's presidential nomination. A few months later, again facing candidates who were better known than he was, he again was victorious.

To win the nomination he had carefully orchestrated a behind-the-scenes campaign against his rivals, pointing out their deficiencies, explaining why one could win here but not there, while another could win there but not here. Each time he made sure to leave the impression that he alone could win all places vital to the party's election.

To win the election he had to refrain from making statements that could possibly alienate any of the groups that made up the Republican

---

*Stephen Douglas was one of those men who also ran at a young age. Like Lincoln, he ran for the state legislature at age twenty-three (but he won). He, too, was very, very ambitious. It was almost inevitable that he and Lincoln would come to clash later in life. Lincoln was probably more popular than Douglas, but senators then were chosen by state legislatures. And in the state elections of 1858 the Democrats were able to win control of the legislature because of gerrymandering, leading to Douglas's victory, even though Republicans won more votes.

Party's unwieldy coalition of businessmen, abolitionists, and Know-Nothings. He did. When asked on one occasion to denounce the Know-Nothings, he refused.

From a position of no political power in 1855 he had succeeded in moving in five short years to a position of supreme political power. On March 4, 1861, at noon, just two weeks and six days past his fifty-second birthday, he was sworn in as the sixteenth president of the United States. ("His ambition was a little engine that knew no rest.")

Lincoln came into the presidency in very much a weak condition, receiving just 40 percent of the popular vote, with the Republicans in both houses in the minority.* But the secession of the South put him in an extraordinarily strong position, immediately reversing his fortunes. For one thing, the secession left the Republicans in control of both houses, an exceptional turn of events given the results of the election just a few months before, in which everything had been left muddy, with power divided geographically between Northerners and Southerners, and politically, between Democrats, Republicans, Know-Nothings, and others. Now the situation would be far clearer and far easier to control.

For another, the crisis, though terribly complicated, gave Lincoln a rich opportunity, a chance to take decisive dramatic action. While Americans had hardly had time to come to know Lincoln, they had instinctively turned to him for guidance once Fort Sumter was fired on. (Lincoln was careful to arrange things so that the South fired the first shot; it put the United States in the superior moral position.) As every president ever after was to discover, in a genuine crisis Americans willingly surrender power to the president, allowing him to do much that they otherwise would not. Buchanan probably would not have known what to do with the power if he had had it. Lincoln did.

In the four months after his inauguration he made it clear that he would use force, if necessary, to keep the Union together, and on his own authority increased the army by twenty-two thousand, the navy by eighteen thousand, and called for a draft of an additional forty

---

*The vote in the presidential election was split four ways, between Lincoln, Stephen Douglas (representing Northern Democrats), John Breckenridge (representing Southern Democrats), and John Bell (candidate of the Constitutional Union Party).

thousand. No president before had ever dared act so boldly, but Lincoln was able to because the people were behind him. So sure was he of public support that he actually suspended the writ of habeas corpus, leading to the prompt arrest of over ten thousand rabid Southern sympathizers, which is said to have helped keep the border states in the Union.* It was, to be sure, an unprecedented grab for power, but because Lincoln did it so openly, hiding nothing, and making sure that the arrests were, as much as possible, made in public, people by and large sustained him in his actions. He deliberately delayed the convening of Congress for four months to give him time to do what he thought necessary to save the Union. But he afterward sought retroactive approval for his actions and Congress gave it to him.[2]

Unlike Buchanan Lincoln hadn't acted sneakily or underhandedly. And as long as events went his way he wouldn't have to. But, of course, events did not go his way. What had been expected to be a short, romantic war—at the First Battle of Bull Run at the end of July 1861 women in hoopskirts had ridden down from Washington in their carriages to watch—quickly had turned into a grim, long, unending and almost unbearable slaughter, the likes of which no one on earth had ever seen, some battles ending in the death or mutilation of tens of thousands. At Gettysburg, hailed as a Northern victory, there would be twenty-five thousand Union casualties. Eventually in consequence there would be draft riots and bread riots. Lincoln was to remain remarkably calm in the face of this evidence of social dissent, but as the war went on and on and on, he became increasingly desperate.

A measure of his desperation was the issuance of the Emancipation Proclamation in January 1863. Lincoln, of course, wished to free the slaves; about that there was never any question. But he had declined to do so as long as he did out of fear of the effect on the border states, where slavery was still regarded positively by many whites. Finally, however, he had felt he had no choice but to act. If he didn't, he feared, England, seeing no moral difference between the North and the South, and hurting because of its inability to obtain Southern cotton, would agree to Southern demands for official recognition of the Confederacy. He did not, in any event, agree to free all the slaves. To keep the border states,

---

*These were the states along the Mason-Dixon Line, such as Kentucky, which could easily have gone either way in the Civil War; they remained loyal to the Union because of Lincoln's deft maneuvering.

he freed only those slaves who lived in the South over whom he had no de facto control. The eight hundred thousand or so slaves remaining in the border states were to continue in slavery. It was an artful compromise, but it *was* a compromise, a "sophistical contrivance wherewith we are industriously plied and belabored," as Lincoln once remarked in another context, a contrivance that left Americans "groping for the middle ground between the right and the wrong." But Lincoln was in no position to offer anything more in 1863. Events were against him.

Lincoln was, by 1864, so divisive a figure that he could not even count on his own reelection. In fact, days before the balloting was to take place he was sure he would not win. Pressed, he behaved very much as other presidents had and would when their reelection prospects were threatened. He compromised, cajoled, and caved. Weak among Democrats in the border states, he decided to encourage the nomination of Democrat Andrew Johnson of Tennessee as his running mate, though Johnson was in no sense qualified for the position. Fearful that Grant might suddenly materialize as a possible opponent, he declined to bring him east as commander in charge of the army until he had received assurances Grant had no intention of running. "No man knows, when that presidential grub gets gnawing at him, just how deep it will get until he has tried it," he recalled afterward. "And I didn't know but what there was one gnawing at Grant." Told by the governors of half a dozen states that there was a fair chance he would lose the election unless he gave tens of thousands of soldiers furloughs to go home and vote, he ordered his generals to send the boys home. Short of campaign cash, he allowed his campaign manager to require all federal employees to kick back 3 percent of their salaries for his reelection drive. He won with 55 percent of the popular vote.

No president had ever exercised more power over the country. Jackson, derided as a dictator by his critics, hadn't even come close. No one had. And yet even Lincoln felt that he was at the mercy of events. As he plaintively explained in 1864, "I claim not to have controlled events, but confess plainly that events have controlled me."*

The means he resorted to to try to gain control were extraordinary.

---

*Elsewhere in the same letter in which he made this statement he made another, equally trenchant: "[M]y oath to preserve the constitution to the best of my ability, imposed upon me the duty of preserving, by every indispensable means, that government—that nation—of which that constitution was

Suspension of the writ of habeas corpus. The waging of all-out war. The delay, for political purposes, in putting Grant in charge of the army. But it was not for these things that he would be remembered, of course. It was for ending slavery and winning the war, both of which he finally was able to achieve, though not nearly as quickly as he wanted and at far greater cost.

Of all the presidents he was forgiven for his transgressions. As was widely recognized after his death, when he was turned into a mythological hero, most of the extreme measures he undertook were helpful in advancing the country's fortunes, not just his own. It was recognized that the only way to wage the war and win it was by going outside the normal bounds of presidential conduct.

Other presidents would bend the rules and be called to account. But not Lincoln. He alone was recognized as having had cause.

Many worried about the precedent he had set, that his presidency would mark a turning point, that ever after presidents would feel free to exercise power as he had, that the country, in the process of freeing the slaves, had saddled on itself a tyrannical, power-mongering, power-crazed presidency. But after four years of Lincoln, four long years of a president doing damn near what he wanted, what the people got was a weaker presidency, weaker in some ways than it had ever been before. And for the next half century, until the arrival of William McKinley (though Grover Cleveland breathed some life into it), it was to remain weak. The power that Lincoln had concentrated in the presidency would shift to the Congress. While the people loved Lincoln, they were not eager for another.

---

the organic law. Was it possible to lose the nation, and yet preserve the constitution? By general law life *and* limb must be protected; yet often a limb must be amputated to save a life; but a life is never wisely given to save a limb. I felt that measures, otherwise unconstitutional, might become lawful, by becoming indispensable to the preservation of the constitution, through the preservation of the nation. Right or wrong, I assumed this ground, and now avow it." Basler, *Collected Works of Abraham Lincoln*, vol. 7, pp. 81–82.

8. The Birth of Industrial Capitalism

How the Industrial Revolution led to the creation of vast wealth and vast opportunities for the greedy, forcing U. S. Grant to tolerate corruption as a means of hanging on to power

It was an accident that Andrew Johnson followed Lincoln, for it was long supposed that it would be Grant who succeeded him. Grant, who had routed the Confederates at Vicksburg. Grant, who had whipped Robert E. Lee. Grant, who seemed to be all that an American hero should be. Grant, the silent, a little gruff, a little primitive hero who dressed in a slovenly way and enjoyed chomping on a wilted cigar. Ulysses Simpson Grant, a child of poverty who had pulled himself up by his bootstraps, overcoming drunkenness and repeated failure to become the nation's commanding general, and now, finally, in 1869, its president.

He knew nothing of government, nothing of finances, nothing of world affairs. And he seemed to have no interest in learning. Beginning his day around ten in the morning, he would quit by five. Probably no president ever worked less. Nonetheless he was very ambitious.

He showed his ambition in a way no president before him ever had—or had ever had to—by his willingness to endure repeated scandals and the way he chose to endure them. Through his two terms Grant would come under relentless attack for his friends' greed and perfidiousness, prompting people to wonder how he managed to survive. To be able to go on day after day despite the headlines, despite the investigations, despite the betrayals of friends. Astonishing.

Grant learned that there was only one way to go on. And that was to ignore a lot of the corruption—to tolerate it, to live with it. Living with it wasn't easy. For people would always wonder if the reason he let the corruption go on was because he himself was corrupt. But as

Grant learned, there was no way to root it all out. There was simply too much of it. And too many people were implicated, people who were powerful, people he was close to, people he needed. So he did what he could, forcing out those who had committed the most egregious crimes, and then he looked the other way.

Grant's administration would, of all of the administrations in American history, be the most corrupt. But from his day forward many presidents would learn as he did that they, too, had to be willing to put up with corruption. It was often the price of power. For corruption would become endemic now that the country was in the throes of industrial capitalism and the opportunities for illegal wealth became extravagantly abundant.

Before the war, certainly before the 1850s, Americans had never had much money. George Washington, though wealthy, had had to borrow money to finance the trip to his own inauguration. Two generations later John Tyler, also wealthy, had had to sell a slave to finance the trip to *his* inauguration (as vice president). The country had been so cash-poor that in the 1830s Nicholas Biddle, head of the Bank of the United States, had had to go begging in Holland for emergency funds when the bank's reserves ran low. Lincoln's career as a politician had nearly been killed because of the inability of the state of Illinois to float ten million dollars in bonds for internal improvements. Twenty years later the country was still short of capital. When the great railroads began to be built in the 1850s American bankers had to turn to Europe for the financing.

The war changed all that. After the war money seemed to be everywhere. Money in amounts that before the war nobody had ever even dreamed of. In 1860 the federal government had collected just $56 million from taxes, tariffs, and land sales. In 1865, the last year of the war, it collected $333 million. A year later, $558 million. And it wasn't just the government that was accumulating giant pools of cash. So was private business. In 1860 the insurance industry had earned a grand total of $6 million. In 1865 it earned $25 million; in 1866, $40 million; and by 1870, more than $100 million. Because the war had been extremely costly, the government had had to go deeply into debt, in 1865 selling $2.2 *billion* in bonds (in 1860, by comparison, it had sold just $64 *million*). That, in turn, had led to the development of Wall Street as a major center of wealth and power. And *that* had led, virtually overnight, to the creation of a whole new industry of financiers, speculators, and brokers. And as the economy grew, so did the take-home pay of the daily worker. In 1860 the average laborer earned $1.60 a day; by 1866, $2.60.[1]

With the abundance of money came a new danger: temptation.

Americans had, to be sure, always been susceptible to corruption. Every president had had to confront allegations of fraud, of unscrupulousness—of greed. Washington had had to face the fact that speculators had swindled millions of dollars out of unsuspecting Revolutionary War veterans. Jefferson had been warned that his postmaster general, Gideon Granger, had possibly paid bribes. In the years following the War of 1812, the Monroe administration had been repeatedly implicated in fraud. First had come the news that corrupt middlemen were making huge sums at the expense of veterans. Then had come the scandal involving Richard Johnson, the chairman of the House Military Affairs Committee. Monroe himself was guilty of a profound conflict of interest.

In the administration of John Quincy Adams a treasury auditor had embezzled $7,000 to cover a bad gambling debt. In the Jackson administration the collector of the customs in New York had made off with $1 million, and then his successor, under Van Buren, had made off with some more. In the Taylor administration there had been the Galphin family controversy. For decades the Galphins had tried, and failed, to persuade the federal government to assume the debts owed them by the British from before the Revolution. After the Taylor administration agreed to pay, it was disclosed that Taylor's secretary of war, George Crawford, stood to personally benefit from the transaction. As the family's attorney he would receive one hundred thousand dollars in legal fees. The secretary of the treasury and the attorney general, who had both approved the settlement, insisted that they were unaware of Crawford's involvement, but Taylor wasn't sure and decided to dismiss his entire cabinet over the incident.*

In the Pierce administration the territorial governor of Kansas had secretly bought up some Indian land and then lobbied, unsuccessfully, to have the capital moved there, so he could cash in on the rise in land values. And then had come Buchanan, whose secretary of war had

---

*Taylor never got the chance to rectify matters. Just at the time that he decided to dismiss the cabinet he died. As mentioned earlier, he passed away after eating a bowl of cherries and cream on a hot summer day.

awarded two sweetheart military contracts to a pack of his predatory friends. Even the Lincoln administration had been implicated in money scandals. Lincoln had had to dismiss his secretary of war, Simon Cameron, for issuing supply contracts without competitive bidding, in express violation of the law. As a result of Cameron's sloppy practices the government had paid top dollar for shoddy blankets, tainted pork and beef, knapsacks that came unglued in the rain, uniforms that fell apart, and guns that blew the thumbs off the soldiers who fired them.*

In the Johnson administration, investigators had uncovered a plot to sell presidential pardons to ex-Confederates. Johnson himself was cleared of wrongdoing, but it was discovered that the person behind the scheme was his son's girlfriend, who sold the pardons for a hundred dollars apiece (fifty dollars down, fifty dollars on receipt of the pardon).[2]

But all that was as nothing compared to what was to come during the Grant years.

The war had unleashed the latent economic strength of American capitalism, leading to the development of gigantic new enterprises (such as the transcontinental railroad) and encouraging a new way of thinking. While Americans had always been tempted by money and had indeed worshipped the almighty dollar, never before had there been the opportunities for making money that there were now. Before, government had been too small for many people to earn huge sums off it, legally or illegally. So had business. The few in the country who had earned giant fortunes had earned them, by and large, as Washington had, through land speculation. Just a relative handful had been able to enrich themselves through business wizardry, through the manipulation of money and goods. But now, suddenly, there seemed literally thousands of ways to make fortunes and thousands who were making them—and quickly.

As news of these vast new fortunes spread, many ordinary Americans began dreaming that they, too, could achieve great wealth. No longer would young men be satisfied with safe, stable jobs. Tens of thousands would now hunger after riches. James Garfield, an ordinary

---

*Cameron was a corrupt machine politician from Pennsylvania. He is credited with the saying: "An honest politician is one who when he is bought will stay bought."

schoolteacher in Ohio at the time, would gamble twice with his hard-earned savings on investments in get-rich-quick schemes. The attitude seemed to be, *If Jay Gould could make himself rich, why not me?* The very bigness of things seemed to invite such thinking. When enterprises were small, profits were small and uninspiring. Now that enterprises were big, profits were enormous and enticing. It was as if, as Vernon Louis Parrington put it, all America had become a "Great Barbecue" at which everyone wanted to feast.

It was not just the business types who wanted to cash in, but also the politicians and the bureaucrats. The more they heard about the big money being made in private enterprise, the more *they* wanted to make big money. As a first step, members of Congress saw to it that their own salaries were increased. In Monroe's day a congressman was paid $6 a day; by Pierce's, $8. In 1855, under Buchanan, members had begun receiving an annual salary, $3,000, a substantial increase over what they had earned before, but all in all still a modest amount.

Then the war had come. In 1865 they increased their salaries to $5,000. And then, in 1873, in what came to be called the "Salary Grab," they increased it again, to $7,500, with the increase retroactive to 1871. The salaries of other officials were also dramatically increased at this time. From Washington to Johnson presidents had earned $25,000 annually. Grant's salary as president was doubled to $50,000. The salaries of the vice president, members of the cabinet, and justices of the Supreme Court were increased to $10,000. The increases for everybody but Grant would be rolled back in the wake of public outrage.* But for the determined there were other ways to cash in on America's newfound wealth—illegal ways.[3]

Grant himself was not very interested in money. He lived simply. The son of an old American family of modest means, he was thrifty by nature; as a teenager working on his father's Ohio farm, he managed to save up a hundred dollars. But he was very ambitious. Although he received a meager education at a local village academy, he used family connections to win an appointment to West Point. There, despite his lack of adequate schooling, he succeeded in graduating in the middle of his class, excelling in mathematics and horsemanship. He then went into the army and fought

---

*Grant's raise could not be rescinded. The Constitution forbids the salary of the president from being reduced during his term in office.

in the Mexican War, where he was cited for bravery. Over the next ten years he remained in the army, moving from one western post to another, doing as he was told, and eventually becoming a captain. But because there were no wars to fight, opportunities to move ahead in the army were slim. Frustrated, Grant, thirty-one, decided to leave the army and to try to make his way as a civilian.

It was a terrible decision. Grant was simply unsuited to civilian life. Over the next seven years failure marked his every move. First he tried farming—and failed. Then, he became a real estate agent—and failed at that. Then he became a tax collector—and failed at *that*. Finally, broke and out of luck, he moved to Galena, Illinois, where his father had opened a modest little store. Grant became a clerk. It was said that he drank too much. Probably he did.

Then the Civil War came—and Grant's fortunes changed dramatically. Although he had trouble attracting attention at first—the War Department did not even bother to answer his letter requesting a commission—he quickly rose through the ranks. Volunteering to serve with an Illinois regiment, he was made a colonel, then, within two months, a brigadier general. In the early months of the war he showed great bravery, continuing to fight fearlessly in one battle even after his horse was shot out from under him. Eight months into his service he conceived a daring plan to launch a combined land and naval attack on two Confederate forts in Kentucky. The Union military commander in the region rejected the plan; Grant insisted on pushing for its approval. Finally the commander gave in to Grant's importuning. The plan worked perfectly. After a fierce battle, with the loss of two thousand lives on each side, the Confederates surrendered. Nearly fifteen thousand enemy soldiers were captured.

During the remaining years of the war, through his victories in Mississippi, Tennessee, and Virginia, Grant proved to be the ablest general the Union had, repeatedly winning against overwhelming odds. In many of his battles Grant faced the staggering loss of thousands of his soldiers. Each time he grimly fought on until he won. At the opening of the Wilderness Campaign in the spring of 1864, he sent a wire back to Washington that summed up his relentless approach: "I propose," he wrote, "to fight it out on this line if it takes all summer." On a single assault during this campaign, he lost more than seven thousand soldiers. Still he kept going. Lincoln, grateful that he finally had a commander who would fight—Gen. George McClellan, who had attended West Point with Grant, often had seemed unwilling to fight—praised him publicly, though some sharply condemned Grant as a butcher.

*　*　*

As president it wasn't his enemies whom Grant had to worry about. It was his friends. *He* may not have been money mad, *they* were. And as they dreamed up one crooked scheme after another to get rich, they repeatedly got him in trouble.

The first to be drawn into scandal was Abel Rathbone Corbin, Grant's brother-in-law. Fixing the mess Corbin created, however, was easy. Six months into Grant's first term, Corbin joined in an insidious plot with Jay Gould to corner the gold market. Corbin's role was to persuade the government to withhold its gold so the price would rise. For a time the government did withhold its gold, and the price did rise, sky high. Corbin received $25,000. Then Grant intervened, selling off enough federal gold to break Gould's control. End of story.

The second to get into trouble was Gen. Orville Babcock. Fixing the mess Babcock created proved almost impossible.

Babcock, a West Point graduate, had been in the army with Grant, serving with him in the difficult days of the Wilderness and Spottsylvania campaigns, among others. Grant subsequently had hired him in the White House to serve as his chief of staff (the position was formally known as secretary to the president). Babcock was a very energetic, very well-organized assistant, and Grant liked him immensely. Grant probably spent more time with Babcock than with any other member of his administration, and gave him an office right outside his own.

Babcock loved Grant, loved working with him in the White House, but Babcock, a newlywed, had a great love of money, too. And in 1869, just a year into the administration, he found a way to make much more money than was due him as Grant's assistant. It was through his friendship with John MacDonald, the supervisor of the Internal Revenue Service at St. Louis.

Though MacDonald was a snake of a man, very conniving, Babcock liked him and agreed to join him in a plot with midwestern liquor distillers to evade the government's steep taxes on whiskey. MacDonald promised that with Babcock's help the scheme could net millions in kickbacks from the distillers. Cheating on the whiskey tax had been widespread in the Midwest since the Lincoln administration, which had dramatically jacked up the tax to help pay for the war. But MacDonald's nefarious plan was fantastic, an utterly brazen swindle, and would, over time, come to involve literally hundreds of people. But for five years the conspirators managed to keep a lid on their activities and to keep raking in the boodle.

Eventually the authorities caught on to the loss in revenues and began an investigation. It was a very preliminary one, however, and when they reported to Grant they did not yet know of Babcock's involvement. Grant, unafraid at this point that the scandal could hurt him or his assistant, publicly announced that he wanted the officials to get to the bottom of the mess. "Let no guilty man escape," he said. The Whiskey Ring (as it came to be called) must be crushed!

Then he found out about Babcock.[4]

Grant himself was very honest, and his honesty was not in question. But he found it hard to let the investigation go forward. He had trusted Babcock, had stood side by side with him in battle, had given him high positions in both the army and the White House. To let the government go after Babcock would seem disloyal. Worse, perhaps, a government probe would undoubtedly reflect badly on the administration, might sink it in public contempt, might sink Grant himself; already there were charges that some of the profits from the scam had gone into his reelection campaign. So Grant decided to do what no president before him ever had (but which several presidents afterward would). He tried to derail an official investigation: to obstruct justice.

He could not put a stop to the probe altogether. Things had gone too far for that. His own secretary of the treasury, Benjamin Bristow, was pressing hard for a thorough search for the truth. But Grant could sabotage the investigation, and that is precisely what he now proceeded to do by depriving the attorney general of the chief means of obtaining evidence against Babcock. He simply told the attorney general that in this case, unlike in any other federal investigation, no witness was to be given immunity in exchange for testimony.[5]

Babcock was indicted anyway, in December 1875.

The trial began in February in the federal courthouse in St. Louis, and things immediately went badly for Babcock. Prosecutors, hobbled but not entirely crippled by Grant's order, presented so much hard evidence against the president's chief of staff that his attorneys decided not to dispute it. Instead, like all good defense lawyers with a weak case, they argued that the evidence shouldn't be admitted.

Babcock worried that he might be convicted, but acted confidently. He had some of the best attorneys money could buy. The Republican press was for him, daily running touching stories about the ordeal to which the government was subjecting his pregnant wife and children. Even the judge was for him, repeatedly ruling in favor of his lawyers' motions. (The judge apparently hoped to be appointed to the U.S.

Supreme Court and thought Grant might just do so if he helped cover up the administration's fraud.)

But the question hanging over the trial from its inception was whether Grant would testify for Babcock and to that question there was for some time no answer. Babcock's attorneys repeatedly wired requests to the president to testify and repeatedly Grant put them off. Finally, when it appeared that Babcock, despite all of the favorable press he was receiving, despite the friendly rulings of the judge, despite the handcuffs the president had put on the prosecutors, might actually be convicted, then finally Grant decided he would testify. At an emergency meeting of the cabinet in Washington, he told his astonished listeners that he planned to take the very next train to St. Louis to appear as a personal witness on Babcock's behalf. The cabinet, objecting strenuously, forced Grant to back down. But the president insisted anyway on giving a deposition for the defense, which he did shortly afterward from the White House itself, sitting through more than five hours of questions.

Grant's deposition marked a turning point. The trial that had been going badly for Babcock suddenly began to go well for him. Grant didn't just testify in his deposition that Babcock was a fine fellow. He went further. He said that they were so close that if Babcock had done something wrong then he perforce would have known about it. And he knew nothing.

As expected, given what the president said—given that Babcock's conviction would be tantamount to the conviction of the president himself—the jury acquitted Babcock on all counts. That night he was given a hero's welcome by a crowd of four hundred, who stood in the cold Missouri winter to hear him give a celebratory speech lasting more than half an hour.

Grant was delighted with Babcock's acquittal and told him he could return to his old job in the White House. But when Babcock went back to his old desk right outside the president's door there was a huge public outcry.

Babcock would have to go. Even Grant finally saw that. The country didn't believe in Babcock's innocence. Too many other people implicated in the scandal had also been indicted for prosecutors to have made a mistake about his involvement. (Eventually 230 people would be indicted; 110 would be convicted, including four government officials.) But the president insisted on letting Babcock keep a second position he held as the superintendent of public buildings and grounds.

* * *

Grant would undoubtedly have handled the Babcock corruption scandal better if it had been the only scandal he faced, but it wasn't. Almost every month somebody or other associated with the administration was charged, indicted or implicated in scandal. The cabinet was especially rotten. *Four* cabinet secretaries were implicated in one sorry scheme or another; three resigned.

Secretary of the Interior Columbus Delano resigned after it was disclosed that his son had taken bribes in awarding government land grants. (As many as eight hundred land grants were subsequently discovered to have been fraudulent.)

Treasury Secretary William Adams Richardson resigned after Congress learned that he had let a private contractor keep *half* of the tax moneys he collected from delinquent taxpayers, earning an astonishing profit of two hundred thousand dollars. This was an especially grievous offense. Congress had authorized the secretary to hire a private agent to collect taxes from hard-to-find deadbeats; instead, the secretary had allowed the agent to collect the taxes owed the government by the railroads.

Secretary of War Robert Belknap resigned after it was learned that he had shaken down the franchise owners of Indian-trading posts for kickbacks amounting to tens of thousands of dollars. There was in this case, however, the suspicion that it was not the cabinet secretary himself who was guilty so much as his first and second wives. The first Mrs. Belknap apparently concocted the scheme, the payments going directly to her until her death. Then they went to her sister, a well-known spendthrift, who, curiously, became the second Mrs. Belknap.

The fourth cabinet official implicated in fraud was George Robeson, the secretary of the navy, who was suspected of receiving three hundred thousand dollars in kickbacks from a Philadelphia supplier, who'd also given him a house. The money was found in Robeson's bank account, but Grant let the secretary remain in office.

Shocking as all that was, it was just what went on in the cabinet. The rest of the Grant government was also suffused with corruption. It was as if greed, like a cancer, had simply taken over. The customs houses, always susceptible to corruption because of the vast sums that passed through the hands of the collectors, became more corrupt than ever. In New Orleans the collector of customs was convicted of malfeasance. In New York the assistant to the collector was implicated in bribery, and the collector himself implicated in fraud.

The federal courts, supposedly the bulwark against corruption, themselves became involved in corrupt practices, leading to the resignation of five federal judges. Overseas, the U.S. minister to Great Britain, Robert Schenck, allowed a British mining company to use his name to raise capital. When the secretary of state ordered Schenck to withdraw his support of the company he agreed to do so, but delayed long enough to give his friends time to sell their stock before the news broke and the stock declined.[6]

Grant, reeling, even had to face the fact that both of his vice presidents were implicated in fraud. Schuyler Colfax, his first vice president, was a very typical politician, deal making, compromising, and appealingly friendly. An old pol, in short. Just what Grant wasn't. Which was good, because Grant was in need of someone who knew his way around politics, knew his way around Washington. Colfax, Speaker of the House, knew his way around as well as anybody. So in a way they were a good fit. But Colfax, too, it turned out, was corrupt. Like more than a dozen other members of Congress, he had gotten involved in the Crédit Mobilier scandal.[7]

Crédit Mobilier was the creation of Oakes Ames, another congressman, and the last person in the world one would expect to become involved in financial hanky-panky. A member of a prominent Massachusetts family, Ames had plenty of money and owned a thriving construction business. But like many at the time he was greedy and wanted more than he had, and to get it was willing to do things no congressman had ever done before.

Crédit Mobilier itself was a shady operation. On paper its purpose was to help the Union Pacific build a main part of the transcontinental railroad. In reality it was little more than a dummy construction corporation through which the directors of the Union Pacific paid themselves inflated prices for the work that was done. Because they made a percentage from every mile of track that was built, the more the line cost, the more money they made. By using Crédit Mobilier as a middleman they could jack up the costs tremendously, giving themselves added hefty profits. But it wasn't Crédit Mobilier itself that got Ames, Colfax, and the others into trouble. It was Ames's decision to bribe key members of Congress in order to obtain their help in winning huge government subsidies.

What Ames did was to give the congressmen stock in the company along with the money to buy the stock. The money ostensibly was in the form of a loan. But nobody had to worry about paying it back; because the company was so hugely profitable, the loan could be paid back out of the dividends the stocks paid in a single year.

For a time it seemed to be a brilliant scheme. Some of the leading members of Congress took Ames's stock: Colfax, Henry Wilson (Grant's second vice president), James G. Blaine, and James Garfield. It was as if they were all members of a secret club, the Ames Club. And it was all very cozy, though only a little profitable, most members receiving just a few hundred dollars. But the scheme depended on the club's being secret, and one day it suddenly wasn't. The unmasking of the club came about as a result of a lawsuit filed against Crédit Mobilier by a disaffected partner who had been disturbed by rumors that Ames was sitting on a block of unsold stock. Actually Ames wasn't sitting on it at all: It was the stock that he was using to bribe members of Congress. But the partner, who didn't know this, thought he was being cheated somehow. Ames tried to straighten out the mess by telling the partner the truth about the secret stock in a series of seriously indiscreet letters (in one he named the members he had bribed). But by then the cat was out of the bag.

In the fall of 1872 the *New York Sun* exposed the scam in a front-page story about Ames titled, THE KING OF THE FRAUDS. The subhead read: "How the Crédit Mobilier Bought Its Way into Congress." Schuyler Colfax, then in the final months of his term as vice president, at first claiming that he didn't know anything about Crédit Mobilier, insisted he was innocent. Then he suddenly "remembered" that one day he had received a thousand dollars in the mail and that he had used this money to buy some three thousand shares of stock in the company. Others had similarly lame excuses. Henry Wilson said the stock had gone to his wife, who'd received some $800 in dividends. On learning that the money was tainted, he said, he'd returned it. James Garfield admitted he'd received a dividend check for $329, but insisted he hadn't known Crédit Mobilier was connected with the building of the Union Pacific, an incredible claim.

Oakes Ames himself insisted all along he'd done nothing wrong, that all he was guilty of was helping his friends get in on a good deal. Two congressional investigations in 1873 uncovered all of the gritty, disturbing details, but in the end the House could bring itself to censor just Ames and one other legislator. All the others got off without even a reprimand.[8]

Grant professed not to believe that the people in his government accused of wrongdoing, people he'd fought with in the war, smoked cigars with, gotten drunk with, were guilty. But it was impossible to believe they had all been framed and Grant really didn't. He simply couldn't get rid of everybody who was tainted. For he'd have to admit he had been grossly irresponsible in giving them all jobs. And that would make *him* look bad. And it would weaken his grip on power. *Let the bastards get my friends, and they'll soon be thinking they can get me, too.* And Grant very much wanted to remain in power. After his first term he ran for a second. And then, after taking three years off, he ran for a third. (On his third try he lost, Americans by then tired of the Grant scandals.)

So Grant tolerated the corruption.

It was a dirty business. A man had to hold his nose to avoid smelling the stench. But for an old soldier like Grant, who had learned to live with the knowledge that he had sent tens of thousands of young men to their deaths, living with corruption was relatively easy.

The scandals were very much a product of Grant's own failings as a leader. But not entirely. They were, to a certain extent, the result of that vast new change that had overtaken the American economy. Grant, the first president to preside over an administration filled with people who wanted to use their government connections to become rich, stumbled badly in dealing with the corruption. But his reaction to the scandals was predictable. Determined to hang on to power he repeatedly agreed to look the other way when evidence of corruption was put in front of him. Other presidents would learn to do the very same thing. It would come to seem easier than making war on one's own friends. And by saving them, it would often seem, one could save oneself.

9. The Birth of Machine Politics

How the arrival of political machines corrupted politics, forcing Rutherford B. Hayes to tolerate massive vote fraud in order to gain power

---

Down through the early 1860s the story of presidential ambition follows a simple narrative. It is, by and large, about what the presidents themselves are willing to do to gain power and to keep power. But beginning in the late 1860s, the story line changes. Now it is not only what they themselves are willing to do but what they are willing to let their supporters do. First, U. S. Grant lets his friends get away with fraud. Then, Rutherford B. Hayes lets his friends get away with stealing votes. The presidents pretend that they are not letting anybody get away with anything. But they are only pretending.

That they know what their friends are up to and pretend they don't is the price they have to pay in this new world for their ambition. It is a steep price, measurable in the diminished respect they command as presidents. But they willingly pay it. For it is, at the moment, the price that has to be paid for the privilege of living in the White House.

The political parties had been established, in theory, to help average people express their will through the ballot box. But after the Civil War the parties became huge machines run by ruthless bosses. And the machines dominated politics, and very often nearly controlled things completely. In some states the only voices that mattered were the voices of the machine politicians. The most notorious of all the machines was Tammany Hall, which operated in New York City for a time with near impunity. There Boss Tweed had more power over things than the mayor. But there were many other machines. And they were nearly as powerful in their domain as the Tammany bosses were in theirs.[1]

What made the machines powerful, in part, was the huge increase in the number of immigrants, whom the bosses found easy to manipulate. Give a foreigner help finding a job and he'd be ecstatically grateful—or at least grateful enough to vote for the candidates the bosses supported.

What also made the machines powerful was the growth in government payrolls that occurred in the aftermath of the war. At all levels—local, state, and federal—government payrolls simply exploded as government services expanded to keep pace with the increase in population, the growth of the cities, and the movement west.* In 1850, for instance, there had been just twenty-six thousand federal employees. By 1870, there were more than fifty thousand; by 1880, more than a hundred thousand. As the number of government employees ballooned so too did the power of the bosses. For it was the bosses who did the hiring. Furthermore, the more people they hired, the more money they could raise. For each employee was required to kick back between 2 and 6 percent of their salary. And that made the bosses even more powerful. (The bosses referred to the kickback as an "assessment," which made it sound almost like a tax. In a way, it was, for nearly every employee had to pay it.)

Along with the power came arrogance. Earlier generations of American politicians had on the whole been rather timid in their attempts to control elections. And living as they were in the shadow of the Founding Fathers, they were reluctant to tamper with things. The bosses seemed eager to control elections and confident that they could.

It wasn't just that they had a lot of money and power. It was also the war. For the bosses, as for people generally, the Founding Fathers were a distant memory, as was the Revolution.† The war that had meaning for them was the war *they* had fought in, the Civil War, and

---

*More than half the people employed by the federal government worked for the post office. As the country moved west post offices had to be established in every community. See Bernard Weisberger, "What Made the Government Grow," *American Heritage* (Sept. 1997).

†The last veteran of the American Revolution died in 1867.

the lesson of the Civil War was that victory goes to the side that is the most ruthless.

It had been a hard war, and it hardened the men who had fought it. They had done things they had never thought anyone would have to do: burn down whole cities; order tens of thousands of men to their deaths; destroy billions of dollars' worth of property. And it changed them—and it changed the country. Business was rougher and so was politics.

The bosses played politics roughest of all.

The candidates the parties put forward in the election in 1876 happened to dislike boss politics. Samuel Tilden, the Democratic nominee, had been a crusading reformer. As the governor of New York he had even gone against Tammany Hall, the Democrats' own machine, and bested it. Rutherford B. Hayes, the Republican nominee, had also been a reformer. Going into the election he issued a statement demanding civil service reform and an end to corruption.

But by 1876 the candidates almost didn't matter. It wasn't the candidates who ran the campaigns. It was the bosses. And the bosses in both parties decided that this year they would win even if they had to steal votes.

Rutherford B. Hayes, while personally honest, had always been eager to be a winner. But as a young man Rud, as he was known, had been more concerned with a rather simpler matter: finding himself. He even suffered a nervous breakdown as a teenager, roaming the rural hills of his native Ohio aimlessly for months.

He was extremely close to both his mother and his sister, and would write them long, loving letters whenever he was away. But the greatest influence on his life was his father, who had died two months before Rud was born. The stories his mother told about his father, the stories his relatives told, the stories neighbors told, led Rud to believe that his father had been nearly godlike, an inspirational figure. In fact the father had been more of a dreamer than a doer, a man who tried to become rich but who never quite succeeded. Although he made money running a whiskey distillery, his great ambition was to make a fortune as a land speculator, and he never did. Shortly after he got into the land business in Ohio, the boom ended and prices plummeted. The son didn't measure himself against the real father, however, but against the fictional, mythical man he'd always heard about. And that inspired him to reach higher and higher and higher.

A second great influence seems to have been an uncle, John Noyes, or, as he was often referred to, the Honorable John Noyes. It was the title that was critical. Uncle John had been a member of Congress. He, too, hadn't been all that the family pretended he was. While serving in Washington he had become an alcoholic. But Rud, once again, seems to have been inspired by the myth. And Uncle John seemed especially mythical. He had, after all, been a politician in far-off Washington, the center of power. And he had a title.

Rud began to figure himself out at Kenyon College, where he had the opportunity to get out from under his mother's influence. Then he went on to Harvard Law School. Harvard would seem to have been just the place for Hayes, for he was bright and inquisitive. But after Harvard he stalled. He settled in Lower Sandusky at the invitation of a relative and went absolutely nowhere for five long years, years he subsequently regarded as a complete waste.

His sister, Fanny, to whom he was very close, implored him to move to Columbus. Hayes, in a blow for independence, moved instead to Cincinnati. There he finally had a chance to prove himself and went all out. To ingratiate himself with the local establishment he aggressively joined five community booster groups. Then, in an extraordinary effort to build contacts with the leaders of the two main religious groups, he joined both the Episcopal and Presbyterian Churches, alternating his attendance from week to week. As if all that weren't enough, he began giving speeches, in order, he confided, to blow "my own trumpet." The world is a harsh place, Hayes noted, and to get ahead you have to "push, labor, shove."

When the war came he joined the Union army and suddenly found a place where he could excel. Although he never achieved national renown, he was appointed a major general after beating the Rebs on several obscure battlefields in the mountains of West Virginia.

There was an inevitability to his career in politics. With a Harvard law degree, a record as a war hero, and the example of Uncle John, he quickly came to the attention of the Ohio political establishment. Before he'd even been mustered out of the army, he was elected to Congress.

He accomplished nothing during his single term in the legislature. But because he was intelligent and dependable, party leaders liked him and ran him for governor. Given the party's predominance in Ohio, he won. When his two-year term was up they ran him again. Again he won.

All agreed he could have won a third term, but he chose not to try. Politics was tiring, he complained. To be elected governor of Ohio twice was enough. So he retired, returning to the full-time practice of law. But he couldn't quit. Six months after leaving he was back, again as a candidate for a seat in Congress. This time, however, he lost. Not because people disliked him but because Republicans all across the country were having a hard time living down the Grant scandals and an economic depression.

Following the election Hayes again took himself out of politics, declaring that he really was out, "definitely, absolutely, positively." But he simply could not *stay* out. Once again, in 1874, he ran for governor. As he explained in his diary, "a third term would be a distinction—a feather I would like to wear." For no Ohioan had ever served a third term as governor. Despite a national trend against Republicans, Hayes won easily.

It was only natural that he would then decide to seek the presidential nomination in 1876. Ohio was a critical state. And as a three-time governor he could brag that he was in a position to win the state in the presidential election. In addition, despite his reformist leanings, he remained on good terms with the bosses.

Hayes pursued the nomination relentlessly but stealthily. Instead of announcing his candidacy early on, he decided to hold back to allow some of the more prominent Republicans to destroy each other in the months leading up to the convention. He even came out publicly in favor of one of the leaders of the reform wing of the party, Benjamin Bristow, Grant's crusading treasury secretary. Bristow didn't have a chance, but by coming to his aid Hayes would be helping to beef up the reform wing, which was where his own strength lay.

The most serious threat to his candidacy came from James G. Blaine, the former Speaker of the House, who at the time was the favorite of the party establishment. Blaine was far better known than Hayes, and with his connections to the bosses, far likelier to succeed. But Hayes proved to be the more adept politician. Made aware of some letters proving that the Union Pacific had paid Blaine $64,000 for some worthless railroad stocks while he was Speaker, Hayes secretly arranged to have the letters published in a New York newspaper.* Blaine's candidacy promptly died, the voters figuring he must have done something untoward in return for

---

*These were the so-called Mulligan Letters, named after James Mulligan, one of Blaine's business partners. Mulligan found the letters in the office files and leaked them.

the railroad's generosity. Even though Blaine had the backing of the bosses, Hayes, not Blaine, would be the party's nominee,

It was a monumentally important presidential election. For it was the first the Democrats had a chance of winning in nearly a generation. Not only were the Republicans being blamed for the bad economy and Grant's scandals. They no longer controlled the South. Reconstruction was ending, and the South was turning solidly Democratic. Only three Southern states remained in Republican hands: Florida, South Carolina, and Louisiana. And all three of these states were about to be taken over by the Democrats.

The Republicans did what they could. As they had before and as they were to do again and again throughout the nineteenth century, they encouraged party speakers to wave the bloody flag to win votes.* A vote for the Democrats, Americans were told, was a vote for the Southerners who had brought on the worst war in our history. "Every State that seceded from the Union was a Democratic State," Robert Ingersoll, the Republican Party's most prominent speaker, told audiences. "Every ordinance of secession that was drawn was drawn by a Democrat." *Remember that!*

But on election day, November 8, it appeared that Samuel Tilden had won. Although it was premature to call the election on the basis of the reported popular vote—Tilden was just a quarter million votes ahead, out of more than eight million cast—the electoral college vote appeared to give the Democrats a landslide victory.

Based on the estimated popular vote, Tilden seemed to have received 203 electoral votes to Hayes's 166. When Tilden's state victories were shown on a map, it looked as if he had swept most of the country. Not only had he won every state in the Old South, he had also won the border states and even several states in the North. Of all the states worth winning the most important was New York; whoever won New York, Hayes had said many times, would win the election. And Tilden had won New York.

In a normal election that would have been that. Tilden would have gone on to the White House; Hayes would have stayed in Ohio. But

---

*"Waving the bloody shirt" was a reference to a stunt pulled by Republican senator Ben Butler; during the impeachment trial of President Andrew Johnson, Butler dramatically stood up and waved a shirt stained with blood.

this was not to be a normal election. The bosses of the Republican Party refused to let Hayes lose.

There was only one way to keep him from losing, and that was somehow to steal the election. So they decided to steal it.

It was not the Republican politicians who first realized how it could be stolen. It was the editors of the *New York Times*, then an avowedly Republican organ. Sitting in their offices late on election night, they heard that the Democrats were not confident of having won Louisiana, South Carolina, and Florida, the three Southern states still headed by Republican governors installed under Reconstruction. The editors then did some quick arithmetic. These three Southern states had a total of 19 electoral votes. If those electoral votes could be taken from Tilden and given to Hayes—a real possibility since the states were under Republican control—Hayes could win the election by a single vote, 185 to 184. (166+19=185; 203–19=184) There was no proof that Hayes had won those states. Every indication was that he had lost them. But in Wednesday's morning edition, contrary to most of the other papers in the nation, the *Times* announced that Hayes *had* won both Louisiana and South Carolina, with Florida still in doubt.

Over at the Fifth Avenue Hotel, the chairman of the Republican Party and one of its leading bosses had gone to sleep convinced that Hayes had lost. In the morning an aide, accompanied by one of the editors of the *New York Times*, persuaded the chairman that the Republicans could claim Hayes had won if the party took the three Southern states still in Republican hands. Within hours the chairman issued this bold statement: "Dispatches received at these headquarters report that Louisiana, Florida, South Carolina, Wisconsin, Oregon, Nevada and California have given Republican majorities. There is no reason to doubt the correctness of these reports and if confirmed the election of Hayes is assured by a majority of one in the Electoral College."

There was in fact *every* reason to doubt that the Republicans had won Louisiana, Florida, and South Carolina. But these were the states they had to win. For these were the states that had the critical 19 electoral votes—19 votes that would decide whether the nineteenth president of the United States would be a Republican or a Democrat; 19 votes that would decide if things were to remain as they were and had been for years or would change. And, most important, 19 votes that would determine which party would be in power the next four years, and which was to control federal patronage.

But how to get them? Never in American history had the electoral votes of a single state been switched. And now not just one or two, but three states would have to be switched. And every electoral vote from these three states would have to go to Hayes. With just 166 votes from the other states, he needed all 19 to reach the winning number, 185. Tilden, on the other hand, needed just one of these contested votes to win.

The keys to victory were held by the three states' Republican governors, who controlled the returning boards that certified the election results. If the boards threw out the votes of white districts, which were overwhelmingly Democratic, leaving the elections to be decided by the black districts, which were overwhelmingly Republican, the Republicans could take the three states.* Of course thousands and thousands of white votes would have to be disqualified. But under the election laws passed during Reconstruction, the returning boards were authorized to invalidate the votes of an entire district if it could be proved that a single black person had been prevented from voting in the district. And blacks *had* been prevented from voting—thousands of blacks. All three states had black majorities, and yet more whites voted. To be sure, lots of blacks didn't vote because they hadn't yet gotten into the habit of voting. But many, many blacks didn't because they weren't allowed to.

---

*From Reconstruction to the New Deal, blacks in the South remained predominantly loyal to the Republican Party. In the states with black majorities they dominated the delegations sent to Republican national conventions. Republicans had to do very little to keep black support; the blacks had nowhere else to turn. When Teddy Roosevelt in 1906 unfairly cashiered an entire black regiment in Brownsville, Texas, after a few of its one hundred and sixty members became involved in a violent brawl, blacks nationwide expressed outrage; but at the next election, in 1908, they once again voted overwhelmingly Republican. Party bosses like Mark Hanna favored black delegations at the national conventions because they were easy to control and repeatedly turned back attempts to limit their voting power. (Under the rules then in effect a delegation's voting strength was determined solely by the population of the state; that meant that blacks from southern states cast as many votes at a convention as whites from similarly-sized northern states even though the southern states never went Republican in the general election. Today the Republican Party gives delegations from states that vote Republican greater representation at its conventions.) The southern delegations remained a powerful force in the Republican party for years. In 1912 they contributed to William Howard Taft's convention triumph over Teddy Roosevelt. (Taft was able to control the southern delegations by seeing to it that they were dominated by federal officeholders he had appointed.)

So, under pressure from the party bosses and out of a healthy instinct for self-preservation, the returning boards began throwing out the votes of white Democratic voters, throwing them out by the thousands, often on the flimsiest of grounds. By the time the board in Louisiana was through, it had discarded more than thirteen thousand white votes. By early December three states that previously seemed to have gone for Tilden now were going for Hayes.

In effect both parties cheated. The Democrats, by stopping many blacks from voting, cheated first. Then the Republicans, trying to steal the election back, themselves cheated.

The members of Congress eventually had to decide which results to approve, the initial board returns or the amended ones. For months they haggled endlessly; finally, just two days before the inauguration, they hammered out a deal that made Hayes president.

Hayes watched what was done in his name—and said nothing. Not one word complaining about the process. In saying nothing he showed just how truly ambitious he was. Say nothing, and he could become president of the United States.

He himself did not steal any votes. But the bargain by which he became president severely hampered his free exercise of the powers of his office. In exchange for the presidency he had to sell out the two groups in Louisiana and South Carolina that had been the most loyal to the Republican party: carpetbaggers and blacks. He had to agree to let the Democrats take over the governments of the two states, forcing the carpetbaggers to flee. And he had to agree to withdraw the federal troops remaining in the states' capitals, troops on whom blacks had relied for protection. The Democrats in Louisiana promised in return to respect the constitutional rights of the blacks and to protect them from violence. But blacks soon after lost their right to vote in Louisiana as they had earlier lost their right to vote in other states taken over by Democrats. Shortly thereafter, more than thirty blacks in Louisiana were killed in clashes with whites.*

---

*The bargain Hayes agreed to also required the Democrats to support government subsidies for the building of new railroad lines in the North. The Republicans, for their part, agreed to approve federal subsidies for internal improvement projects in the South.

Hayes actually had always intended to withdraw the troops from the two states. But because he was forced to do so as part of the electoral bargain that put him in office, he was unable, when the Democrats took over the states, to use the possible return of the army to threaten the Democrats when they began to break their promise to protect blacks. He protested. His protests were ignored. And so Reconstruction came to its ignominious end.

Hayes always insisted that he had won the election fair and square. But if he did, his victory raises questions about the fairness of the elections held afterward in the South. For if it was right in 1876 to throw out the votes of whites in districts where blacks were prevented from voting, it should have been right to do so in the elections that took place subsequently. And yet this was not done. Hayes never once said it should be.*

As President Hayes initially fought the machine bosses in his own party, even insisting, over the objections of Senate leaders, on the seating of a cabinet dominated by reformers. At great risk to his standing in the party, he even challenged the boss of the bosses, New York senator Roscoe Conkling, who ran the biggest machine in the country. After a commission reported that the New York Custom House controlled by Conkling was rife with corruption, Hayes demanded that the place be cleaned up. When Chester Arthur, the head of the Custom House, refused to make real reforms, Hayes fired him.

Hayes even took on the explosive issue of assessments, the lifeblood of the machine. He issued an outright ban against them by executive order. This was a marvelous move, and it got everybody's attention, invigorating the reform movement as nothing before ever had. But it was theater, a bit of play acting. Hayes, it turned out, had no intention of banning assessments at all. In the off-year election of 1878 he expressly permitted them as long as they were "voluntary," which was as good as lifting the ban outright. For an assessment to be legal, all the party hack sent out to collect it had to tell an employee was that it was "voluntary." The employee, knowing who buttered his bread, then "volunteered" to pay.

Hayes really couldn't go further than he did in his little war on the

---

*The outcome of most of those elections would not have been different if blacks had been permitted to vote. But in at least two elections—in 1884 and 1916—it probably would have. In these two elections—two stolen elections, in effect—blacks probably would have voted Republican in sufficient numbers to give the Republican candidates victory.

machine. Go further, and he risked destroying the very party that had put him in power. For the party depended on assessments to survive. And he very much wanted the party to survive. Like millions of other Republicans who had fought in the war, he believed in the Republican Party—even if it was now beholden in so many ways to the bosses. Besides, destroy the party and he'd destroy his own administration. Without the party the administration would collapse.

Hayes did not run for reelection. The bosses had made him promise in 1876 not to run to give one of their own favorites a chance in 1880. After he left office he went back home to Ohio. There he remained a local celebrity of sorts. But history forgot him. He simply hadn't done anything worth remembering. Even his responsibility for the disreputable way he came to power didn't seem worth remembering: For he himself hadn't stolen votes. He'd just passively let them be stolen, enabling him to escape the notoriety that he probably deserved.

Like most of the presidents between Lincoln and McKinley, he was too weak to bother with, one of "the lost Americans," as Thomas Wolfe characterized them, "whose gravely vacant and bewhiskered faces mixed, melted, swam together. . . . Which had the whiskers, which the burnsides: which was which?" He was weak because the bosses wanted weak presidents. For the weak could be controlled. To win, therefore, a man perforce had to give up his principles, had to cave. Next to the power of the bosses, principles were nothing.

Vote stealing was not a new phenomenon in American history. Aaron Burr had tried to steal the election of 1800. And Andrew Jackson had claimed that the presidency *had* been stolen from him in 1824. And there had been innumerable individual instances of vote stealing in nearly every presidential election since 1840, when Thurlow Weed, the boss of the Whigs in New York, reportedly bought votes to help elect Tippecanoe.* But in none of these elections, deplorable as the circumstances surrounding them were, was there anything compara-

---

*Chester Arthur, one of Weed's protégés, was to remark that Weed in the 1850s resorted to the use of street-gang ruffians to "break up the lines" of voters hostile to the party's candidates. "On one occasion," recalled Arthur, "these ruffians were provided with awls, which they prodded into the flesh of the majority, thus dispersing them." At other times, he said, Weed simply stuffed the ballot boxes.

ble to what happened in 1876. It was the first election to be stolen out-right.*

Once the bosses saw how easy it was to steal an election, they were encouraged to keep on doing it. So in 1880, just four years later, they stole another.

---

*In 1844 the Whigs are said to have raised twenty thousand dollars to buy votes in Philadelphia; the Democrats claimed to have intercepted a letter lay-ing out the scheme. There were rumors that the Whigs in Boston had raised a hundred thousand dollars to buy votes there. (See Charles G. Sellers, *James K. Polk: Continentalist* [1966], p. 155.) In 1868 Tammany Hall Democrats in New York supposedly stole as many as fifty thousand votes in New York City, enough to win the state for Horatio Seymour; one man is said to have voted twenty-seven times. See Thomas C. Reeves, *Gentleman Boss: The Life of Chester Alan Arthur* (1975), p. 48; John Niven, *Martin Van Buren* (1983), p. 471.

10. The Story of Chester A. Arthur

How a party hack fired for corruption from his post as head of the New York Custom House tried to rehabilitate himself by becoming the vice president of the United States

---

In 1880 James Garfield is selected as the Republican nominee for president. An extremely talented man—he can simultaneously write Latin with one hand and Greek with the other—he is also extremely ambitious, too ambitious. To gain power he now oversees one of the dirtiest campaigns in American history, secretly condoning the stealing of votes in both Indiana and New York. To help finance campaign operations he cuts a corrupt deal with the railroads—the biggest special interest in the country—agreeing to give them the right to veto any of his appointments to the Supreme Court in exchange for tens of thousands of dollars in campaign contributions.

But just three months after Garfield is inaugurated president, he is shot by a deranged nobody, Charles Guiteau. Two and half months later, Garfield is dead. Chester Alan Arthur, his vice president, is now president. It is a shocking turn of events. Nobody can believe Chet Arthur is president of the United States.

It is no wonder they wonder.

If an ambitious man was after political power in nineteenth-century America there were plenty of places he could find it. But if he was after more, after wealth and position as well as power, real wealth and real power, there was only one place: the New York Custom House, where the government collected tariff revenues levied on three-fourths of the goods imported into the United States. Three officials ran the Custom House: the collector, the surveyor, and a naval officer. All three were powerful, but one was more powerful than the others and by a substantial margin: the collector. His was the job really ambitious men craved.[1]

Of the thousands and thousands of posts available to those with political connections in the federal government, there was no other like it. For one thing, it paid more than any other job: more than fifty thousand dollars some years, more than the members of Congress made, more than the justices of the Supreme Court made, more even than the president himself made. For another, it came with immense patronage power. The person who held the collector's job had the power to give a thousand other people jobs, more jobs than were controlled by any other single individual in the entire federal bureaucracy, except the president. And that was not all.

Because employees had no recourse if one day the head of the Custom House decided to fire them, his control over them was nearly total, making them dependent on him in a way ordinary citizens in a democracy are seldom dependent on anybody. He could levy assessments on them for political contributions, forcing them to kick back to the party coffers between 2 and 4 percent of their annual salary, which brought in a huge amount of money, often as much as fifty thousand dollars. And he could make them work on behalf of machine-backed candidates at election time. And because he could raise huge campaign funds at a snap, and because he controlled an army of campaign workers, candidates for offices high and low became dependent on him, too.

With all that power, inevitably, came corruption, the corruption typical of nineteenth-century big-city political machines. Businessmen complained that the port charged them extortionist rates to store their goods in warehouses if the goods weren't immediately picked up from the docks after being unloaded. Government auditors charged that the place was overstaffed: A third of the work force may have been unneeded. Many who appeared on the payroll never even showed up for work; they got paid because somebody—the collector, one of his friends, or a political big shot—was owed a favor. The operation was so corrupt that many of the collectors were fired for fraud. The collector under Andrew Jackson—Samuel Swartwout—fled to Europe with his million-dollar boodle.

Many people in politics hungered for this lucrative, powerful, and corrupt post. Only one made it his life's ambition: Chester Arthur.

And that in a way was odd.

For it had seemed for a long time that all Chester Arthur was interested in was good causes, and one in particular: black rights. As a college student in the late 1840s he had championed abolitionism. Subsequently he had joined with Free-Soilers to help establish the Republican Party in New York State. As a young lawyer he had even helped win a lawsuit that brought about the end of segregated streetcars in New York City.

So it was odd for Chester Arthur to decide on a career as the holder of the most powerful and corrupt patronage post in the country—odd, but by the middle of the 1860s not unexpected.

By then Arthur had changed. The onetime idealist, son of a minister, had developed an insatiable appetite for power and wealth. He married a rich woman from Virginia. He moved into a two-story brownstone on a fashionable street in New York City. He hired servants. He ate lavish meals. He wore fine clothes.

His friends in the Republican Party were rich. He wanted to be rich like them. He even changed churches to be like them. Though he had been raised a Baptist, he now joined an Episcopal Church. His rich friends were Episcopalian.

And this man who had once upon a time devoted himself to social causes turned to another: improving the fortunes of the Republican machine. He became so devoted to the party that he started spending more evenings with political friends than with his family, not returning home until two or three in the morning. Soon he was spending all of his evenings with his political friends and none with his family. His wife eventually began to think of leaving him. (By 1880 she had decided she would, but then suddenly and unexpectedly, she fell ill and died.)

After the Civil War Arthur served in a variety of political jobs. But there was only one he really coveted: Custom House collector. Become the collector, even for just a few years, and he could earn enough money to become rich for life. And for years he lobbied for it.

The collector was appointed by the president, but it was not the president in the second half of the nineteenth century who ordinarily decided the matter. It was the party bosses in New York State. That meant the Republican Party bosses when Republican presidents were in office, and one boss in particular, Roscoe Conkling, the state's unctuous senior U.S. senator—the boss of bosses.

Arthur made it his business to know them all.

* * *

They were a querulous bunch. Conkling especially was egotistical and insufferable; highly intelligent, he grew impatient with people who were not. He could be withering in his scorn. One verbal jousting match on the floor of the U.S. Senate almost ended in a duel. "Let me be more specific," Conkling had shouted at one point in the course of a debate. "Should the member from Mississippi, except in the presence of the Senate, charge me, by intimation or otherwise, with falsehood, I would denounce him as a blackguard, as a coward, and a liar; and understanding what he said as I have, the rules and proprieties of the Senate are the only restraint upon me."

Arthur, his eye always focused on the ultimate prize, made it his business to get along with the bosses—Conkling included—Conkling especially. Even when they weren't on speaking terms with one another they spoke with Arthur, who was on good speaking terms with everybody. To achieve his ambition he had to submerge his own ego. Week after week, year after year, he did so, eventually earning a reputation as a man who went along to get along.

And that, given Arthur's ambition, was essential. For it meant, highest of compliments in his world, that Arthur was sound. If the bosses wanted something done, Arthur could be trusted to do it. And to do it even if the request was unseemly, even if it involved breaking the law.

The bosses had once made the mistake of installing a collector who was not sound, Moses Grinnell, who had declined to be as pliable as the bosses wanted. They had gotten rid of him quickly.

The bosses did not have to worry about Chester Arthur. "Mr. Arthur is a gentleman," Boss Thurlow Weed was to say of him fondly some years later. "I have known him for twenty-five years as Whig and Republican, and a more loyal party man and truer Republican cannot be found anywhere."

Soundness was especially valued by the bosses in the late 1860s and 1870s, when Arthur began campaigning for the Custom House position. For by then the stench of political corruption had become so pronounced that reformers were threatening to clean up the place; only a sound collector could be trusted to stop them.

Honest men had long held the Custom House in contempt. But they had lacked the facts needed to stir the public to action. Then, slowly, they had begun to accumulate the facts—damning facts. An investigation showed that the Custom House was run by incompetent political hacks for the benefit of the machine.

The fact of corruption became so obvious that eventually it was not only wild-eyed reformers who began pushing for change but prominent politicians, too. And the change they wanted was profound: the establishment of a professional civil service, composed of people who obtained their jobs by merit and then were allowed to keep them, irrespective of the party affiliation of those in power. In 1871 even President Grant, a beneficiary of the machine, felt compelled to join the effort, at least in name.

A presidential commission endorsed several reforms; Grant then ordered that they be implemented. Two reforms were especially dangerous to the continuing power of the bosses: One was the requirement that open positions go only to candidates who could meet certain competency qualifications, which could limit the ability of the bosses to hand out plum jobs to party workers, many of whom were barely literate. (One of the competency qualifications included the ability to read, write, and speak English.) The other reform that threatened the bosses was a ban on the practice of political assessments.

The bosses were afraid of reform but convinced that they could survive it if they put the right man in charge of the New York Custom House, the institution that sat like an industrial-strength engine in the center of their creaking political machine. Through clever manipulations a devious collector could get around the changes, making them so innocuous that nothing really changed at all.

But who?

At the end of 1871, after the incumbent was exposed as a swindler and forced out, that was the question that faced the bosses. The answer, all quickly agreed, was Arthur, dependable Chester Arthur. In November, President Grant named Arthur collector of the New York Custom House. At last Arthur had his dream job. He was forty-two years old.

The first task he faced was to find a way around those damn competency exams. If he couldn't hire who the bosses wanted hired they would be angry, and Arthur never wanted any of them angry with him. He wanted to remain the collector for a long, long time. It paid better than any other job he was ever likely to have. And it gave him entrée into the best homes in America.

A later generation of clubhouse politicians found that they could hire incompetent individuals simply by exempting the applicants from the civil service. But of course the difficulty with that approach was that it still limited the power of the bosses to hire who they wanted. They couldn't exempt everybody.

But they could rig the tests so that anybody they wanted passed could pass. And that is what Arthur now proceeded to do. As an initial step he installed three friends on the board established to administer the entrance exams. "Sound" men all, they could be counted on to rig the tests the way the party wanted. Next, he excluded from the board's proceedings anyone who took the reform cause seriously. This had been necessary because a reformer had started showing up at their meetings. Arthur simply told the man he couldn't attend.

Then he and his minions designed the test to be easy, so easy a grade-school student could answer them. Unfortunately, many of the applicants lacked even a grade-school education. But that did not pose much of a problem either. If they gave the wrong answer, they could still be passed.

"Into what three branches is the government of the United States divided?" To this question one applicant answered: "The army and navy." He was given a passing score of 70.0 and was hired. Another answered, "Publick stores, Navy Yard." He was given a score of 69.6. He was also passed and hired.

"By what process is a statute of the United States enacted?" To this question another applicant answered: "Never saw one erected and dont know the Process." Passed and hired.

More of a threat to the bosses than the entrance exams was the ban on assessments—if it were to be implemented. But it wasn't. In 1872, just as in years past, a party hack opened an office next to the Custom House at election time and collected the employees' kickbacks. Every one of the thousand people employed at the Custom House paid except for Silas Burt, an old friend of Arthur's, who had turned reformer. He flat out refused. This prompted an outburst from Arthur. "He said," Burt afterward recalled, "that every person who accepted a position under the government made a contract, as solemn in its implication as if plainly expressed, to give a part of his salary to sustain his party in power and that to repudiate such a contract was not only ingratitude but mean[n]ess." Burt still refused to pay.

Given the state of politics in 1872, Arthur would have been justified in firing Burt. But he didn't. Arthur believed in being loyal, even if that meant being loyal to a friend who had become a reformer. As time went on, though, the brooding conflict between Arthur and Burt grew wider and wider. One day Arthur, who never liked to express anger at anybody, got angry with Burt when the latter expressed concern about a corrupt official. "You are one of those goody-goody fellows," Arthur said scornfully, "who set up a high standard of morality that other people cannot reach."

The bosses had picked Arthur because they thought he could find a way to keep doing things the old way despite the new rules, despite the popular impulse to clean up the system and stop corruption. And he had. But then in 1873, he went too far.

The trouble began one day when the Custom House opened a case against an old respected importer named Phelps, Dodge, and Company. An informer was said to have reported that the company substantially undervalued a recent shipment, resulting in a huge loss to the government in tariff revenues. As a penalty Custom House officials decided to fine the firm a whopping $271,017. This was the largest fine that had ever been levied in such a case, but company officials were led to believe they were lucky. It was hinted that they could be made to pay a fine equal in amount to the value of the shipment: $1.7 million. So they paid up quickly.

Then they found out the truth. The local U.S. attorney revealed in a letter to the company that in fact only a part of the shipment had been undervalued, and a small part at that, a part worth less than $7,000, costing the government less than $2,000 in lost tariff revenue. Further, it seemed that another part of the shipment had actually been overvalued, more than offsetting the part that was undervalued. So the government, in effect, hadn't really lost any money at all. If anything, it had received more than it was owed.

Outraged, company founder William Dodge demanded an investigation. At a congressional hearing he testified that he had been bullied into settling the case. "We paid the money in ignorance of the fact of the amount we owed to the Government. We never had a bill of specifications; we have not got one to-day; we have got simply the list of the vessels on which the goods were imported." It sounded incredible. Dodge agreed that it was. "We settled, and this has become the biggest case on record. It is known the world over. We look back upon it and think as you gentlemen, think, no doubt, that we were fools."

How had it happened? Better yet, why? It was because of greed.

Under the law, an old law passed in the early days of the Republic, any fine or forfeiture received by the Custom House was split between the informer, the three top officials at the Custom House, and the government. The law was designed to encourage the collection of revenues. Its actual effect was to encourage officials to file bogus claims against good companies in the expectation that a substantial part of the money collected would go into the officials' own pockets. It was this law, known as the moieties law—a moiety is the part received in a settlement by the individual parties—that made the post of collector as lucrative as it was. Arthur's actual salary as collector was small: some six thousand dollars a year. But with the moieties he earned many times that amount, one year earning $56,000. From the Phelps, Dodge settlement alone Arthur received nearly $22,000.*

It was a scam. The testimony before Congress revealed that the Custom House officials knew, as the U.S. attorney had indicated in his letter to Phelps, Dodge, that the undervaluation involved a negligible amount, and knew in addition that part of the shipment had been overvalued, resulting in that net gain to the treasury. Nevertheless, they had decided to put the squeeze on the company because of the windfall in moieties that they themselves would receive.

The figure of $271,017 in fines had been arrived at arbitrarily. The officials had considered pressing for even more. Under the law, as they had hinted to Phelps, Dodge, they could technically have demanded an amount equal to the value of the shipment, or $1.7 million. At the key meeting where all this was discussed, Arthur had wanted to demand the full $1.7 million. But he had finally relented when others pointed out that Phelps, Dodge, would probably go to court to challenge the higher amount, and that in court the Custom House would probably lose; the facts would come tumbling out and the jury would side with the company.

The scandal involving Phelps, Dodge, resulted in the passage of a new law abolishing moieties. Because of the new law Arthur's income fell dramatically the next year. Instead of earning $56,000 he earned just $12,000.

---

*It is worth noting that Arthur got to keep every penny of the money he earned. There was as yet no income tax.

But Arthur escaped public blame for the scandal. By denying early on that he knew anything about it he created the impression that underlings were responsible. A judge testified to the contrary at the congressional hearing, stating that Arthur had been present at that key meeting where the amount of the fine was determined. But by the time the judge testified the press had lost interest; only a few newspapers noted that Arthur was as deeply involved as anybody.

In 1875 President Grant, still beholden to the bosses, reappointed the bosses' hand-picked choice as collector: Arthur. By then Arthur had held the post for four full years, longer than any other collector in decades. But he would not get to keep it another four years. For in 1877 Grant was replaced by Rutherford B. Hayes and Hayes was committed to civil service reform. Like Grant, he, too, was indebted to the bosses and was elected with the money they raised from assessments. Unlike Grant, he actually seemed intent on seeing civil service reforms implemented even if that meant undermining the very people who had helped bring about his election.

Arthur tried to hang on; despite the salary decrease, he was still earning more money than he probably could otherwise and the position continued to give him power and respect. At times it seemed as if he just might succeed.

The great threat to his job was facts, facts that seemed often to be going against him now. One investigating body, known as the Jay Commission, found that the Custom House was overstaffed by at least 20 percent and that corruption was rife: businessmen testified that they had had to pay officials bribes to recover their merchandise; others confirmed that the Custom House was filled with people hired only because they were friends of the bosses. A second investigation demonstrated the corruption was widespread; when senators loyal to the bosses demanded to see the evidence for this, investigators obliged immediately by dropping a wheelbarrow full of documents off at the Senate door.

But Arthur was wily and found ways to get around the facts.

There are many techniques a wily politician can use to hang onto his job when pressure builds to give it up. Arthur seemed to try them all.

If you can't fight them, join them. When the Jay Commission reported that the Custom House was overstaffed, Arthur cleverly agreed; by agreeing he hoped to confuse the public and defuse the issue. But he denied the

Custom House was overstaffed by 20 percent. He said that after a thorough investigation of his own he had found that maybe 12 percent of the work force could be cut. But certainly not more than that.

Blackmail. One of the chief charges against Arthur was that he had put people on the payroll at the behest of the bosses. This was undeniably true. But it so happened he had helped not just the bosses but President Hayes's own secretary of the treasury, John Sherman. At the very same time that Arthur was under attack for hiring the bosses' friends, Sherman had written Arthur to ask for help in hiring *his* friends. Arthur let it be known that he would release these letters if the attacks on him continued.

Stonewall. President Hayes at first seemed willing to leave Arthur in place if Arthur agreed to cooperate with the reformers. When it became clear Arthur wouldn't, Hayes decided to fire him. But Arthur refused to go, and under the prevailing law Hayes lacked the power to order him to go. Month after month Arthur simply brazenly hung on. To counter charges that he was a scoundrel, Arthur revealed that the president had offered him a post as consul in Paris if he resigned; the implication was obvious: If he truly was a scoundrel, no offer would have been made.

Eventually he lost his fight. When Congress adjourned in the spring of 1879 Hayes, under the authority granted him in the Constitution, appointed an interim replacement. By the time Congress returned, the Senate, which had stood by Arthur for months at Conkling's insistence, finally relented. Arthur was out.

That should have been the end of Chester Alan Arthur. It certainly seemed as if it was. It wasn't. A year later Arthur was back, this time as the Republican Party's vice presidential candidate. It was the greatest resurrection in American political history as of that time and was to remain the greatest until Richard Nixon sprang back to life in 1968 after he, too, had been left for dead on the battlefield of politicians who once had stood high and then fallen.

How had it happened?

It came about entirely through Arthur's own determined efforts. No one else wanted him as vice president. No one had ever thought of Arthur as vice presidential material. But Arthur wanted to be vice president, and wanting it he made it happen.

Wanting the vice presidency was unusual. For Arthur didn't want it as a steppingstone to the presidency as others did, but for itself. The presidency didn't interest him. He felt, rightly, that he simply wasn't qualified for the office. Anyway, all he wanted the vice presidency could give him. And that was just one thing really: respect. After his humiliating dismissal as custom house collector respect was what Arthur now yearned for most deeply. Not money, not power, simple respect.

There were many offices that could give Arthur respect, but none were within his grasp. Only one office was: the vice presidency. And that was because of a fortuitous set of circumstances that happened to come about on the afternoon of a single day—Tuesday, June 8, 1880—the day the Republican Party met in convention to nominate James Garfield for president. And that day, quite on the spur of the moment, Arthur decided to go after the vice presidency once the opportunity presented itself.

That morning Arthur was a failed and discredited political boss. By the end of the day he was on the party's national ticket.

From the time Garfield was nominated it was clear the second position would go to a New Yorker as it had in 1876; New York, the most populous state, was critical to the Republicans. But as late as two that afternoon it had seemed likely the prize would go not to Arthur but to Levi Morton, a fellow New Yorker. Morton was Garfield's own choice, and Garfield had told him he could have the nomination if he wanted it.

Until this moment Arthur had expressed to no one any desire to be named to the national ticket. But then, because of a mix-up in the ranks of the Garfield forces, it suddenly seemed possible that he could be. Not knowing that Garfield had already decided on Morton, one of Garfield's managers told Roscoe Conkling that New York could name whoever it wanted.

It was then that Arthur struck. On hearing of the offer, he decided on the spot to go for the nomination and instantly began rounding up support in the delegation, which would meet just a few hours later to pick a candidate. Several New Yorkers immediately backed Arthur, including Boss Tom Murphy, who had served as Custom House Collector prior to Arthur. But one man's opinion mattered more than anyone else's—Roscoe Conkling's—and so Arthur went to him.

And Conkling told him no. Conkling was still smarting from the defeat of his own favorite candidate for president—Grant, who had lost to Garfield after the convention deadlocked—and at this point was so dejected he wanted to have nothing to do with Garfield.

"What, sir, you think of accepting?" Conkling asked.

"The office of the vice president," Arthur retorted, "is a greater honor than I ever dreamed of attaining. A barren nomination would be a great honor. In a calmer moment you will look at this differently."

"If you wish for my favor and respect," Conkling responded, "you will contemptuously decline it."

Arthur, who had long been regarded as Conkling's chief protégé. Arthur, who had never once before gone against Conkling. Arthur, who had in fact spent a lifetime doing what Conkling and the other big bosses had wanted him to. Suddenly this Arthur had to decide whether to be his own man, as he had never been before, or be Conkling's.

The impulse to accommodate Conkling ran strong in Arthur. He owed almost everything to the senator: His appointment as custom house collector. His wealth. His position in society.

At this moment of decision, however, Arthur did not hesitate. He wanted to be on that ticket. And he would be. And he would be on it with or without Conkling's support.

What suddenly became clear to people as they watched Arthur maneuver for power on this summer day was that he had long been misunderstood. He had not been subservient all the years he served the bosses because it was in his nature. He behaved in a servile manner because to get ahead he'd had to, and he'd been able to do so because his ambition was so powerful that he would subordinate even his ego to serve theirs. And now his ambition demanded that he behave differently, independently, even if that risked crossing the very man before whom he had been fawning for years.

"If you wish for my favor and respect," Conkling had warned him, "you will contemptuously decline it." But Arthur no longer was wishing for Conkling's favor and respect. He had a bigger wish.

That afternoon he campaigned for the votes of his fellow New Yorkers and won them. That evening he was named the Republican nominee for vice president of the United States.

It was to be another corrupt election. For the only way the Republicans could win was by stealing. Now that Reconstruction was over, the Democrats completely controlled the South, giving them at the outset of the contest 138 electoral votes. To win they needed just 47 more (138 + 47 = 185).

The likeliest place they could find them was in Indiana and New York, both of which had gone Democratic the last election. And that in fact was where the Democrats concentrated their efforts.

And that was just what the Republicans wanted. Because both states were nearly evenly divided between the parties, they were ripe for vote fraud. As the Republicans deduced, they could win the presidency by buying a relatively small number of votes in each state. In other words, an election in which millions cast their ballots could be decided by a few thousand stolen votes in Indiana and New York.

That fall the Republicans raised hundreds of thousands of dollars from Wall Street and assessments in their campaign to win the two prized states. How much was raised in all it's impossible to say. Stephen Dorsey, the secretary of the Republican National Committee, bragged that the party raised more than $400,000 for Indiana alone, although some historians say this was an exaggeration. But clearly a lot was spent; afterward party officials in New York would remember that in 1880 they were assessed at a higher rate than they had ever been before. Even judges had to pay. "Please let me know when and where to pay it over," one judge wrote Arthur, who personally handled the accounts of judges. "I am ready at any time."

However much was raised, it was never enough. Arthur, who served as the chairman of the Republican Party in New York State, kept imploring partisans to raise even more. "I am in serious trouble," Arthur complained to the campaign's chief fund-raiser, "if something is not done at once."

And what was all this money that was raised, raised for? Some undoubtedly went for legitimate functions, for banners, headquarters, newspaper ads, and so on. But a lot undoubtedly went for buying votes. The Democrats at one point intercepted a Republican telegram referring specifically to the buying of votes.

Ten million ballots were cast in the election. Garfield and Arthur won by a mere seven thousand. Four years earlier Indiana and New York had gone to the Democrats. This time they went to the Republicans, deciding the election.

It was obvious to reporters covering the campaign that it had been stolen. And it also was obvious that Garfield and Arthur must have known it was stolen, even if they hadn't been part of the stealing. But there was no hard evidence.

Then, on February 11, 1881, at a big Republican bash held at Delmonico's, the famous New York steak house, Arthur himself seemed to incriminate the party. He did not admit to fraud. But neither did he deny it. And in not denying it he seemed to confirm that there had been fraud.

"I don't think we had better go into the minute secrets of the campaign, so far as I know them," he began, "because I see the reporters are present, who are taking it all down; and while there is no harm talking about some things after the election is over you cannot tell what they may make of it, because the inauguration has not yet taken place. . . . If I should get to going about the secrets of the campaign, there is no saying what I might say to make trouble between now and the 4th of March."

What kind of trouble he did not say. But he did not have to. Everybody could guess. And then, as if he could not help himself, he began to talk about the campaign in Indiana. "The first great business of the [national] committee," he noted, "was to carry Indiana, and Mr. Dorsey was selected as the leader of the forlorn hope to carry Indiana." He added sarcastically, "That was a cheerful task." After the guests had stopped laughing, he went on: "Indiana was really, I suppose, a Democratic State. [But] it had always been put down in the book as a state that might be carried by close and careful and perfect organization and a great deal of—"

Of what? The obvious answer was money. But someone in the audience shouted "soap," and that prompted hysterical laughter. Everyone knew it wasn't soap that had won the election. It was bought votes. "I see the reporters here," Arthur again reminded himself, "and therefore I will simply say that everybody showed a great deal of interest in the occasion and distributed tracts and political documents all through the country." Again the audience began laughing. When they finally quieted down Arthur finished: "If it were not for the reporters I would tell you the truth, because I know you are intimate friends and devoted adherents to the Republican Party."

And then he sat down.

Until that dinner Arthur had always been discreet. But on that night, perhaps because he drank too much, perhaps because he was surrounded by friends, perhaps because he was simply feeling boastful and confident, he forgot to be discreet. He had not actually disclosed any secrets. But he had admitted that there were secrets. And that was damaging enough.

* * *

After Garfield died in September 1881 and Arthur suddenly became president, a cry could be heard around the country: *Chet Arthur, president of the United States! Good God.* Things had been bad enough with Grant, Hayes, and Garfield in office. The spoils system had gone wild. Even Hayes and Garfield, putative reformers, had had to bow in the end to the power of the bosses, giving in to demands for the continuation of assessments. For the bosses ruled. With Arthur in office . . . well, it was evident that Arthur, a boss himself, would also do what the bosses wanted.

The country was in for a surprise.

While Arthur did indeed fill his cabinet with machine politicians, he refused to knuckle under completely to the bosses, refused even to name Conkling secretary of state as Conkling demanded.* And then in 1883 Arthur came out in favor of genuine civil service reform, the Pendleton Act, which outlawed assessments and created a class of protected employees. The people demanded reform, and Arthur accommodated them, in part because it was politically wise to do so, in part because he genuinely wanted to. By then he wished to be remembered as more than a machine politician.†

---

*Arthur did offer to name Conkling to the Supreme Court. The Senate approved the nomination, but Conkling declined to accept it. It would have meant giving up his lucrative private law practice.

†It was not a great victory for reform, however. Assessments continued to be levied—and would be on into the next century. In the 1930s reformers were still complaining about them; it was learned that in several states employees of the Works Progress Administration were required to kick back 2 percent of their paychecks to the Democratic Party. And while some federal jobs were protected under the newly formed civil service, most were not, just 10 percent or so.

But it was a first step, and as the years went by, assessments declined, and more and more jobs *were* covered. Not because the politicians had suddenly been persuaded of the goodness of civil service reform. But because it was in their own interests. When the Republican lock on the presidency finally was broken, and the two parties began alternating control of the White House, the politicians in both parties realized that the only way they could assure the continuation in office of their own supporters was by getting them listed on the civil service rolls. The more the presidency changed parties, the more people were added to the rolls.

He did not run for election when his term was up. He was too ill. Just a year after becoming president he had come down with Bright's disease, leaving him barely able to function some days. A year and a half after he left the White House, Bright's killed him.

Today, when Arthur is remembered it's usually in connection with his efforts on behalf on civil service reform. In 1997, in a story about the Pendleton Act, he was even praised on the front page of the *New York Times*.* But the rub is, Arthur seldom *is* remembered. Like so many of the other bewhiskered presidents of the nineteenth century, he's become one of Wolfe's lost men of American history.

The system did not really change. The bosses were to hold on to their power well into the twentieth century, aspiring presidential candidates

---

*The story ran after it was revealed that Vice President Al Gore had made campaign fund-raising phone calls from his office during the 1996 election. A provision of the Pendleton Act prohibits federal employees from soliciting campaign contributions on government property. Gore insisted that the law did not cover phone calls.

†Teddy Roosevelt was to owe his elevation to the vice presidency to New York Boss Tom Platt, who wanted to get Roosevelt out of New York. (Platt reassured party regulars that in the vice presidency Roosevelt could do no harm; of course, that was true only if he stayed in the vice presidency, which he did not. It was the worst political miscalculation of the century.) William Howard Taft was to owe his renomination as president in 1912 to the bosses, who were adamant about denying the nomination to Roosevelt, though he'd won all the primaries. Woodrow Wilson was to owe his elevation to the governorship of New Jersey to Boss James Smith (whom Wilson subsequently betrayed), and his elevation to the presidency to the bosses of Illinois and Indiana; at the Democratic national convention in 1912 Wilson won the nomination from Champ Clark by secretly bargaining for their support while publicly denouncing Clark for accepting the support of Tammany Hall. FDR was to owe his nomination as vice president in 1920 to the boss of Tammany Hall, whom he had nurtured for years and years, and his nomination as president for a third term in 1940 to a phalanx of urban bosses. Harry Truman was to owe his political career to Missouri boss Tom Pendergast, a crook who later went to jail. Hubert Humphrey was to owe his nomination as president to Chicago boss Richard Daley. Humphrey's nomination in 1968 would be the last, however, to be dictated by a boss. After that the media consultants took over.

kowtowing to them for years and years.† And even after the bosses had finally gone, their legacy of ruthlessness and arrogance remained. Ever afterward politicians would seek to manipulate the system cunningly and sometimes illegally, if not by stealing elections outright, then by other means pioneered by the bosses.*

---

*In 1888, after hearing that he had been elected, Benjamin Harrison was said to have commented, "Providence has given us the victory." That had provoked derisive laughs from Pennsylvania boss Matt Quay. "Think of the man," said Quay. "He ought to know that Providence hadn't a damn thing to do with it." But then Harrison was deliberately kept ignorant. Only the bosses would know what really went on, "how close a number of men were compelled to approach the gates of the penitentiary to make him president," as Quay observed.

Harrison was to get an idea of what had gone on, however, when he tried to form a government. "When I came into power," he later lamented to a small group of friends, among them Theodore Roosevelt, "I found that the party managers had taken it all to themselves. I could not [even] name my own cabinet."

In a rare display of independence—for both him and American politicians in general—Harrison boldly refused to appoint New York boss Platt as secretary of the treasury. Platt withdrew his support from the administration. Four years later Harrison lost New York—and the election.

11. The Arrival of the Immigrants

How the great flood of immigrants to the United States in the last quarter of the nineteenth century prompted Grover Cleveland to do what no president ever had: pander to immigrants as a means of keeping power

After Lincoln all of the presidents had been weak. None won in a landslide. None could control either his party or Congress. None could count on his party controlling Congress. None could claim a personal mandate. So they did as their predecessors had done in like situations: They compromised. And compromised. And compromised.

The public grew sick of its compromising presidents and became eager to find somebody—anybody, almost—in whom they could place their trust. It was a measure of their desperation that they finally turned, in 1884, to Grover Cleveland. For nobody really knew much about him.

Three years before his nomination he had been mayor of Buffalo. Then, after just a year as mayor, owing to a quirk in the politics of the day, the Democrats in New York State being without a leader, Cleveland suddenly was nominated governor. Just a year after *that*, in the quickest rise of any politician in American history, he found himself president of the United States.[1]

But what little people knew about Cleveland they liked. And what they knew was that he was a leader. A man willing to fight. And by 1884 they were desperate for a fighting leader again. Hayes, Garfield, Arthur—not one of them, whatever his virtues, had seemed like much of a leader, and not one of them had ever seemed like much of a fighter.

Cleveland's chief characteristic was his earnestness. The son of a minister, he had been raised to be earnest. Not perfect. Earnest. Sincere.

Straightforward. A boy who did the chores his parents gave him because he was supposed to.

He was always interested in politics. Just after reaching his majority he joined the Democratic Party and immediately tried to ingratiate himself with the leaders by doing all the little chores *they* wanted done. Whatever they wanted. If they needed an elderly voter to be escorted to the polls, Grover was there to help. If they needed banners put up, Grover put up the banners. It was boring work, but it was the kind of work you have to do to get ahead in local politics. And Grover very much wanted to get ahead. He was an extremely ambitious young man.

The extent of his political ambitions became evident in 1862, when at age twenty-five, after passing the bar, he lobbied to be named an assistant district attorney in Buffalo. Taking the post meant giving up a job in a private firm that was already earning him a thousand dollars a year. In the prosecutor's office he would make just five hundred. But Cleveland didn't worry about the loss in income. He wasn't interested in money or security. He wanted a career in politics. And as an assistant prosecutor he would be in a position to have such a career: to run for district attorney in a few years, and after that, a seat in Congress.

Just three years later, at twenty-eight, he ran for district attorney. And he worked very hard to win, giving speeches everywhere, day and night. And he won a great deal of attention, most of it positive, from people who were impressed with his earnestness. His earnestness, that quality he had inherited from his father, made him very appealing. For in the postwar political environment earnestness was in short supply, most men in politics feeling that they couldn't afford to be earnest. A man had to be willing to cut deals with the machines, and a man who did that had to be calculating, not earnest.

He was popular in the wards where he was known best, where his personality, his honesty, were known. But in the suburban parts of the district, where he was not well known, he found it very difficult to become popular. Giving speeches there didn't help because Cleveland's chief asset, his earnestness, was not a quality that came through in his speeches. He simply wasn't a very good speechmaker. He was the kind of man who impressed people by his actions, not his words. So he lost.

And the loss affected him profoundly. Shook him up. For the first time in his life he had faced a reversal of fortune.

The Cleveland who emerged from this ordeal was a vastly changed Cleveland. Where once he had seemed uninterested in security, he now seemed consumed by it. And for the next ten years his every step was geared toward achieving security. Financial security. He didn't want to become rich. Money per se never interested him. Well into his forties he lived in a modest one-room apartment. But he wanted the security that came with having money. So he devoted himself to earning money by working extremely hard as an attorney in private practice.

In the middle of this period of his life, he ran for sheriff of Erie County, serving three years. It was an extremely quirky move, and none of his friends could quite understand it. Most sheriffs were political hacks. But being sheriff met two of Cleveland's deepest needs. First, foremost, and most simply: It paid well. Because sheriffs were allowed to pocket substantial fees in payment for their work, he was able in three years to save tens of thousands of dollars, helping him win the financial independence he very much prized. Second, it brought him back into politics, his first great love, giving him the chance to meet people, make speeches, and become better known.

He made a good name as sheriff, displaying the earnestness that was so very much a part of who he was. When his office was given the responsibility of executing a man found guilty of murder, Cleveland personally conducted the hanging. To make somebody else assume the disagreeable responsibility, he decided, would be dishonorable, essentially dishonest. A little later another man was put on trial for murder. Cleveland felt the man was guilty but also felt strongly that he shouldn't be put to death because he was obviously deranged. Intervening in the case, Sheriff Cleveland asked the court to spare the man's life. But the man was sentenced to death anyway. Cleveland thereupon decided, once again, that he would personally carry out the execution. That would be the principled thing to do.

It was becoming clear to people that Cleveland was the kind of man who did what he felt he should do. And that made him a very strange bird in politics. Most people in politics at the time did what the bosses told them to do.

After leaving the sheriff's office, he returned to the full-time practice of the law, ever eager to accumulate enough money to achieve financial independence. Again, he worked very hard, usually putting in twelve-hour days, six days a week. Because he was very good at detailed legal work he was very successful. But it was plain to his friends that he didn't have his heart in his legal career. He worked hard at it simply to be able to obtain that financial security he craved. And no more. When lucrative posts began to be offered to him, jobs as the legal counsel to big corporations, he turned them down. He even turned down the opportunity to become the regional general counsel to the New York Central Railroad, a position that would have earned him fifteen thousand dollars more a year than he currently was making. He simply didn't need the extra fifteen thousand by then. For he finally had accumulated enough to be secure, about seventy-five thousand dollars, which in those days was a huge sum.

For Cleveland, having fulfilled one ambition, there remained yet one other that he hadn't. But as the years passed this ambition seemed as if it never would be—not as long as he remained the stubbornly earnest man he had been. For there simply didn't seem to be a place in politics for a man with Cleveland's conscience. In fact, if anything, politics seemed to be getting dirtier and dirtier. In 1881, the same year he turned down the job with the New York Central, the corrupt Republican machine that ran Buffalo put a corrupt city alderman up for mayor. That Grover Cleveland would ever find a place in such a world seemed unlikelier than ever.

It was just then that he got a lucky break. Because average people were revolted by the machine's mayoral candidate, they began clamoring for a candidate who was pure, clean, incorruptible. Cleveland was the obvious choice.

When the reformers came to him he at first put them off, leaving the impression he no longer was interested in politics. Actually, he remained very interested. But he was by now a very prudent man of forty-three, and it took him time to reach a decision. Within a week, however, he made up his mind. He'd run.

Because reform-minded Republicans were disgusted by the machine's choice for mayor they bolted and joined Cleveland's campaign. This guaranteed Cleveland the election.

He was a very courageous mayor, exactly the kind of leader people had hoped he'd be. Getting into one fight after another with the corrupt members of the city council, he dominated the headlines and attracted statewide attention. For here was something no one had seen in a while: a politician willing to take on the machine.

It was Cleveland at his best. Not Cleveland speaking, but Cleveland doing. Vetoing extravagant appropriations that kept the local machine going. Beating up on the bosses. Streamlining the government's operations.

Less than twelve months later he was elected governor.

Again he took on the machines, including the Democrats' own, Tammany Hall. Again he vetoed extravagant appropriations. And the people loved him. They even loved him after he vetoed an extremely popular bill cutting the fare on the elevated railway in New York City. While they wanted the cheaper fare, wanted it very much, they wanted an independent governor who did what he thought best even more. And Cleveland was sure he was doing what was best. The high fare people objected to was included in a contract made with the private company that ran the system. It couldn't be reduced without violating the contract, and under the Constitution all legal contracts are inviolate. So the fare had to remain where it was until the contract expired.

Again he made headlines, this time attracting nationwide attention. Within twelve months he was nominated to the presidency.

It was an extremely dirty race. The Republicans put up James G. Blaine, which prompted the Democrats to reopen the business about those damaging letters he'd written to the Union Pacific, which revealed that the railroad had paid him sixty-four thousand dollars for some worthless stocks. Cleveland was attacked for alleged licentiousness and drunkenness.

The most sensational charge of the campaign was leveled by a Buffalo newspaper, which claimed that Cleveland, a lifelong bachelor, was the father of an illegitimate child born to an impoverished woman of dubious morality. For a time it was all anybody talked about. When the news broke Cleveland was told by his advisers to deny the charge. Instead he publicly claimed that he was the father and even admitted he'd been paying child support.* People couldn't believe his honesty. Forced to choose between a politician who may have taken a bribe and falsely denied it, and a politician who was guilty of fornicating but who admitted it, millions swung behind the fornicator. But Blaine remained the favorite; as the former Speaker of the House he was much better known.

Then, just days before the election, at a public function in New York City, the Reverend Samuel D. Burchard referred to the Democrats as the party of "rum, Romanism, and rebellion," a crude appeal to upright rural

---

*It is beyond question that Cleveland had an affair with the woman, but it is still unclear if he was the father of her child. While the mother claimed he was, and Cleveland provided child support, it is highly possible that he was simply covering for a married friend, who also had become involved with the woman.

Protestants who believed the stereotypes about Irish drinking. Furious, the Irish in the city rallied to Cleveland even though Tammany Hall had instructed them to vote for Blaine (Tammany hated Cleveland). That gave Cleveland New York and the election. (It was a very tight race; he won New York by just 1,047 votes.)*

It was an astonishing victory, not just because his path to power had been meteoric, but because of the way he had come to power. He hadn't stooped. Although politics had gotten terribly complicated, instead of maneuvering this way and that to gain advantage, Cleveland had stood ramrod straight and in doing so had won nationwide support. It was almost as if he were another George Washington. By being himself, his plain old incorruptible, earnest self, he gained in stature and became ever more popular. Instead of playing to the powers that be, the machines, he ignored them. The more he ignored them the stronger he got.

He was to play George Washington through most of his first term. Although he gave in to the bosses' demand for jobs, replacing forty thousand Republican postal workers with Democrats, he mainly remained stubbornly apolitical, doing what was right even if it cost him support. He even took on the veterans, one of the country's most powerful special-interest groups. When members of Congress began passing private bills allowing veterans with bogus claims to receive pensions they didn't deserve, Cleveland began vetoing the bills.

It was a particularly courageous move because he himself hadn't fought in the war. When his draft notice had come in 1863 he'd done what thousands of other middle-class Americans had. He'd hired a substitute, a Polish immigrant, to do his fighting for him, Cleveland paying the fellow a $150 fee. It was perfectly legal; anybody could get out of serving by getting a substitute. And Cleveland had the excuse of having to support his younger siblings. But it looked bad. And it complicated matters. A president who hadn't fought in the war could not hound those who had. But by focusing on the abusers of the system he succeeded in winning public support.

---

*Excepting the three contests decided by the House (1800, 1824, 1876), the closest election in American history had been the election of 1796; the shift of a few hundred votes in New York would have made Jefferson president over Adams. In 1916 the shift of two thousand votes would have made Charles Evans Hughes president instead of Woodrow Wilson.

As the years rolled by Cleveland continued to play George Washington. At the end of his third year in office he even took on the high protective tariff, which nobody in American politics had touched for a quarter century. He got nowhere; Congress was too beholden to the special interests who prospered behind America's tall tariff walls. But at least he tried.

Then, as his first term was winding down and he was preparing to run again, Cleveland came face to face with the great force of immigration. He folded.

It had been fun all those years playing George Washington. But now there was an election to be won. And this time it couldn't be won simply by being a Washington. While the Washington image would be helpful, it wouldn't get Cleveland the support of the immigrants. And that year he desperately needed their support because the Republicans were making calculating moves to lock up the immigrant vote. So Cleveland decided to risk the Washington image he had been working hard all these years to perfect in order to play hardball politics—hardball immigration politics.

They had always been a factor. And they had always been controversial, always exploitable, both parties engaging in the exploitation, but differently. The Democrats (and before them, the Jeffersonians) sought to enlist the immigrants in their cause, naturalizing them, registering them, and then seeing to it that they voted the party line. James Polk's forces in the election of 1844 registered thousands in New York City; on the day of the election they registered 497, so many that the only way people could get in and out of the courthouse to sign the necessary documents was by climbing through the windows. James Buchanan's campaign in 1856 naturalized thousands in Philadelphia, most of them illegally. In the election of 1868 Democratic presidential candidate Horatio Seymour supposedly bought the votes of up to fifty thousand immigrants in New York City. (As a result he won the state, but still managed to lose to Grant.)

The Republicans (and before them the Whigs, and before *them*, the Federalists) usually ran against the immigrants, against their perceived differentness, against their perceived power, against their willingness to work for lower wages. Always against. When the immigrants started voting Democratic, the Whigs came out against immigrants even getting the vote. As Henry Clay put it, "this constant manufacture of American citizens out of foreign emigrants" is a great "evil."[2]

Sometimes the immigrants brought out the best in Americans. Usually they brought out the worst. It had been the fear of French immigrants in the 1790s—French revolutionary–loving immigrants supposedly—that had led to the passage of the Alien and Sedition Acts. It was the fear of Irish immigrants beginning in the 1840s that prompted widespread xenophobia and the creation of the Know-Nothings, whose whole appeal was to people's prejudices and fears. As early as 1844 the precursor of the Know-Nothings, the American Republicans, swept the city elections in New York. The following year they took control in Philadelphia. Passions inflamed, the nativists there then rioted, waging wholesale war on the city's newcomers.

But important as the immigrants had been through most of American politics in the nineteenth century they had not affected the outcome of a single presidential election. In the three elections in which they had played a prominent role, Polk's election in 1844, Buchanan's in 1856, and Seymour's in 1868, their votes had almost certainly not been decisive. But all that was to change after 1880. After that they would affect every presidential election.

In part their power was simply a function of their numbers. And their numbers were growing tremendously, faster than ever before, accounting for the most remakable cross-oceanic population shift that had ever occurred. In 1880 nearly 500,000 immigrants entered the country; in 1881, more than 650,000; in 1882 more than 750,000. So many immigrants poured in that in little more than a generation the foreign population of the country doubled, from three million in 1870 to nearly six million in 1900.

But it wasn't just that so many came but that so many came to one place, New York—and then stayed there. For that one state was so large, so important, possessing thirty-six electoral votes—more electoral votes than any other state—that it frequently decided presidential elections. Almost always if a candidate won New York he could be certain he won the election. Between 1880 and 1912 *every* winning candidate won New York. So every candidate tried hard—very hard—to win New York. And to win New York they had to win over the immigrants and win over, specifically, the immigrants of New York City, who made up over half the city's population.

Because of New York the dynamics of American politics changed. Earlier in American history the Republicans and the Whigs had discovered that there was a big constituency in the country against immigrants (and Catholics) and the parties had played to it. Now they had to modify their approach drastically. They simply had to abandon smear attacks on the immigrants of New York City. The last president who had attacked them had been Hayes and Hayes had lost there.

Hayes had tried a tricky approach. In the past the anti-immigrant, anti-Catholic politicians had always been "againers," against this, against that. Hayes's tack was much more subtle and devious. Instead of sounding a negative note, he sounded a positive one. He did this by coming out in favor of public schools. Of course, this was meant to be taken as an attack on the Catholics, who liked to run their own schools, and was understood as such, but it was a coded attack and had a veneer of respectability. Still the fact was Hayes had lost New York and hardly anybody afterwards could afford to. (Hayes had won the election, of course, but he had achieved victory through fraud.) By the end of the century both Democrats and Republicans would be printing tens of thousands of campaign pamphlets in Polish and German in an open appeal for immigrant votes.*

There yet remained one immigrant group the politicians could safely campaign against, and both parties did: the Chinese. Like the European immigrants in New York, the Chinese were often hated. Hated for supposedly taking the jobs of whites, hated for supposedly driving down the wages of railroad workers, hated mostly for just being *Chinese*. Throughout the West, where the Chinese mainly settled, there was a huge backlash against them. In 1885 some two hundred had been driven out of Tacoma, Washington, at gunpoint and forced to flee to Portland, Oregon, leaving the city of Tacoma without a single Chinese resident. The next day white vigilantes burned down

---

*The Republicans were to resume their attacks on immigrants in the late 1890s, when for the first time in American history more immigrants began arriving from southern Europe than northern Europe: Poles, Italians, Jews. The shift in the origin of the immigrants so inflamed public opinion that the Republicans adopted a proposal to restrict immigration severely. It cost them heavily in New York City, but was so overwhelmingly popular in the rest of the state that they were able to win New York in several presidential elections. To help shore up support in New York they frequently put a New Yorker on the ticket (most notably, Teddy Roosevelt). When Roosevelt became president he made numerous efforts to gain Catholic and Jewish support, and in 1904 became the first Republican president to win a majority of the Irish vote; in 1907 he named the first Jew to the cabinet (Oscar Straus, secretary of commerce and labor). Lewis Gould, *The Presidency of Theodore Roosevelt* (1991), pp. 140–41; 258–59.

what remained of the city's modest Chinese settlement. And what happened in Tacoma was repeated all over the West.

But because there were far fewer Chinese immigrants than European immigrants, the politicians felt free to exploit peoples' fears and prejudices. The politics of Chinese immigration was in fact almost exactly the reverse of the politics of European immigration. While the politicians discovered that they could *not* attack the European immigrants (because it would cost them New York), they found that they almost *had* to attack the Chinese (for it was the only way to win the West). And they did. In 1880 Garfield let loose with a broad swipe at the Chinese, comparing their emigration to an "invasion."

Because nothing was simple in American politics anymore, the politics of immigration became very complicated. It was not just that there was now a "good" immigrant group and a "bad" immigrant group, though that certainly made things difficult; a politician had to be sure that in making the case against the "bad" immigrants he wasn't somehow impugning, by implication, the "good" ones. A politician couldn't simply rail against all immigrants, he had to make distinctions, sometimes very fine ones. It was also that the politics of immigration was full of traps. A candidate had to worry about not only what he said about the immigrants, but what his opponent *said* he said—or what his opponent said he *should* have said on an occasion when he had said nothing, absolutely nothing! The Republicans had found themselves in trouble in 1884 not because of what they said about the immigrants, but because when Reverend Burchard popped off about "rum, Romanism, and rebellion," they said nothing. Because immigration was so emotional an issue immigrants could be easily manipulated into thinking that a candidate had slighted them when he hadn't intended to at all. So it was a very, very complicated business.*

* * *

---

*It was emotional for good reason. Most immigrants, the Irish in particular, faced terribly painful discrimination: social discrimination, job discrimination, housing discrimination. Signs hung in the windows of Boston shops bluntly warned: "No Irish Need Apply." In 1881 Oxford Professor Edward Freeman joked during a public lecture in the United States that "the best remedy for whatever is amiss in America would be if every Irishman should kill a Negro and be hanged for it." When Irish leaders objected, Freeman replied blandly that he had not intended to hurt anyone's feelings. It was a sign of the times that he believed he could get away with the "explanation."

How complicated was to be seen in the election of 1888, when Cleveland ran for reelection against Benjamin Harrison. Cleveland was already having problems because he had taken on the fight against the tariff, which had weakened his support in the business community. Then the Republican Senate, just two and half months before the election, robbed him of the one victory he'd had in foreign affairs. The senators voted down a treaty he'd signed with Canada (the Bayard-Chamberlain Treaty) that settled a long simmering dispute over fishing rights, a dispute that had become so heated the Canadians had begun seizing American sailing ships.[3]

This put Cleveland in a desperate position. While only a few people gave a damn about the arcane fishing issues that lay behind the Canadian-American imbroglio, by rejecting the treaty the Republicans put Cleveland in the position of defending it, and that could be politically fatal. For he wouldn't simply be defending a fishing treaty, he would be defending a treaty made with Canada. Or more to the point, a treaty made with a country that was part of the British Empire. And in the late nineteenth century nothing could be more deadly to a politician running for national office than to do that. For the Irish hated the British, and the Irish were the key to winning New York, and New York was the key to winning the general election. (Cleveland had won the election in 1884, after all, only because at the last minute the Irish had shifted to his side in the wake of Reverend Burchard's slander.)

Now, in 1888, he needed their votes again. Desperate to get them, he inaugurated a new era in American presidential politics: He openly pandered for the immigrant vote in order to attract their support.

It was a new Grover Cleveland. The old Grover Cleveland never pandered. But the old Cleveland had never had to. In fact, the only way he could win was by not pandering. For pandering would seem at odds with the image he was selling of himself as a kind of reincarnated Washington. But now, on the eve of his reelection, the Republicans had forced him to confront the force of immigration. And in confronting it he had to face a fact about himself.

He really wasn't as much of a Washington as he had led people to believe. He was a politician. And a politician who very much wanted to avoid losing. Losing was what he had done as a young man of twenty-eight. And it had shaken his world. He did not want to lose again. To be sure, he no longer was the insecure person he had once been. He was now the president of the United States. And he had money in the bank. But losing now in a way would be even more devastating. For he had more to lose: the highest office in the land. And that seemed intolerable.

So he pandered, making sure to pander in a subtle enough way that he did not suddenly seem to be trading in his Washington image for another. For he still needed to be regarded as a Washington as well. This would be tricky but not impossible. The Washington image was so well established by now that Cleveland could make a number of political moves before people caught on to the change in his behavior.

It wasn't good for the country for Cleveland to be pandering. It undermined the presidency, cheapening it. And it politicized foreign policy. In effect, Cleveland was putting his own interests above the country's, showing just how deep his ambition to succeed really was.

The Republicans had expected Cleveland to defend the treaty. Instead, in his message to the Senate following its rejection, he took another tack. And it was brilliant, politically brilliant.

Instead of defending the treaty, which would have cost him greatly in the Irish community, he attacked Canada for the depredations on American shipping that had given rise to the treaty in the first place, which put him in the enviable position of attacking a member of the British Empire. Then he went on to demand that the Republicans give him extraordinary emergency powers to deal with the issue, including the right to suspend all trade with Canada. Cleveland explained that if he was to have any hope of curbing Canadian perfidy, he needed to be able to back up his tough talk with the threat of action.

Cleveland suspected that the Senate wouldn't give him the authority he demanded. And he didn't really want it. What he wanted was to be turned down. By turning him down the Republicans would make themselves look pro-British. And that would make Cleveland appear anti-British. And that would appeal to the Irish.

It was a simple political trap but the Republicans fell into it. Just as Cleveland had hoped, the Republicans, unwilling to let him start a trade war with Canada—a trade war that would appeal to American patriotism and redound to his credit—turned down his request for emergency powers. And just as he had hoped, this put him in good stead with the Irish. In the days following his message to the Senate he received hundreds of letters from Irish families congratulating him on his "devotion to old Erin."

Then there was yet another turn in the campaign, yet another flap involving immigration politics, this time to Cleveland's disadvantage. And it cost him the race.

Although he won more popular votes than Harrison, Harrison won more electoral votes, a freak occurrence so uncommon that the same situation has occurred just three times in all of American history (in 1824, 1876, and 1888).

It had been unlikely that Cleveland would win reelection. He had taken on the protective tariff lobby. He was a Democrat from New York running at a time when the country was electing Republicans from Ohio.* And he had barely won his first election to the presidency. Worse than all that, he was a Democratic president up for reelection. That put history against him: No Democrat had been reelected president of the United States since Andrew Jackson, and that had been fifty-six years earlier.

But in the end just one state stood between Cleveland and victory: his own, New York. And he lost New York because at the last moment one group thought to be leaning heavily his way turned against him: the Irish.

How had he lost the Irish vote? Just two weeks before the election the Republicans released a letter that Sir Lionel Sackville-West, the British ambassador to the United States, had written to Charles F. Murchison in September. Murchison, who described himself as a naturalized Englishman now living in California, had asked which candidate for president would be good for Great Britain. The ambassador indiscreetly suggested that that would be Grover Cleveland. And when that got out, the Irish swiftly abandoned Cleveland, probably contributing more than any other single factor to his defeat.

Grover Cleveland had come close to doing in 1888 what no other Democrat had done since 1832 and what no Democrat actually would do until 1916. And that was to win reelection. But he lost because of a dirty trick. The Englishman who wrote Sir Lionel Sackville-West was not really an Englishman at all. His real name was George Osgoodby.

---

*All the presidents elected between Andrew Johnson and Teddy Roosevelt came from Ohio except Cleveland. And all of them were Republican except Cleveland.

And he was a California Republican. Cleveland was done in by a Republican dirty trick—the first in American history to alter the outcome of an election.*

He took the loss well, though his friends thought he simply may have been putting up a brave front. It almost appeared that he was taking it too well—as if he were hiding his true feelings.

But he would be back. As he was leaving the White House to make way for Benjamin Harrison, his wife† remarked to a butler, *Don't change anything. We'll be back.* Cleveland already was plotting his return.

Four years later he was again selected as the Democrats' nominee for president. In a rematch with Harrison, Cleveland won, making him the first (and so far only) president to serve nonconsecutive terms.

---

*Dirty tricks had been all but unknown in American politics. The first ones hadn't been played until 1844. And those were almost innocuous. One involved a newspaper story put out by the boss of the Whigs, Thurlow Weed, about a traveler named Roorback who on a visit down South supposedly came across some of Democrat James Polk's slaves cruelly branded with Polk's initials (a president could own slaves but he couldn't brand them like cattle—*that* was considered inhumane). Another involved an attempt by the Whigs to demoralize leading Democrats; to persuade the Democrats that the campaign was going badly the Whigs sent out a letter to that effect supposedly written by some high party muckety-muck. Still another involved the printing up of phony ballots; the ballots mixed up the names of Democratic and Whig electors, which was intended to confuse voters. All in all pretty tame stuff.

But then the war had come along, and then the bosses, and then the special interests, and then the immigrants. And the dirty tricks suddenly got much much rougher. In 1880 the editor of the *Truth*, a New York scandal sheet, published a letter Garfield had supposedly written to H. L. Morey, the head of the Employers Union of Lynn, Massachusetts, endorsing the right of a corporation to hire the cheapest labor available, including Chinese labor. Garfield had not written the letter. It was a forgery. But it threatened to undermine his candidacy in the West. See Charles G. Sellers, *James K. Polk: Continentalist* (1966), pp. 152–53; Allan Peskin, *Garfield* (1978), pp. 505–11.

†In his second year in office, he had married; more on her later.

12. The Media

How the revolution in mass communications forced Grover Cleveland to manipulate the media and lie about his health in an effort to maintain control

Lucky as Cleveland was to win reelection, his timing was terrible. Just a month before his inauguration a financial panic struck. The Philadelphia and Reading Railroad—one of the nation's leading lines—went bankrupt. Wall Street started on a steep decline. And the country's gold reserves began running dangerously low. As spring turned to summer, businesses began failing at an alarming rate. In Colorado half the industries in the state closed down. As the price of silver collapsed, mines shut down. In New York City, the financial powerhouse of the country, the banks all but stopped giving out cash. Even the rich found that the banks wouldn't give them their money. One fellow with twenty thousand dollars on deposit at a bank was told he would be allowed to withdraw just fifty dollars.

And then, at the height of the crisis, Cleveland found out that he had cancer.

Grover Cleveland had always enjoyed good health, despite his portly frame, despite the 250 pounds he carried on his five-foot eleven-inch body. But on Friday May 5, 1893, just a few months into his second term, he noticed a growth the size of a quarter on the roof of his mouth. A doctor who examined it thought it might be cancerous. A lab test subsequently confirmed that it was. And Cleveland was told it had to come out—and come out quickly before the cancer had a chance to spread.[1]

No one knew what the reaction of the country would be to the news that the president was ill, but it could only contribute to the feeling that things were out of control. It had been just a little more than a decade since Garfield had died, and the last thing anyone wanted was to see another president die. It wouldn't just be the fact of death that could prove unsettling, but that Cleveland might suffer a long lingering death precisely as Garfield had, though for different reasons, of course.

Worse perhaps than the fear that the president might die—worse in Wall Street's eyes for sure—would be the fear that the vice president, Adlai Stevenson, would take over. Stevenson believed in a policy of soft money, believed, that is, in flooding the market with silver coins to inflate the currency and to help out debtors. The money men back East would be fearful of a Stevenson presidency, so fearful the markets would be likely to perform even worse than they already were, deepening the financial panic, maybe deepening it so much as to plunge the nation into a full-blown depression.

He never should have been put on the ticket with Cleveland. For Cleveland believed in hard money; one of his chief goals in his second administration was to dampen the fears of inflation. But it was *because* Stevenson believed in soft money that he had been run with Cleveland. The Democrats needed Stevenson to win the West, where indebted farmers and the silver mine owners lived.

The Wall Street concerns were paramount. In the great new economy that was developing in the postwar period, Wall Street loomed large. If Wall Street did well, the economy usually did well. It wasn't as if they were directly linked. Sometimes when Wall Street was going gangbusters, the economy was in the doldrums. But there was a relationship there, and presidents had to be careful about Wall Street in a way that they had never had to be careful before about any economic entity.

In the old preindustrial days, presidents had had very little to do with the economy. Even the government as a whole had had very little impact. It could issue currency, set tariff schedules, run a budget surplus or deficit, and deposit government funds, but not much else. The economy pretty much ran by itself. Industrialism changed that, government becoming deeply involved in the economy, especially in the distribution of substantial subsidies to the railroads. During the Civil War the government virtually took over the economy. And that altered the relationship between the government and business, the president and business.

Even after the war ended, the relationship was never the same again. Now they were interdependent, the one affecting the other in myriad and complicated ways. One word from the president about silver or gold and stocks could take a dive or rise, investors gaining or losing millions. In a way they never had before, presidents now could affect Wall Street and business and in turn the economy as a whole. So they had to be much much more careful than ever before; they had to measure everything they did by its potential effect on business.

Cleveland now took all that into account and decided that he had no choice but to lie about his cancer. The operation to remove the cancer in his mouth would have to be secret, utterly secret, so secret as not even to become the subject of rumor, for even the rumor of illness could exacerbate the financial crisis. If anybody happened to ask about his health, they would have to be told a falsehood. They would have to be told that the president was well when in fact he was not.

Earnest Grover, sincere Grover, truth-telling Grover would have to lie. It was an incredible and ironic turn of events, considering Cleveland's reputation for honesty. But he felt he had no choice. He was desperate. Just as desperate as he had been when he'd pandered to the immigrants in 1888. Only then he had stooped to save himself. Now, he told himself, he was stooping to save the country. Doing his duty.

To be sure, there was a great deal of patriotism in his lying. Like Polk, he felt that lying was in the country's interest. Sometimes it is. Perhaps this was such an instance. But it would be naive to think that his decision to lie was only about patriotism. It was also about satisfying his great driving, ambitious need to remain in control of events. Tell the truth, and he'd lose control. Lie, and he had a chance of retaining control. So he lied.

The critical factor in his decision to lie was not the change that had taken place in the economy or the president's new role in it. It was another change, this one so obvious most people tended to overlook it. But it was profound. It was the change in the media and the way the media were beginning to cover the presidents. For the first time they were beginning to report on the state of the president's health. And because they were, presidents now felt compelled to lie when the news about their health adversely affected their ability to control things.[2]

In the old days the presidents hadn't had to lie when they fell ill, because the papers rarely reported on their illnesses. Washington in his second year as president caught pneumonia and nearly died. The press largely ignored the story. Jefferson in his first administration suffered from nonstop diarrhea and during both terms faced recurrent bouts of dysentery, malarial fever, migraines, and backaches. The press never said a word. Madison suffered from bad nerves and anxiety and during the War of 1812 almost died from some kind of "bilious fever," probably dysentery. The press skipped that story as well.

And so it went. Jackson was especially prone to illness: abscesses, coughs, vomiting, stomach cramps, malarial fever, and possibly even lead poisoning. At the height of the bank war he collapsed and was confined to bed. Polk suffered from gallstones and diarrhea and worked himself to death. Pierce was an alcoholic. Buchanan was simply old and often feeble; at his inauguration, just prior to taking the oath, he had to take several shots of liquor to steady his nerves after he came down with some kind of unspecified illness, probably food poisoning. Lincoln, while physically strong, suffered from melancholia—depression. All in all a frightening record. Two presidents had even died after falling ill, as had four vice presidents.* And yet hardly a word was ever printed about the president's health. A line here, a line there, and that was usually all.

So what changed? Why did the presidents' health suddenly begin to assume importance? It was because they became celebrities.

Reporters had never really written about the presidents as celebrities. They had not even thought to write about them that way. Before the war—*the* war—when presidents had gone on vacation reporters hadn't followed after them; the presidents had simply gotten on their horse or climbed into a carriage and ridden off. Jefferson spent months at Monticello without ever seeing a reporter. John Quincy Adams spent seven months at home in Massachusetts without ever seeing a reporter either; no editor had thought to send a reporter up to talk to the president about his extended vacation from the capital. Buchanan had gone home to his mansion in Pennsylvania and complained because he was a little lonely. Again, there were no reporters.

And then the war came. While Lincoln found that he still could travel about alone, reporters now became a presence in his life. The White House in a way even became an unofficial beat, reporters hungry for war news hanging around for a scoop (though the telegraph office was usu-

---

*The two presidents were William Henry Harrison and Zachary Taylor. The four vice presidents were George Clinton, Elbridge Gerry (both of whom served under Madison), William King (who served under Pierce), and Henry Wilson (who served under Grant).

ally a better place to wait for one). But nobody yet thought to ask the president for an official newspaper interview. Newspapers didn't yet interview presidents; the presidential interview hadn't been invented.

But things were changing. In a way they hadn't been before, presidents were becoming bona fide celebrities. Not of course as they were to become in the twentieth century, but celebrities nonetheless. Everything about the presidents now seemed to be of concern. What kind of food they ate, what kind of clothes they wore, how they wore their hair. When Benjamin Harrison ran in 1888 the papers reported his shoe size ($6\frac{1}{2}$), his hat size ($7\frac{1}{2}$), the price of his shirts (twenty-seven dollars a dozen), and his favorite sports (fishing and baseball). What their wives were like was also of interest, the kind of clothes *they* wore, the kind of food *they* ate.

Even where the First Family vacationed became of interest. And when they went on vacation now a whole press contingent followed. When Cleveland went on vacation, which was usually to a place on Buzzards Bay in Cape Cod, reporters reported on his every movement. Who came by to visit. How he relaxed. What he did all day.

And then he got married! To a twenty-four-year-old named Frances! A young woman whose guardian he had been since her birth! After a secret engagement! And the press went crazy. Every detail seemed momentous. *Marrying at his age? To a woman so young? To a woman he'd known since birth? And nobody knew a thing about it until the official announcement?* Fascinating! When Cleveland and his new bride went on their honeymoon, the press followed, Cleveland complaining of the "newspaper nuisances."

It was a curious time for presidents suddenly to become celebrities. For the presidency itself had never been less powerful or less interesting. Hayes, Garfield, Arthur, Cleveland, Harrison—these were not terribly colorful, exciting people. But as the bourgeois middle class had grown, as average people had become concerned with status and had looked about them for models from whom they could take their social cues, they had quite naturally turned to the presidents.

A turning point of sorts was reached in 1887 when Mrs. Fanny Gillette and White House steward Hugo Ziemann put out *The White House Cookbook*, a runaway bestseller, which included, in addition to the presidents' favorite recipes, all sorts of advice on correct social etiquette, very Victorian social etiquette: "A gentleman should not bow from a window to a lady, but if a lady recognizes him from a window, he should return the salutation." Or: "Cleanliness is the outward sign of inward purity. It is not to be supposed that a lady washes to become clean but simply to remain clean." Why such stuff was included in a work passed off as the White House cookbook was something of a mystery. But that the presidents were now linked in some way to such matters was telling.

Editors had quickly sized up the public's appetite for gossipy news about the presidents and begun to feed it. So great was the interest in the presidents that they could no longer go anywhere without reporters following them. Which in turn led to another development—the creation of the White House press corps.[3]

Just yet the corps did not quite exist. For nobody covered the president full-time. One day a reporter might report on the president's activities, the next day on the secretary of state's. But inevitably, as things became more formalized and specialized, certain reporters began focusing on the president exclusively.[4]

Presidents often liked the new attention they were receiving. It was flattering to be followed around all the time. It made them seem extra important. But the added coverage also came at a cost. Just as increased newspaper coverage of the Grant scandals had riled Grant, increased newspaper coverage of his successors' personal lives riled them. Not always, but on occasion and for the same reason: control.

The more the newspapers wrote the more control they had over the national agenda and the president's life, and the more control *they* had, the less control *he* had. And their agenda was quite different from his. They wanted to write about the things that sold newspapers. Often that would be innocuous stuff, but sometimes not. For what the newspapers liked most was the dramatic and the most dramatic news usually wasn't good news but bad news, and that usually wasn't the news the presidents wanted people reading.

Newspapers had been powerful for a good long time by now, since the late 1830s at least, when Benjamin Day, the publisher of the *New York Sun* and the elder James Gordon Bennett, publisher of the *New York Herald*, had invented the penny press, revolutionizing the business. Before Day and Bennett papers had sold at best a few thousand copies a day; by 1842 Bennett's *Herald* was selling fifty thousand copies a day; by 1860, seventy-seven thousand.

But after the war, as Grant had discovered, the newspaper had become something more, almost a fourth branch of government, and more powerful in some ways than the other three branches of government.

It wasn't just that the newspapers helped shape public opinion. It was something more than that. The newspapers defined the terms of the national debate about things. If they said something was a fact, it *was* a fact. If they thought something wasn't a fact, it *wasn't*. Grant had felt that his friends had been railroaded. But the newspapers hadn't seen it that way. And the newspapers' view had prevailed over the president's. It wasn't just an opinion that the administration was corrupt. It was now considered a fact. The newspapers said it was corrupt and so it *was* corrupt.

The press could not of its own authority pass laws or wage war. But it could help get laws passed, and it could help get wars started. (It was to have its war in 1898, when press baron William Randolph Hearst used his paper to whip up public support for an invasion of Cuba. When the artist Frederic Remington, sent to Cuba to produce pictures for Hearst of the looming war, sent word that "there will be no war" and things were quiet, Hearst wired back, "You furnish the pictures and I'll furnish the war." And he did.)

An early indication of the press's new power had come in 1872. That year the Democrats, searching for a candidate to run against the ever-popular Grant, nominated a journalist to be president, Horace Greeley, publisher of the *New York Tribune*. A first. The campaign, however, was a disaster. Greeley made a terrible candidate. He simply talked too much. And then tragedy struck. His wife became deathly ill and Greeley became despondent. "I disagree with you about death," he wrote to a friend who had expressed a fear of death. "I wish it came faster. . . . I wish she were to be laid in her grave next week, and I to follow her the week after." On October 30 his wife *did* die. Five days later Greeley lost the election. Three weeks after that *he* died.[5]

It would be a long time before another newsman would be nominated for president,* but there was no stopping the increasing power of the press. As time went by it would grow more and more powerful.

Its power was evident in all sorts of ways, quirky ways. In 1875 Boss Tweed, the corrupt boss of New York's Tammany Hall, the most corrupt boss there probably ever was, had been accused of swindling the government out of millions of dollars and been arrested and held in Ludlow jail while awaiting trial. Then, to the everlasting embarrassment of the

---

*Warren Harding was a newspaper publisher; John Kennedy served as a war correspondent for the International News Service in 1945. He covered the Potsdam Conference and the founding of the United Nations.

authorities, he had somehow managed to escape. The police looked everywhere but couldn't locate him. Then came word that he had been arrested in Cuba, a Spanish official spotting him after seeing a newspaper drawing of Tweed done by cartoonist Thomas Nast. As an isolated incident it was simply a delightful little story. An anecdote. Something to talk about over dinner. Which people did. But of course it was symptomatic of something larger. A change in the society. The rise of a newly powerful institution in American life.[6]

The newspapers themselves did not quite knew what to do with their new power. There was no handbook to follow, no rules, nothing to help editors decide what could and couldn't be done. So the editors winged it. Tried everything. Had fun. One day James Gordon Bennett, Jr., who took over the *Herald* from his father, even tried a hoax. Just to see what would happen he ran a front-page story claiming that the wild animals at the city zoo had escaped and were roaming loose: TERRIBLE SCENES OF MUTILATION. A SHOCKING CARNIVAL OF DEATH. Readers naturally became horrified and frightened and for hours they remained off the streets.*

Journalist Lincoln Steffens, by accident, made people think that the city of New York had been hit by a crime wave. Steffens began innocently enough. After finding out about a certain crime not reported in court papers, he ran a story about it and readers went wild. The other papers in town, relying solely on the courts as their source of crime news, missed it—were beaten, scooped. So Steffens leaned on his friends in the Police Department for more scoops and started running more and more crime stories. Suddenly it seemed as if the city had been overrun by criminals.

Teddy Roosevelt, the police commissioner, convened a secret meeting of the police board to find out what was happening. Told that crime was not up, that just the *reporting* on crime was up, Roosevelt became angry, and called in Steffens, a friend, and asked him to please, please pull back on the crime coverage. Steffens did, ending the crime wave. But the newspapers continued to be sensationalistic.

---

*This was, as it happened, the second great newspaper hoax of the century. The first was Benjamin Day's front-page claim in 1835 in the *Sun* that life had been discovered on the moon by an astronomer using a giant telescope in South Africa. According to the newspaper the moon was inhabited by four-foot-tall batlike creatures. Some believed the story, some did not. But the hoax was wildly successful in attracting readers, nineteen thousand a day, then a world record.

Sensationalism sold papers. A famed Indian fighter, William Travers, complained in 1891 that the papers killed more Indians than his troops. "Why, if we killed half as many Injuns as the newspapers do," he observed, "we'd be short of Injuns!"[7]

Powerful as the newspapers were, presidents were powerful, too. While they could never gain back completely the control they lost over the national agenda once the newspapers became rich and powerful, they could gain back some of it by restricting the flow of information. Information was power, and the president ultimately had more information than the reporter, meaning that the president ultimately was in the more powerful position.

It wasn't really ethical for a president to play games with information; in theory a president should give everybody equal access. But in the new environment in which they were now operating, presidents couldn't afford to be particularly Ivory Soap pure. They *had* to play games, had to exercise control. Of course, they didn't figure this out all at once. The situation was confusing. But slowly they did. By Cleveland's day they had mastered most of the techniques of control that are now so commonplace: playing one reporter off against another, leaking certain facts and not other facts, giving off-the-record interviews, sending up trial balloons. It was a fundamental turning point, almost the beginning of modern times. And once again it was desperation that drove them to behave as presidents before them never had.[8]

They tried to keep tight control over as much information as possible, but they were especially fanatical about controlling information about one area of their lives in particular—even if they had to lie to do so. For this one area could often be critical. Most importantly, it was personal. It was news about their health.

It happened to be Chester Arthur who was the first to lie.

When he took office he seemed to be in perfect health. Just fifty-one years old, one of the youngest presidents, he had always been vigorous, always able to keep long hours, and was known for his stamina. But then he'd come down with Bright's, and he suddenly began feeling terrible. Not just a little terrible. A lot terrible. And not just once or twice a month, or once or twice a week, or once or twice a day. But all the time, every day of every week of every month. He found he had trouble getting up in the morning, didn't have an appetite, often felt depressed, and didn't want to work.[9]

Arthur knew he was dying but he decided to keep the news secret. He thought if people knew it would undermine his ability to control the government. The only thing that anybody would talk about would be Chester Arthur's illness. Any chance he had of getting legislation passed, of putting his stamp on the presidency, would be lost. From the moment he disclosed his illness he would become a cipher, the connivers inevitably scrambling to fill the resulting vacuum. And Arthur very much wanted to control the administration, if only to prove that he could be more than a simple machine politician. Thus he felt he had no choice but to keep his illness secret. Open up, tell the truth, and the administration would suffer terribly. *He* would suffer terribly. The country would suffer terribly. Just a year after losing one president people would begin worrying that they would lose a second.

Keeping an illness like his secret, however, wasn't easy. Not with all the news reporters now swarming over the White House. And, as it happened, a writer for the Associated Press managed somehow to find out about the illness almost immediately, just after the diagnosis was made, and sent out a report on the wire. It was full of specifics and seemed irrefutable. According to the story the surgeon general had made the diagnosis, and a New York specialist had confirmed it. But Arthur wasn't about to let a wire service story destroy his administration, so he sent out a close friend to deny that the story was accurate. The president *was* sick, the friend related. But he had a mild case of malaria. Not Bright's disease.

Having decided to lie it was relatively easy to keep the lie going, because nobody had ever heard of a president lying about his health. It simply hadn't been done. So people discounted the AP story and figured the president had malaria. The only real danger was that Arthur would begin to look ill, which would inspire rumors. But for most of his term he was able to limit his public appearances and to conceal the deteriorating state of his health.

Near the end of his administration, however, he had a close call during a trip to Florida. It was one of many trips he took during his last year (which provoked a lot of discussion and criticism), all in an effort to find relief from his ailments. But on this trip he nearly died. In addition to kidney failure, the Bright's was leading to hypertension—high blood pressure—and one day it nearly stopped his heart cold. The secretary of the navy, who was traveling with Arthur, put out the word that the president was ill. But Arthur rallied and was able to travel home to New York, where he hid out in a hotel for a week while he recovered. When he finally returned to Washington reporters asked him about his health and Arthur brazenly lied. "I am feeling perfectly well, as well as ever in fact. I have not been sick at all." *Not at all?* the reporters pressed. Well, Arthur admitted, he had suffered a "slight indisposition."

Arthur was able to keep the lie going by keeping the secret tightly held. Aside from his family, just one or two friends knew, and only one member of the cabinet, the secretary of state. He even declined to tell the man who was organizing his reelection drive. Arthur had no intention of running again. He was too deathly ill. But he let the poor devil keep working feverishly to reelect him right up to the Republican convention.*

It was possible, of course, that Arthur had lied not because circumstances made lying inevitable but because *he* was simply a liar. But then a president known for telling the truth lied, too, when a few years later *he* became ill. That seemed to settle the matter. Presidents felt it was their prerogative to lie about their health.

Cleveland's operation could not, of course, take place in a hospital. Too many people would have to be in on the conspiracy at a hospital—where the president would be likely to come into contact with dozens and dozens of nurses, orderlies, and doctors—for the silence to be maintained. So the operation would have to occur in a place hidden completely from public view. As it happened there was such a place—a yacht.

And so the charade began.

Friday night, June 30, 1893, after the last visitors had left the White House, the president took a train from Washington to Newark, New Jersey. Then he crossed the Hudson, boarded a horse-drawn carriage and arrived at Pier A, off the East River in New York City. There he boarded the *Oneida*, a private yacht belonging to Commodore E. C. Benedict, a wealthy friend with whom he'd gone boating many times before.

Only a dozen or so people in the whole country knew what was

---

*Coincidentally, Bright's disease struck two other prominent Americans. It took the lives of the first wives of both Theodore Roosevelt and Woodrow Wilson.

planned for the *Oneida*: Commodore Benedict, of course; the members of the ship's crew; a dentist; the five doctors who would perform the operation; the president's wife; a couple of friends; and one—and only one—officer of the federal government, Daniel Lamont, the secretary of war, who'd been with Cleveland since his days as governor of New York. At the doctors' insistence Lamont was to be present at the operation in case something went wrong.

That night Cleveland and the others relaxed on the deck of the ship, exchanging small talk and smoking cigars. (Even the president smoked.) They stayed up until midnight, even though Cleveland complained of fatigue.

Since his inauguration on March 4 he had contended not only with the crisis caused by the panic, but with the thousands of office seekers who descended on Washington after every election. One, a Mr. Nibge, had been hounding him day and night. "Oh, Dr. Keen, those office-seekers!" Cleveland burst out bitterly. "Those office-seekers! They haunt me in my dreams." Cleveland had been so beset by office seekers that he had recently issued an unprecedented order. On May 7, two days after he had discovered the cancerous patch in his mouth, he had announced that "the limitations placed upon human endurance oblige me to decline . . . all personal interviews with those seeking appointments to office, except as I on my own motion may especially invite them."*

And, of course, it wasn't just the panic or the office seekers that had worn Cleveland out, it was the diagnosis of cancer. Though he hadn't felt any ill effects from it, the diagnosis itself had been a severe strain. The president was a young man even by nineteenth-century standards. But his father had died young, at age forty-nine. Cleveland at the time of the operation was fifty-six and concerned that *he* would die young, too.

Saturday morning the president had a light breakfast, including toast and coffee. Then he had two teeth pulled to give the surgeons room to cut into his jaw. Finally, the operation itself began. It was 1:24 P.M.

Cleveland, sedated with nitrous oxide and ether, lay in a chair located in the ship's saloon. The chair was anchored to the ship's mast, a tangible reminder that this was the unlikeliest of settings for an

---

*The patronage burden was especially great for Cleveland because with his election the presidency had changed parties. He had followed a Republican.

operation on the president of the United States. That morning one of the doctors had mentioned to the captain, "If you hit a rock, hit it good and hard, so that we'll all go to the bottom."

First the surgeons cut straight into his upper left jaw, separating the jaw from the cheekbone. As they started removing the upper front of the jaw they found the cancer, a "gelatinous mass." Working quickly, they cleared out the contaminated fragments. It was at this point that they could see that the cancer had gone further than they'd suspected, into the sinus cavity above the roof of the mouth. They took out the visibly cancerous sections, but Doctor Keen argued that the surgeons should remove the rest of the upper left jaw and continue cutting up past the sinus and into the bottom of the eye socket. But the other doctors decided that this would be unnecessary and stopped cutting, leaving the president's eye structure intact. At 1:55 P.M., after just about a half hour, the operation ended. Cleveland was given a shot of morphine and put to bed.

When he awoke he was in a feisty mood, obviously still haunted in his dreams by those damnable office seekers. "Who the hell are you?" he asked one of the doctors who happened to be standing over him at the time. "Erdmann," said the doctor, "from Chillicothe." "Oh," said Cleveland, "do you know Mr. Nibge there?" "Yes, he's the druggist." "Well," asked the president, "is he so poor that he needs a job from me?" "No," said Erdmann. "Then he won't get one," snapped Cleveland.

Obviously, the public should have been told about the operation. The risk of the portly president suffering a stroke had been so real that a specialist had been brought in to monitor his progress. But Cleveland wanted the operation kept secret, and it was.

For five days nobody outside the conspirators' circle even suspected that the president was ill. From Saturday, July 1, to Wednesday the fifth nobody but the conspirators knew that the president of the United States had been operated on for cancer. Nobody. Not the members of the cabinet (except Lamont, of course), not the members of Congress, not even the vice president. Nobody even suspected anything. People had been told the president was spending his July 4 vacation aboard the *Oneida*, and they believed it. On July 5 the *Oneida* dropped Cleveland off at his summer home in Buzzards Bay on Cape Cod. All was quiet.

Then on Thursday the president's personal physician, Dr. Joseph Bryant, who'd made the first diagnosis of cancer, held a press conference to declare that the president was recovering from a slight illness. The event was all part of the elaborate plan the conspirators had concocted to conceal the operation from the American public, but it went badly. Doctor Bryant said that Cleveland was suffering from "rheumatism and somewhat from his teeth." When pressed for details the doctor said that Cleveland's knee was lame, and that his left foot was swollen so greatly he had to wear a big shoe. He added that a dentist had extracted a bad tooth. But the reporters sensed something more serious was amiss and began asking penetrating questions. Finally one of them asked if the president had cancer.

Doctor Bryant was not used to lying, but at this moment there was no avoiding it. He told the reporters the president was absolutely free from cancer. He went on to deny specifically that the president had been operated on. Secretary of War Lamont, who was also present, insisted that the president was simply suffering from a sore tooth.

Neither Bryant nor Lamont, however, made very good liars. When the reporters retreated to their hotel to file their stories, half of them were convinced that the president was seriously ill and suspected he did have cancer. But the presidents' men, this day anyway, were lucky. For the sake of consistency the reporters decided it was necessary that they all take the same position on the president's health, and that position, they finally concluded, was that the president was recovering from rheumatism and a toothache. For now at least the conspiracy was holding.

As Cleveland recuperated he received dozens of visitors at his Cape Cod home. It was obvious to all of them that he was in great discomfort. His mouth was stuffed with cotton and he had trouble speaking. Under the circumstances it was difficult for Cleveland to keep completely silent about the operation, and he found that he couldn't. When Richard Olney, the attorney general, arrived Cleveland blurted out the secret and cried, "My God, Olney, they nearly killed me!" But the public remained in the dark. And it remained there through July and into August, even after he returned to the White House.

Then on August 29 the *Philadelphia Press* told all. In a story rich in detail, Elisha Edwards reported that the president had cancer and that he had undergone a secret operation earlier in the summer to have it removed. Edwards had gotten the story from his family doctor, who in turn had gotten it from the dentist who had removed Cleveland's two teeth. It was entirely factual; the dentist, under the impression silence was no longer necessary since Cleveland seemed to have recovered, had confirmed the facts in an exclusive interview with Edwards.

But then a funny thing happened. The members of the conspiracy went on the offensive and denounced the story, one of the doctors afterwards remembering that he lied more at that time than he had in his whole life. And the story was forgotten. Completely forgotten— buried in a hail of lies. The American people had been fooled.

And to no end. It had been thought that lying would help keep the panic from turning into a depression. But in the end the panic became a depression anyway. All the lying was for naught.

Cleveland didn't make a very good liar. He lied only because he felt he had to. He didn't lie again.

Mostly he remained true to himself, putting the country's interests above his own and his party's. And he was very impressive. He fought so hard for the repeal of the Sherman Silver Purchase Act, which required the government to buy millions of ounces of silver a month, that he put the very survival of the Democratic Party at risk. Both farmers and silver mine owners, two key groups in the party's coali- tion, supported the law; the farmers because it helped inflate the money supply, thereby reducing their debts; the silver mine owners because it helped make them rich. But when Cleveland realized that the law was bankrupting the country, dangerously depleting the gov- ernment's gold supply (the government was required to buy the silver with gold), he demanded the law's repeal.

One day the Speaker of the House, Charles Crisp, came out to the White House to protest Cleveland's demand for a speedy vote. "Do you know what this means to me?" Crisp asked. "My people in Georgia are for silver. My political career will be ruined." "Mr. Speaker," retorted Cleveland, "what is your political future weighed in the balance against the fortunes of the country? Who are you and I compared with the welfare of the whole American people?" "Well," Crisp responded, "if you put it that way, I'll consent."

The law was repealed. But the Democrats, divided between hard and soft money factions, split in two, leaving the silverites and newly minted politicians, like William Jennings Bryan, in control. It would take the Democrats a generation to heal the rift and return to power.

But he wasn't always so tough, so true. Although he served during one of the most tempestuous periods of American history, when the struggle between labor and capital was intense, giving him numerous opportunities to demonstrate presidential leadership, he largely passed them all up. He let the trusts run wild. He let the special interests that controlled the Senate sabotage genuine tariff reform. He refused to fight for higher taxes on the rich.

It wasn't that he failed to see the great problems facing the country. He saw them and did nothing. But on two occasions he did pause and take notice. On the first occasion he assailed the corporations, declaring that they were "fast becoming the people's masters." On the second he attacked the trusts, assailing them for their "palpable evils." But both times he spoke when speaking plainly no longer posed a danger to his political career. He criticized the corporations in a message to Congress in December 1888—at the end of his first term, after he had lost to Harrison and was on his way out. He took on the trusts in December 1896—at the end of his second term, when he was no longer in a position to right the wrongs he had identified.

It wasn't that he was a political coward. There simply was only so much he could do, given the powerful forces arrayed against him. Had he done more he very likely would have destroyed the Democratic Party, not just split it in two.

Increasingly, as the country became more complex, presidents would often find themselves in situations like Cleveland's, situations in which they wanted to do good but simply couldn't. Desperate, they would cave in to the pressures. They really had no other choice. In trying to do what they thought they should, they would only succeed in destroying themselves. So they compromised. Again, and again, and again.

By the time he left office he was widely scorned. Among the groups that scorned him the most were reporters. Cleveland had manipulated them as no president before ever had, and they resented it. One of them, once Cleveland was safely out of office, went public with their complaints, pointing out in a national magazine all the detailed ways in which the president had used information to maintain his control over the news.

Cleveland came off terribly in the piece, but he had the last laugh. The reporters who covered him didn't find out for years about the secret operation. Not until 1917—nine years after Cleveland died, twenty years after his administration ended, twenty-four years after the operation—was the lie finally exposed, when Doctor Keen told all in the *Saturday Evening Post*.

The reaction of the public to the news about the Cleveland lie was interesting. People read the story, talked about it, ohhed and ahhed about it, and then they forgot about it. It was as if the Cleveland lie had never been told. A president could lie about his health and get away with it.

The politicians didn't need the Cleveland example to figure that out, however. They'd already figured it out by themselves. In their shrewd way, based on years and years of experience dealing with people and politics, they had calculated that health was an area in which a president could lie. So they lied and lied and lied, telling more lies about their health than about anything else. It was almost as if they felt they had a right to lie about their health because, really, nobody should be asking them questions about it anyway. Their health was *their* concern.

13. World Power: I

How William McKinley caved in to the demand for a war with Spain but then went on aggressively to monopolize the control of foreign policy

---

Through most of the nineteenth century the presidents were pre-occupied mainly with domestic issues. Now they become deeply involved in foreign affairs, as the United States steps on the world stage, transforming itself into a true world power.

Unlike many of the other great changes that took place previously, this one actually enhances executive power. Presidents find that in foreign affairs they can unilaterally commit the United States to vast undertakings, sometimes even involving the country in war.

But there is a catch. To exercise power they often must resort to subterfuge and manipulation. Most important of all, they learn that to have a free hand they must find ways to exclude Congress from the decision-making process.

Compared to Cleveland, McKinley looked like a weakling. Almost anybody would have, Cleveland had been so tough and determined, so willing to act in defiance of public pressure. But McKinley seemed especially weak. He spoke softly. His hands were soft. His mien was soft. And he was very devoted to his wife, Ida, a chronic invalid subject to epileptic fits, and that made him look soft, too. When he served as governor of Ohio he arranged for himself and Ida to live in a hotel directly across from his office so he could be near her in the event of an emergency. Every day at three in the afternoon he'd stop what he was doing, go to the window, and wave to her. On those occasions when she suffered a fit in public, he'd carefully place a cloth napkin over her face to save her the humiliation of being stared at. He remained so devoted to his wife that Mark Hanna, his campaign manager, was to complain jokingly that McKinley "made it pretty hard for the rest of us husbands here in Washington."

He wasn't really soft at all. It was just that he appeared to be soft. That was one of his talents actually. He could push himself very hard and push himself ahead without anybody thinking he was working hard or working to get ahead.[1]

He was born in Ohio, which already told you a lot about the man. Like other Ohioans, he came of people who moved west for opportunity. His father was a small-scale businessman. His mother was a devout Presbyterian; for many years she had high hopes that her boy William would grow up to be a preacher. McKinley had other ideas.

When the Civil War broke out he enlisted as a private. Within a year he was a mess sergeant, but he was unsatisfied with his position. At Antietam he quickly attracted attention by daringly driving a mule team loaded with hot food to the soldiers fighting on the front lines. McKinley thought he deserved a commission for his feat and was disappointed when one wasn't immediately offered. So he quietly went to the commander's brother-in-law and asked to see if it could be arranged that this obvious oversight be corrected. And it was. Shortly thereafter he managed to attract the notice of the commander himself, becoming an aide. McKinley had a sharp eye for people who were obviously going places. The commander was Rutherford B. Hayes.

After the war he returned to Ohio and studied law. At age twenty-four he opened his own law office, and then, just two years later, he was elected county prosecutor. Two years after that he married Ida, then a raving beauty, who also happened to be the daughter of one of the richest men in town. And three years after *that*, in 1875, he helped Hayes win reelection as governor.

It was becoming obvious by now that McKinley had political ambitions, but it wasn't yet obvious just how ambitious he truly was. So far he hadn't had to demonstrate how far he was willing to go to fulfill his ambitions. His victories had come cheaply. But now they were to come at immense cost. He and Ida had already lost one child and now, in 1876, they lost another, Katie. A little while later Ida's mother died, and Ida began showing the symptoms of invalidism that were to dominate her life. McKinley, however, pushed on with his own plans for a political career. Five months after Katie died, he ran for Congress and was elected.

It appeared for a time that probably all he was going to be was a congressman, and for the next fourteen years that was all he was. But behind the scenes he was working aggressively to become a power in the Republican Party. In 1880, just four years after he was elected to Congress, he became the temporary chairman of the state party convention, then four years later, the permanent chairman.

By 1888 he had succeeded in becoming so powerful that people were beginning to think of him as presidential timber. At the national convention that year, some delegates even pushed for his nomination, but McKinley wisely protested that he wasn't running; he didn't have a chance anyway, and by taking himself out of contention he won the loyalty of candidate John Sherman, whom McKinley was pledged to support.

His immediate ambition was to be elected Speaker of the House, and the following year he waged a fierce battle for the post. He lost, but he was given the chairmanship of the Ways and Means Committee, the most powerful committee in Congress. As chairman he was then able to champion the passage of a high protective tariff, which put him in good stead with the businessmen who controlled much of the party.

What was so fascinating about McKinley was that even as he was becoming a power in the party, nobody was becoming jealous of his power. He just wasn't the kind of man who inspired jealousy. So it seemed as if his way was clear to run for president. But he decided that he would have a better chance if he were first to be elected governor. As the governor of Ohio he would have instant credibility as a presidential candidate—Hayes had, after all, been governor of Ohio—and by winning election as governor he could demonstrate his vote-getting abilities. So in 1891 he ran, winning handily.

It would later be said that Mark Hanna made McKinley president. This was only half true. As the party boss of Ohio Hanna played a key role in McKinley's ascension, but it was McKinley who pushed McKinley first, McKinley who waffled on the issues so as to avoid alienating important blocs. If McKinley hadn't exercised extreme shrewdness in his handling of these issues in the early nineties, he never would have been considered suitable for the presidency in 1896, when Hanna's help finally proved to be critical.

To his credit McKinley demonstrated consistency on the tariff, even after the country seemed to reject a high tariff in the elections of 1892. But he waffled badly on the silver issue. In 1890 he supported the Sherman Silver Purchase Act. In 1891 in his race for governor he seemed to come out against silver. In 1893 he again reversed course and seemed to come out for silver. By 1896 nobody knew where he really stood, which was just what he wanted. By being ambidextrous on the issue, he could appeal to all corners of the party in his drive for the nomination.

After he won the nomination the party came out four-square for gold. For any other politician with McKinley's record the party's stand might have proved an embarrassment. After all, McKinley just three years earlier had seemed to be for silver. But McKinley was oleaginous. He could get away with a flip-flop. And he never publicly displayed the least discomfort about his obviously inconsistent positions. Inconsistency was inevitable in politics, he thought.

His inconsistencies, however, made him appear weak. Speaker Joe Cannon commented that McKinley had his ear so close to the ground that it was full of grasshoppers. But he wasn't weak in the sense that he could be pushed around or controlled. In his deliberate and steady path to the White House, he'd showed an iron will. In a way, by folding on the silver issue he was showing his strength. Or at least his ambitiousness, which was a kind of strength. Ambitious to the core, he wouldn't let anything get between him and the White House, not even principle. What McKinley truly wanted, McKinley usually got.

Two days before his inauguration McKinley had a brief conversation with Cleveland, one of those conversations leaders inevitably seem to have that afterward appear so tragically ironic. Cleveland frankly told McKinley that he'd left him a war to fight in Cuba. McKinley responded that he hoped he wouldn't have to go to war. His single greatest hope, he said, was to get through the next four years without one. Cleveland nodded in agreement.

McKinley hated war. He remembered the bodies piled up on the Civil War battlefields and didn't want to see those same kind of gruesome piles again, and certainly not on his watch. And he organized his cabinet as if he didn't expect to. To appease veterans he put in a political hack as secretary of war, Russell Alger, the head of the GAR— the Grand Army of the Republic. He put Ohio Senator John Sherman in as secretary of state, though Sherman at seventy-three was utterly feeble by then. (The Sherman appointment was part of a little old-fashioned political bargain. By putting Sherman in the cabinet McKinley opened up a place in the Senate for Ohioan Mark Hanna.)

Having selected his cabinet McKinley settled into his job and prepared to concentrate on economics. The country had been through hell the last four years; the Cleveland depression, the worst in American history, had demoralized people and McKinley wanted to focus on recovery. His first act was to convene a special session of Congress to enact a new tariff law. The Democrats had lowered the tariff (a little) just when the depression had hit. The Republicans wanted to raise it again. A high tariff, they claimed, would bring back good times. By July a new tariff had been passed; McKinley, pleased, headed off for vacation. If everything went as planned, the economy would come roaring back and he could take it easy.

As things turned out the economy did come back, the depression ending about as quickly as it had begun four years earlier, and McKinley and the Republicans reaped the rewards. Everything had worked out as planned.

But if McKinley felt that everything was in control it wasn't to remain so. The media, that great vast new powerful force in American history, was about to join with another, equally as powerful: imperialism. Together they would rock his presidency, forcing him to do what he did not want to do, which was go to war. Strong as he was when he wanted something, he wasn't strong enough to hang on to peace. So he caved. Cave, and he at least could remain somewhat in control of things. Defy the media, defy imperialism, and Congress would take control.

America as it neared the end of the century was a much more exciting, more vibrant, more cosmopolitan place than it had ever been before. The mixture of cultures, the growth of the media, the development of industry—all had worked marvelous changes in the country. But in one way it was very much like the old America. It was inward looking, self-absorbed, and ignorant—almost defiantly ignorant—of the world outside, the world across the Atlantic and the Pacific. Little attention was even paid to the countries on its own doorstep, Mexico, Santo Domingo, Colombia, Nicaragua, or even Cuba. Grant had tried to annex Santo Domingo; the Senate turned him down. Americans simply didn't want to be part of the outside world. They especially didn't want to do anything that might involve them in unnecessary conflicts with the European powers. Most of their ancestors after all had come to America to escape from

Europe, to get away from history and start fresh. The expectation that one could create a new Eden had been a central part of the American dream. As a result no president had yet even dared venture abroad during his term of office. When Cleveland on a fishing expedition had gone past the three-mile limit, it was considered something of a dramatic moment; as far as anybody knew, no other sitting president had ever done so.

But by the 1890s Americans had become fascinated with the outside world, fascinated in particular with the Great Powers—England, France, Germany, and Russia—and what they were doing around the globe to increase their power. Americans were both repelled by the empire builders' conquests—the division of Africa into European spheres of influence, the division of China into European commercial zones—and yet also quite a bit jealous, millions wondering why the United States didn't have an empire. Slowly the idea became prevalent that the United States should.

Imperialism, however, wasn't one thing. To different people it meant different things. To businessmen with hopes of establishing new markets it meant profits. To jingoists it meant a chance to go to war. To the Teddy Roosevelts it was a way to project American power. To Christian missionaries it was an opportunity to redeem the world's oppressed classes. To others it was a way to get new land now that the American frontier had supposedly closed. So it was a complicated, contradictory thing. But common to all of the various meanings attached to it was a growing, fervent belief that it was America's destiny to take a leading role in world affairs, which itself grew out of a fervent belief in American nationalism.[2]

Where all these feelings first crystallized was around the Sandwich Islands—Hawaii. Exotic Hawaii. Most Americans had never heard of the place and certainly couldn't find it on a map. But it sounded interesting, and annexing Hawaii would, people were told, extend U.S. influence deep into the Pacific. That was appealing. It instantly could turn the United States into a Pacific power.

Behind the high-flown imperialistic campaign was a small group with a profound self-interest in annexation: the sugar lobby. If Hawaii were made a part of the United States, the white American sugar planters who in the last generation had established gigantic sugar-growing operations on the islands would immediately become eligible for a federal subsidy worth millions.

But little was heard about the sugar lobby's pecuniary stake. What excited people was the imperialistic, nationalistic rhetoric. At the height of the campaign the islands' white planters overthrew Queen Lili-uokalani, with the help of a rogue American minister, who on his own authority had deftly dispatched a small contingent of U.S. Marines to the scene at a critical moment. The planters then proclaimed Hawaii an independent republic and applied to Washington for annexation. The American minister wired the State Department, "The Hawaiian pear is now fully ripe, and this is the golden hour for the United States to pluck it."

The Harrison administration was fully behind the acquisition and in 1892 prepared to "pluck it," but just then Harrison was voted out of office, replaced by Cleveland. Cleveland, believing the queen had been cheated out of her kingdom by conniving businessmen, scotched the plans for annexation and repudiated the involvement of the American minister in the coup against the queen, which the Harrison administration had covered up.*

Hawaii would not become part of the United States—not yet, anyway. (It finally was to be annexed a few years later under McKinley's aegis.)[3]

Cuba was a lot closer than Hawaii, ninety miles off Florida, and in the middle 1890s in the throes of a colonial rebellion against Spain. Cleveland watched what was happening on the island and for several years pressed Spain to settle the conflict. But he had other problems, the silver crisis, the tariff fight, the depression, and so mainly ignored what was going on. When he wrote up his final message to Congress, he included a sharp warning to Spain that the United States would not remain indifferent forever to the rebellion in Cuba, that with us peace was not a necessity, but then he had second thoughts and dropped the warning. McKinley could deal with Cuba; it would be his headache.

The imperialists very much wanted Cuba. Cuba was the key to the Caribbean. And the Caribbean was key to Central America. Which in turn was key to South America. The imperialists didn't want to take over all the Americas, but they very much wanted the United States to dominate; currently the European countries still exercised enormous influence, Germany and Great Britain in Venezuela, Spain in Cuba.

Imperialists were, however, in a somewhat awkward position. In

---

*Cleveland refused, however, to help the queen return to power. Southerners in the Democratic Party wouldn't allow him to replace a white regime with, as it was put in those days, a "colored" one. The whites remained in control.

Hawaii they had sided with the oligarchs against the indigenous people; now in Cuba they were siding *against* the oligarchs and *with* the
indigenous people. No matter. The point was to expand American
power. Anyway, they didn't really want to help the Cubans become
independent, totally independent. They simply wanted the Spaniards
thrown out. Then the United States could step in, exercise benign control over the island, perhaps making it a protectorate.

The Cuban colonial war was a journalists' dream. There were good
guys and bad guys. And the press lords of the day, William Randolph
Hearst and Joseph Pulitzer, made the most of the situation, competing
daily to see who could come up with the most sensational evidence of
Spanish tyranny.

It was practically a game, almost. You'd open the Pulitzer paper, and
there'd be a story about some poor Cuban woman who'd been raped by
a Spaniard. Then you'd turn to the Hearst paper and read how the Spanish authorities had decided to starve out the rebels holed up in this or that
part of the island. It was exciting. Fun, almost. Bad news that curiously
made you feel—not good, but alive! Most importantly, it was bad news
that prompted an emotional response. While the reader picked up some
facts from the papers the facts were incidental; what was key was the
emotionalism, the outrage, the welling feeling that *something must be
done.*

Of course the situation in Cuba was actually horrific. The longer the
war went on, the harsher the Spaniards became. Finally they adopted an
atrocious policy, the creation of unsanitary concentration camps to hold
tens of thousands of relocated rural peasants. This was supposedly done
for the peasants' benefit so that they wouldn't become infected with the
rebels' propaganda. But of course it was a terrible infringement of their
liberties, and completely demoralizing, which only turned the Cuban
people even further against the Spaniards. Which, naturally, became yet
another day's story in the papers, which were becoming so sensationalistic that their coverage got a special name: yellow journalism, which
became a byword for distorted, exploitive, distorted reporting. (The
name was an allusion to "the Yellow Kid," a character in a cartoon that
ran in Pulitzer's *New York World*.)

Both papers became so caught up in the story that they lost all pre-
tense of objectivity. Hearst at one point even staged the daring rescue
of an imprisoned young Cuban woman, Evangelina Cisneros, whom
the paper had turned into a cause célèbre. Cisneros, the paper had
repeatedly blared in front-page stories, had been jailed because she

had declined the lustful advances of a depraved Spanish officer. When the authorities refused to free her, and American officials proved incapable of arranging her release, Hearst decided simply to do it himself. "It is," the paper bragged, "an illustration of the methods of new journalism."[4]

Successful as Hearst and Pulitzer were in inflaming public opinion, they were not singularly responsible for bringing on the war with Spain. Americans by the millions wanted a war, wanted to project American power, wanted to flex their national muscles. So it was really the two forces that were drawing America into the war, the media *and* imperialism.

Given the forces driving the country toward war not much was needed to trigger American participation, but then suddenly, in bewilderingly quick succession, there were not one but two triggering events that made war all but inevitable. First, in early February 1898 the papers got hold of a letter written by the Spanish minister to the United States, Dupuy De Lôme, in which the minister referred to McKinley as a weak vacillating politician; the Hearst paper headlined the story: WORST INSULT TO THE UNITED STATES IN ITS HISTORY. Then, on February 15, the *Maine* blew up, killing 264 people. A month later an American board of inquiry ruled that the explosion had been triggered by an exterior cause; Americans naturally leaped to the conclusion that the Spaniards were to blame.* McKinley asked Congress for an emergency defense appropriation of fifty million dollars and got it in two days. In April the United States and Spain went to war.

McKinley had not wanted war, had not expected war, and he had done everything he could to avoid it. For a full year he had worked secretly to put pressure on Spain to settle the issues giving rise to the war: to close the concentration camps, to end the policy of starving out the rebels, to give the Cubans autonomy, if not exactly total independence. Spain, for its part, had refused to make reforms. Pushed to give the Cubans autonomy, Spain had come up with a plan that left the real control of the island in the hands of a governor appointed by Madrid;

---

*What caused the *Maine* to blow up has never been answered definitively. Adm. Hyman Rickover, after a thorough investigation, reported in 1976 that the explosion was caused by a malfunction in the ship's own engines. Most experts agree.

both the Americans and the rebels had immediately grasped the emptiness of the gesture and rejected it. Spain had never even closed down the camps, though it promised to relax its control. It was apparent the Spanish didn't intended to reform; De Lôme, in his infamous letter, had said as much. What Spain needed to do, he had counseled his superiors back in Madrid, was "to carry on a propaganda [campaign] among the Senators and others in opposition to the [Cuban rebels'] junta, and to try to win over the [Cuban] refugees."[5]

It may not have mattered what the Spanish did. For the forces moving in the direction of war were simply greater than the forces summoned against them. Trying to stop the prowar forces was like trying to turn back nature. It simply couldn't be done. For there simply was no force in the society strong enough to resist the twin forces of media and imperialism—not even the presidency of the United States. McKinley repeatedly tried. Repeatedly he failed.

Not everybody in the country was prowar, of course. There were pockets of opposition to imperialism in every community. On Wall Street there was substantial opposition, many businessmen fearing that the war would possibly destabilize the economy, derailing the recovery and sabotaging profits. But more people were for war than against it, many millions more. Even Christian leaders. "You have no idea of the pressure on William from religious peoples," McKinley's brother complained.

So war came.

It was, of course, a turning point. Never before had a president seemed to be pushed into war the way McKinley had been. Before, presidents had taken the country into war because they believed it was the right thing to do, the necessary thing to do. And they had resisted calls for war when they felt war was not right or necessary, as Jefferson had resisted calls to answer British attacks on American shipping with war, as Lincoln had resisted calls to start an unprovoked war with Britain in order to divert attention from the slavery crisis. But McKinley had no real choice. As one irate senator explained in March, when McKinley seemed to be stalling for time, "[I]f he doesn't do something, Congress will exercise the power and declare war in spite of him. He'll get run over and the party with him!"

Unprepared for war, the country largely botched the effort at the outset. The army was put in the incapable hands of Gen. William Shafter, who was so fat and so sick he couldn't ride his own horse; to get around in Cuba he had to be carried on a wooden door turned sideways. The soldiers were put in the wrong uniforms, thick woolen uniforms instead of the light airy ones needed for fighting in tropical Cuba at the height of summer. Training camps were set up in parts of the South where there had recently been an outbreak of malaria. The canned beef used to feed the soldiers was often rancid. Because there weren't enough ships available to transport the soldiers to Cuba, several regiments, including Teddy Roosevelt's Rough Riders, fought with others to get passage. In the ensuing melee the Rough Riders had to leave their horses behind in Florida.

It was then that McKinley took control of the war effort himself. He saw that Secretary of War Alger was incompetent, so he bypassed him, establishing the White House, for the first time in history, as the operational brain of military administration. He enlarged the White House staff, had a half dozen telegraph lines routed directly into the building, and then made all the important military decisions himself. Not even Lincoln had ever taken such detailed control of the military.

McKinley even began taking control of the media. Discovering the great secret about the press—that it was far far easier to control reporters by keeping them close than by keeping them at a distance—he held them close so they could be watched over as they did their watching. He even gave reporters space in the White House to write their stories. It was just an old wooden table and some chairs located in a hallway, but it was reserved for them, a place where they could gather at any time of the day or night. He also began the practice of holding official White House briefings for the press. At ten o'clock every night, in time to meet the morning papers' early deadline, his private secretary, John Porter, would tell reporters what had happened during the day and then field their questions. Porter was, in effect, the first White House press secretary. Not surprisingly, McKinley got very very good press coverage.

Alger still caused trouble. At war's end most of the men in the army fell sick and nearly died from rampant malaria and Yellow Fever because Alger refused to bring them back quickly enough; it took a public circular from Roosevelt to draw attention to the problem, saving the army from near disaster. As it was, 5,000 died of disease; only 298 died in action.

The soldiers generally performed well despite adverse conditions, the Rough Riders especially (even without their horses). But only the navy could be said to have showed itself truly capable. Fortunately for both the army and the navy Spain was even less prepared for war than the United States. In three months it was over.

William McKinley, the gentle president who hadn't wanted war, who had in fact done all he could to avoid it, and who had looked weak in putting off war and then had looked weak in caving in to the demand for it, emerged from the conflict in a strangely powerful position. It wasn't just that he had led the country to victory, though that of course was part of it. The war had made the presidency count again in a way it hadn't since Lincoln. In the crisis Americans naturally had turned once again to their president for leadership. And McKinley had provided the leadership people craved. Once he had taken control of the levers of war, the military effort had gone more smoothly. Almost all the mistakes were due to Alger.

Imperialism had cost McKinley control of events, but the war had given him the opportunity to gain back control. By war's end he had become almost imperial in a shocking way. Although the war lasted just a brief time, it was long enough to give him a chance to exercise real power. Once he had it he refused to give it up. There simply was no turning back. McKinley would remain in charge, would keep power concentrated as much as possible in his own hands. Almost symbolically he refused to trim the White House staff back to prewar levels. Now that it had been expanded, it would stay expanded. The more people he had on the staff, the more power he had: He was the first president to figure that out.

It wasn't immediately apparent at the end of the war how much more powerful he had become. For he seemed the same, looked the same, talked the same way he always had. He was still silky smooth, so smooth he was said to have had the best relationship with Congress of any president ever. To help cement his relationship with Congress, he even included a few senators in the peace delegation sent to Paris to negotiate a formal end to the war—a shrewd and unprecedented gesture. (Significantly, he also included a journalist, Whitelaw Reid, a prominent Republican who had run for vice president with Harrison.)

But there really was a newly powerful McKinley, which became obvious when he submitted the peace treaty to the Senate for approval. Instead of waiting for the Senate to act, which could be dangerous given the opposition of a small but vocal crowd of anti-imperialists, he used all the powers at his command to get the treaty ratified quickly, bargaining jobs for votes, promising the fence sitters control of their state's patronage if they gave him their support, threatening others that they would lose patronage if they blocked him. And he kept up the pressure right up until the day of the vote, in February 1899. The final tally showed just how important it was that he had used his patronage power liberally: The treaty passed by one vote.

Even more telling than McKinley's decision to trade jobs for votes was his decision to have the vote taken in February at all. For in February the odds were less in favor of the treaty than they would be in March, when the term of a newly elected Senate was due to begin. In the lame-duck Senate in February the Republicans had a thirteen-seat majority; in the new Senate in March they would have a twenty-seven-seat majority. Getting the treaty past the Senate in March would be far far easier, but waiting until March entailed a substantial risk to McKinley's power.

To get a treaty in March out of the Senate he would have to call the new Senate into special session. Once convened, the Senate would probably remain in session for months and months. And that would mean that the senators could oversee the president's administration of the country's new colonial empire in Cuba, Puerto Rico, and the Philippines. By getting the lame-duck Senate to approve the treaty in February, McKinley ensured himself a free hand in the colonies. For, when the session ended in early March, the new Senate would not convene until December. Between March and December, McKinley would have supreme authority in the colonies under his power as commander in chief. What he wanted to get done would get done. Nobody would be in a position to challenge him. It was therefore very important to McKinley that he get the treaty through the old Senate. Get it through the Senate then and he could have what he always craved: real power.

In domestic affairs McKinley continued to behave very much as presidents had for decades. He was slow, methodical, and restrained. He had to be: He simply lacked the kind of power in domestic affairs that he had in foreign affairs. So it was almost as if there were two McKinleys. The weak and accommodating McKinley, and the strong and dominating one.

This was odd but not unexpected. For it was in foreign affairs, matters involving war and peace, that presidents had always taken the most liberties. It was in those matters that they *had* to. In foreign affairs the government often had to act quickly, decisively, and with energy, and that gave the president an extraordinary incentive to assume vast powers. For only the president could act with dispatch; Congress by its very nature couldn't. Often Congress wasn't even in session. In even years it met for just four months, in odd years for six, seven, maybe eight months. Only the president, in effect, was in session year-round. Only he could respond to events as needed.

So it wasn't just McKinley who had a split personality—most presidents did. Even the strictest of the strict constructionists had felt compelled in foreign affairs to act boldly, sometimes almost recklessly, in the exercise of their powers. The most famous case, of course, was Jefferson's decision to strike a deal with Napoleon for the Louisiana Purchase. But there were many many others: John Tyler had secretly promised Texans that as soon as they signed a treaty of annexation with the United States he would, as commander in chief, personally guarantee the protection of their republic, even if that meant going to war with Mexico. He had, of course, no right to make any such commitment. (Congress alone under the Constitution has the right to make war, as Tyler himself had always admitted.) He especially had no right to guarantee the safety of a place that had as yet not been legally absorbed into the United States. Only Congress could annex Texas, and Congress had not yet done so for the very reason that annexation carried with it the clear risk of war. But Tyler felt he *had* to act. Act, and the country would obtain a huge tract of land (and Tyler would get the credit). Do nothing, and Mexico might yet awake from its torpor and try to take Texas back. The danger of war was obviously very great. A year later, in 1846, the United States and Mexico actually went to war, when James Polk, acting on his own authority, infamously sent the army into an area long claimed by Mexico, in effect launching an invasion.

Even the weakest of presidents, Millard Fillmore and Franklin Pierce, had acted boldly in foreign affairs. Fillmore on his own authority had sent Commodore Matthew Perry to open up Japan and issued orders to use force if the Japanese showed disrespect. The mellow Pierce had sent a warship to Nicaragua to go after the residents of Greytown, whom he held responsible for an attack on the American minister to Central America. The president subsequently vigorously defended the commander's decision to level Greytown when the residents refused to apologize. And virtually all the presidents had used the military in police-style actions to go after pirates, smugglers, slave traders, and the like.[6]

So there was plenty of precedent for McKinley's imperial pretensions in foreign affairs. But now he was to do what no president before him ever had, and that was to go to war abroad without congressional consent, without even consulting the leaders of Congress. McKinley hadn't gone power crazy, though he loved power. The fact was that imperialism had changed the world in which presidents operated. To succeed at imperialism a country had to react quickly to swiftly moving events. Delay, and other countries would take advantage of the opportunities created by events. So a president felt great pressure to go it alone, to act without first getting Congress's say-so. To take control of events, they had to.

When the United States conquered the Philippine Islands, it was unclear exactly what they were good for because nobody really had ever given them much thought. Thanks to Commodore George Dewey, they had just kind of dropped into our lap. About the only thing we were sure of was that we didn't want to give them up.

But by 1900 Americans had figured out what the Philippines were good for—business. Not because people thought they could make much money in the islands themselves, but because the islands could be used as the gateway to the place where one *could* make a lot of money: China.

The big news coming out of China in 1900 was the Boxer Rebellion, and it was very frightening news. First came the report that the Boxers had massacred Christian missionaries, then that they had gone on a rampage killing any Westerners they could find. Then came news that the Boxers had taken control of Peking. And *then* came reports that they had murdered all the foreigners in Peking. The reports did not happen to be true; while the Boxers had indeed taken over Peking, just a small number of foreigners (including two diplomats) had been killed. But what Americans read in their newspapers made the rebellion sound just horrible, the triumph of the barbarians. When the Western European powers announced that they were going to send in an international force to retake the city, Americans cheered. The barbarians would be taught a lesson.

The Europeans' interest in Peking was obvious. It was mostly their people, after all, who had supposedly been killed. And it was they who had the most to lose. They had all sorts of profitable concessions in China: railroads, ports, custom houses. So they would go in, of course.

What Americans should do, however, was not very clear. There weren't many Americans in Peking, and in all of China there weren't any American concessions. None. In the scramble to divide up China, the United States had been left behind. But there was a danger in standing on the sidelines. It would almost permanently guarantee that Americans would never obtain a foothold in China, almost guarantee that American business would be excluded from one of the world's burgeoning new marketplaces. The McKinley administration had recently taken the position that nobody should have concessions in China, that all foreign powers should be treated equally there, and that they should respect Chinese sovereignty. In the words of Secretary of State John Hay, China should be an Open Door. Which was a very idealistic policy at the same time that it was entirely self-serving, since it was in our interest to have a door opened that formerly had been closed to us. But the European powers hadn't been much impressed with the Open Door policy, and it was obvious that if they sent in an army to put down the Boxer Rebellion that they would demand even more concessions.

So as the news continued to come in from China—all of it very bad news—McKinley was under intense pressure to join the European force. And to do so quickly. It would do the United States no good to join the force after Peking had been retaken. To have a say in Chinese affairs the United States had to go in *with* the force. McKinley was thus faced with a very modern problem, one that nearly every president since has faced. On the basis of obviously incomplete reports he had to decide whether to commit the country to international action that could very well end in war, and he had to decide so quickly—in a matter of days—that he could not even consult Congress (which was not in session). Complicating matters, as if they weren't already complicated enough, was the fact that this was an election year, a presidential election year. Mess this up, make the wrong decision, and perhaps the administration would be voted out of power. Make the right decision, and he had the chance to go to the voters in November as a real leader again, a man who'd made history a second time.

It was thus a very typically modern presidential dilemma, and McKinley went about deciding what to do in a very typically modern way. He assembled a small group of advisers in the White House and hashed out the problem, looking at it from every angle, without including the public in the discussion. An elite group would decide a matter of international war and peace.

On June 18, just five days after the first news story had come in that Peking had been taken over by the Boxers, he made his decision to intervene militarily. He issued an order to the navy commander in the Philippines to assemble a force "to act in concurrence with other powers so as to protect all American interests." There would later be arguments about the size of the detachment that was sent. Some accounts said 5,000 went, some 2,500. Either way, it was a substantial, impressive force.

The whole thing could have gone terribly badly. Trying to control events at home was often difficult enough, but trying to do so thousands and thousands of miles away—well, at first glance trying to do that almost seemed foolish. But the fact was a president could almost control events better thousands of miles away than right on his own doorstep. Because in foreign affairs he could act unilaterally. He could, in short, do what needed to be done without worrying overly much about the democratic process, which was always messy and unpredictable.

As it happened things did go just as McKinley had hoped they would. Although the Chinese government took the side of the rebels and declared war on the European powers and the United States, by the end of the summer the international force, numbering some twenty thousand, put down the rebellion and retook Peking. The United States, in consequence, was in a position to influence the way business was to be conducted in China. As McKinley boasted in a phone call to his secretary of war (probably one of the first times a president used a phone to communicate with a key adviser during an international crisis), "It seems to me we have been most fortunate in all our dealings with this most delicate question."

The hard part, it turned out, was not getting the U.S. military into China but getting it out. While McKinley had hoped to bring the men out fast, in time for the election, there were enormous pressures on him to keep them in. As Secretary of State Hay noted, if the troops were pulled out before a new order could be arranged "we shall be left out in the cold." Hay's hope was that the United States could use the troops as leverage to implement his Open Door policy. But now that the troops were in China, McKinley suddenly seemed less committed to the Open Door than he had been before. As Hay confessed to a friend, McKinley "seems to take the other view and to want a slice." In the end the president took a decidedly muddled course. He let Hay issue a second Open Door note, promising to respect the integrity of China, while also authorizing the navy to negotiate the acquisition of a military base. In September, just in time for the election, he pulled out most of the American troops, leaving only enough to keep up a presence. Two months later he was reelected in a landslide, the first Republican president to win reelection since Grant, a generation earlier.

It was McKinley who, under the pressure of imperialism, broke with past practice and sent American troops into a war zone without the authorization of Congress, but it was Teddy Roosevelt who would be remembered as the first imperial president. William McKinley was simply thought to be—dull. But it was McKinley's very dullness that made him interesting in a way. That such a dull, plodding politician acted so boldly was fascinating. Teddy was expected to be bold. But McKinley?

It was almost a paradox—almost, but not really. For McKinley, while dull, was always ambitious. And like all ambitious presidents he wanted to control his destiny. Faced with the force of imperialism, the only way he could do that was by aggressively monopolizing the control of foreign policy.

The very week he was shot, the first week of September 1901, just six months after his second term began, McKinley took a walk along the bridge connecting the United States and Canada at Niagara Falls. He was very careful to make sure he did not go too far, however. In keeping with tradition, he did not want to become the first president of the United States to cross into a foreign country. So, well before he reached the Canadian side, he turned around and went back. He was, however, planning on taking several international trips. To Hawaii, Puerto Rico, and Cuba. He felt that visits to these places couldn't really be construed as foreign travel. And of course he was right. By the time he died, as a direct result of his presidency, these no longer really were foreign places. The American flag flew over all three.

And then in a blink there was Teddy.

14. World Power: II

How Teddy Roosevelt secretly became involved in a revolution in Panama to enhance his chances of winning election as president in his own right

Theodore Roosevelt had been a lucky child. Not because everything had gone right in his youth, though an awful lot of things *had* gone right—son of a wealthy philanthropist, he had grown up in one of the toniest neighborhoods of New York and lived the life of a pampered, privately tutored, rich boy—but because something apparently had gone wrong. And ever after he had always felt he had to prove something. He had to be smarter than others, bolder than others, better educated than others—and, of course, physically braver than others—as if at the core he wasn't sure of himself, even though he gave the appearance of being very sure, cocky almost.

He had, of course, been very sick as a child. An asthmatic, he had had trouble simply breathing. As his father told him, in a famously Victorian way, "You have the mind, Theodore, but you have not the body." The realization that he lacked ordinary physical abilities profoundly affected his psychology. Like many sickly children he felt inferior, felt that there was something deeply wrong with him, something he had to fix, which would seem to explain why he was always busy fixing himself, improving himself, and telling others that *they* should work on improving *themselves*.

It was in college—Harvard, of course—that he came into his own. He gained weight, his asthmatic attacks eased and then virtually ceased altogether, and he became a pugilist. He positively loved fighting, each blow seemingly signifying his liberation from illness. But he almost loved it too much. It appeared as if he needed to prove to himself that he *really* wasn't a sickly child anymore.[1]

Some thought that he was trying to prove even more—that he was fearless, perhaps because he worried that his father had not been; during the Civil War his father had opted to hire a substitute.* And there may have been something to this. His whole life he was to put himself repeatedly in harm's way, as if to say to the world—to himself—*See, see, I'm tough, I can take it, Theodore Roosevelt can take it, Theodore Roosevelt is brave.* Soon after college he went out West to take on the cowboys in the Black Hills.[2]

Whatever had gone wrong in his childhood, whatever it was that prompted him to want to prove himself over and over again, he became an astonishingly driven, astonishingly ambitious man. At twenty-four he ran for a seat in the New York state legislature. At twenty-eight he ran for mayor of New York City. At thirty-one he became a member of the United States Civil Service Commission. At thirty-seven he became president of the Police Commission of New York City. At thirty-nine he became assistant secretary of the navy. At forty he was elected governor of New York. At forty-two he was nominated to be vice president of the United States.[3]

While he was nurturing his career as a politician he was pursuing numerous others. Unhappy with who he feared he really was—a skinny asthmatic rich kid—he had to become somebody else. But because he was never sure exactly who he should become he kept trying to be different somebodies, one day a boxer, the next a cowboy, then a historian, then an entomologist, then a naturalist, then an orator, then a writer, then a philanthropist, then a soldier—he loved soldiering. He said afterwards that his ride up San Juan Hill† through a hail of gunfire was the greatest day of his life. Because he was smart, very, very smart, and very, very talented, he was able—as no other man of his generation was—to excel in all of these roles. His history of the War of 1812, for instance, remains to this day a credible account. But he always gave the impression of being an overachiever.

Given the circumstances under which he became president—that he hadn't been elected, that he was just forty-three years old, the youngest president ever, that he lacked a mandate—he might have

---

*His father always claimed that he chose not to fight because he had family on both sides of the conflict; his wife was raised in the South.

†Actually, Kettle Hill.

been expected to go slow and to stick with the McKinley program, at least at first. But that wasn't Teddy. Teddy wasn't a go-slow person. He was fast, full of get-up-and-go, the American boy wonder. So, when to his surprise he found himself president in the fall of 1901, he grasped the reins of power and charged ahead, just as he had in Cuba on San Juan Hill. "I want you to understand at the start," he told the press within days of McKinley's death, "I feel myself as much a constitutionally elected president . . . as McKinley was. . . . Due to the act of a madman, I am President and shall act in every word and deed precisely as if I and not McKinley had been the candidate for whom the electors cast the vote for President. I have no superstitions and no misgivings on that score. That should be understood."[4]

And act he did.

In February 1902, just five months after becoming president, he filed an antitrust suit against the Northern Securities Company, the big railroad holding corporation formed by three of the titans of American business, James Hill, E. H. Harriman, and J. P. Morgan. McKinley hadn't busted any trusts. TR would bust lots. Best of all he didn't have to ask anybody's permission. He could just do it. So he did it without consulting anybody except the attorney general, and he probably only told him because *he* had to file the cases.

In May the coal miners went out on strike—and stayed out on strike. For months and months and months. If they stayed out on strike much longer the country would freeze during the coming winter. Already in many places coal was in short supply, and the price of course had skyrocketed. Teddy naturally got involved. He called labor and capital to a conference at the White House and told them they had to agree to a settlement. Had to. When the mine owners balked, Teddy told them if they didn't agree to arbitration he would simply send in the military and take over the mines. Legally he probably couldn't, but the threat succeeded in bringing the mine owners around.*

Again Teddy had acted boldly and again he had acted alone. It was beginning to seem as if he could do anything. And then a crisis arose in foreign policy. In Venezuela a tinhorn dictator had run up millions of dollars in debts. Great Britain and Germany, which were owed the

---

*When Harry Truman, in 1952, during the Korean War, seized the steel industry after the leading producers increased prices, the courts ruled that he had exceeded his power.

money, decided to intervene. The British blockaded the coast and the Germans sent in troops. Roosevelt earlier in his administration had told the Europeans that they could "spank" the nations of Latin America, but he now changed his mind. He told them in no uncertain terms that they had to agree to arbitration. (It was beginning to seem that this was TR's answer to every problem.) Fortunately by then they had reached the same conclusion, but TR claimed privately that he had forced a settlement by threatening the kaiser. True or not, Teddy thought it was true. Again he had acted boldly, again he had acted alone.

Later in the year there was another international dispute, this time between the United States and Great Britain over the boundary of Alaska. Gold had been discovered in the Klondike region, and both countries were intent on staking a claim to the disputed land nearby. Roosevelt pressed Great Britain to turn the matter over to an arbitration commission stacked in favor of the United States, and the British agreed. The pattern had held. TR acting boldly and acting alone had succeeded.

For once a president actually seemed to *be* in control of events. He even seemed to be able to control the media.

By then he was an old hand at media manipulation. As police commissioner he'd succeeded in getting Lincoln Steffens to stop his crime wave. Then, during the Spanish-American War he'd made arrangements to take film photographers with him to Cuba when he charged up San Juan Hill, garnering him brilliant coverage. But as president he excelled at media manipulation.

Reporters loved Teddy. Several went on to write adoring biographies of him. But they always had to keep in mind that *he* loved *them* only so long as they did what he wanted. There was an incident, right at the outset of his administration, in which this was made abundantly clear. It was at the White House, just after McKinley's funeral, when TR called the reporters together to have a heart-to-heart talk about media relations.

It was a strange time to be having such a conversation—an inappropriate time, really. But TR had something he wanted to say that he felt needed saying right away, that couldn't even wait a day. It was a warning, a threat: *If any of the reporters should at any time violate a confidence or publish news that the president thinks ought not to be published, he would be punished by having legitimate news withheld from him*. The reporters objected. TR ignored them and ended the meeting.

It was not an idle threat. When reporters began printing unfavorable news about the administration, Teddy punished them just as he had promised. Then he went further. He singled out the worst offenders and barred them from the White House.

Teddy had a name for the reporters he barred. He called them members of the Ananias Club. Ananias was a liar who dropped dead when Peter rebuked him. The reporters TR rebuked did not die. But they were dead just the same—in his eyes anyway.[5]

Teddy defined his presidency by his media personality, deducing that in the media age, personality counted even more than party. This last insight was brilliant, and it came at just the right moment. For parties just then were beginning their gradual decline in American history, weakened by corruption, civil service reform, and (soon) the establishment of primaries in a dozen states. A president could no longer rely solely on his party to control public opinion, to muster support for his programs, as presidents in the past had. Now a president had to rely on himself to a great degree, and the only way he could do that was by exploiting himself, exploiting his media personality.

Teddy himself was a natural. Always on. Always ready to do anything to give the boys in the press room (for whom *he* built a special office) good copy. Nothing was off limits, it seemed, not even his family. The exploitation of his family was a great innovation. Nobody before him had even thought to exploit their family. But Teddy excelled at it, letting photographers swarm all over the mansion and the grounds to get good pictures of his family at play and work. His daughter, Alice, actually became something of a star. Like future presidents who were to use the media he often chafed at the reporters' intrusiveness at the same time that he was encouraging it. But he never cut off their access. They were simply too important. By giving them access, *he* had access.

"Let me have free access to the channels of publicity," Roosevelt said at one point, "and I care not who makes my country's laws—or what the other fellow does." It was a startling, shocking boast, but it also happened to be accurate.

Teddy made such a colorful, inspiring president that almost everybody was sure he'd win the Republican nomination in 1904 and then, of course, the general election—almost everybody except Roosevelt himself, that is. No matter how hard he worked, no matter how successful he seemed to be, no matter how positive his press notices,

Roosevelt seemed convinced that he was in terrible political trouble. Mark Hanna, still chairman of the party, refused to endorse him, and business seemed hostile.* So Teddy went to work on them both. He tried to coddle Hanna and he made peaceful overtures to business, easing his trust-busting campaign. He even invited J. P. Morgan to the White House.

He didn't exactly become conservative, but he became very prudent, almost too prudent. When people out West began screaming about the great threat Chinese immigration posed to their well-being, he agreed to make it even harder than it already was—and it was already very, very hard—for the Chinese to get through the door. There would be no Open Door policy for Chinese immigrants; not now, anyway, not with an election approaching.

His fears weren't exactly groundless. History seemed against him, after all. No vice president who had succeeded a fallen president had ever been elected on his own. And no vice president had been elected president since Martin Van Buren. While the vice presidency had once been a stepping stone for presidents it no longer was.

So 1904 was not a sure thing. And Roosevelt wanted it to be, *needed* it to be. It wasn't any good being president only by virtue of an assassination. He had to be elected on his own. Had to prove that the years he had spent in the White House since the assassination had been deserved.

And, of course, he wanted not only to win. He wanted to win big.

There was, Roosevelt felt strongly, just one way to have a chance at winning big and that was to get started on the building of the Panama Canal. A president who could get the dirt flying on a canal that linked the Atlantic and Pacific would be unbeatable. Americans were clamoring for a canal.

Getting the right to build the canal across Panama was proving extremely difficult, however. Panama was then part of Colombia, and in August the Colombian Congress, dissatisfied with the ten million dollar offer he had made for the rights to the canal zone, turned him down.

---

*Hanna had always opposed Roosevelt, hadn't even wanted him for vice president. When the convention in 1900 rallied round the Rough Rider, Hanna famously cried out, "Don't any of you realize that there's only one life between this madman and the White House?"

Roosevelt was furious. He wasn't used to losing. And he had had such great luck thus far, getting everything he wanted, and getting it really without even having had to take any extraordinary measures. The most extreme thing he had felt compelled to do was to threaten to militarize the coalfields, and that had been just a threat.

But now apparently he *would* have to take extraordinary measures. Would have to act really boldly—if he was to get the canal quickly. And that was the point. He had to get it done *now*. It wouldn't do to go back to the negotiating table with Colombia, for that would take time, probably a year or more, and even then there'd be no guarantee that Colombia and the United States could reach a settlement. Probably they could, but it wasn't guaranteed. There were no guarantees in international affairs. Things were often out of reach. And TR needed a guarantee. In 1904 he'd be up for election.

The problem therefore was time. There might not be enough time. By rejecting the treaty Roosevelt had negotiated, the Colombian Congress had left the United States back where it was in June 1902, more than a full year before, when talks with Colombia had first begun—which was nowhere. Only then the election of 1904 had seemed far in the future. Now the election was just fifteen months in the future. Fifteen short months. Roosevelt would have to do in the next fifteen months what he had been unable to do in the last seventeen.

A tall order. Even for TR.

All of August he sat and fumed. In September he continued fuming. *Those Colombians, they're dastardly, untrustworthy, undependable, criminal.* But then in October, finally, he got a break. It was the news that Panamanians were planning a revolution to divorce themselves from Colombia. Better yet, they wanted to strike a deal with the United States for a canal. Roosevelt might get his canal after all, and get it in time. There was, of course, no time to lose. Now the election was just thirteen months away.

But to get his canal Roosevelt would have to shade the truth a little. For the truth was, it really wasn't Panamanians at all who were behind the revolution. Yes, yes, they wanted to be free of Colombia. Over the years they had repeatedly tried to obtain their freedom (and over the years the United States had repeatedly helped Colombia stop them). But it was a Frenchman, Philippe Jean Bunau-Varilla, who was arranging things. From an apartment on the eleventh floor of the old Waldorf-Astoria in New York City he drafted a constitution, designed a flag, and generally took care of just about everything. He even personally selected the Panamanians he wanted to lead the revolution. It was all very bizarre, but it made sense if you understood the history.

In the late nineteenth century a French company had poured hundreds of millions of dollars into Panama in hopes of building an Isthmian canal, only the effort had ended in failure. A new company had been organized to take over the project but it, too, failed. Ever since, company officials, among them Bunau-Varilla, had been trying to get somebody, anybody, to take the equipment they had purchased off their hands so they could get something out of their costly misadventure in Central America. Initially they had wanted the full amount they had lost, but after the U.S. Congress declared that the United States would pay no more than forty million, they relented and agreed to the American terms.

Everything then had seemed settled. The Colombians would get ten million for giving the United States the right to build a canal. The French would get forty million for their equipment. But then in August, Colombia suddenly had turned Roosevelt down, meaning we wouldn't get our canal and the French wouldn't get their forty million. (Congress obviously would pay them the money only if the canal was to be built.)

Like Roosevelt, Bunau-Varilla had been under tremendous pressure to find a way to overcome the Colombian logjam. And then he found a way. Start a revolution! That was how he, a Frenchman, had come to be the mastermind behind a revolt in Panama.

It was an unlikely scenario, but by the fall of 1903 Bunau-Varilla had managed to pull almost all the pieces together to get a revolution going by about November. He had just one little problem. After independence was declared, he somehow had to be able to stop the Colombian military from crossing into Panama and crushing the revolt. On October 9 he went to the White House to see Roosevelt.

Roosevelt very much wanted to help, of course, but because of his position as president he couldn't come right out and say so. The White House couldn't be involved in planning a revolution in Panama, since Panama was then still a part of Colombia and we were at peace with Colombia and pledged by treaty to preserve the peace. Intervening would make us look as if we were willing to do anything to obtain our canal, just as the European powers in China had been willing to do just about anything to obtain their concessions, for which we Americans were always chastising them.

So TR couldn't come right out and tell Bunau-Varilla that the United States would help prevent the Colombians from interfering with the revolution. And he didn't. But then he didn't really need to. It was obvious that it was in the interest of the United States for the revolution to succeed, obvious that the U.S. government would use the navy to see to it that it did, obvious that the United States would agree to give the Panamanians the same financial deal previously offered to Colombia (namely, ten million dollars). So of course the United States would help. And as TR later said, Bunau-Varilla would have to have been "a very dull man had he been unable to make such a guess." Dull, indeed, considering that Roosevelt had his secretary of state tell Bunau-Varilla straight out that "orders have been given to naval forces on the Pacific to sail towards the Isthmus"— and that the ships would arrive by November 2.

Not being a dull man, in fact being a very smart, very alert, very energetic man, Bunau-Varilla then immediately relayed the information to his foot soldiers in Panama. On November 3, one day after the arrival of the U.S. warships, the Panamanians revolted. The Colombians attempted to put a stop to the revolt by shelling the city of Colón. The U.S. Navy prevented them.

On November 6, within one hour of receiving notice that a new regime had taken over in Panama, Roosevelt recognized it as the de facto government. A week later he gave it official recognition. Five days after that the United States signed a treaty with Bunau-Varilla, who had made himself Panama's first minister to the United States. Under the treaty we got the canal and the Panamanians got the ten million dollars that the Colombians had turned down as inadequate. (Meanwhile, the French got their forty million.)

What had seemed impossible just a few short months ago had been done. TR had gotten his canal. And he had gotten it with time to spare. In three months he had been able to do what he hadn't been able to in the previous seventeen.

Of course, it wasn't exactly ethical. The United States in effect had started a revolution in Panama. Without the help of our navy, without the strongly implied promise of ten million dollars, Panama would never have revolted. Roosevelt, under self-imposed pressure to secure Panama in time for his own election, had involved the United States in a gigantic international fraud, something no other president before him had ever done.

TR insisted while he was president that he had done nothing unethical in Panama, nothing untoward, even, and he kept to that line the rest of his life. But years later, during an address at the University of California, he was to brag that he was "interested in the Panama Canal because I started it. If I had followed conventional, conservative methods, I should have submitted a dignified state paper to the Congress and the debate would have been going on yet, but I took the canal zone and let Congress debate, and while the debate goes on the canal does also."

In the spring before the election TR grew morose, fell into a deep depression, and became utterly pessimistic about his chances in November. There were no grounds for pessimism, however. After all he *had* gotten the canal going. And everybody else thought he was a shoe-in. But TR did not. And that was what mattered. *He* thought he was in real trouble.

So he did what other presidents sometimes did when they thought they were in real trouble. He became desperate, and out of desperation he hatched an uncharacteristically underhanded plan. Convinced that what he needed to ensure victory was money, lots of money, to fund his campaign, he turned for financial help to the very corporations and businessmen he had been at war with the last three years. To make sure they gave—and gave big—he put his campaign in the charge of George Cortelyou, the very man who had been regulating business. (Cortelyou had been the first secretary of commerce and labor.)

The appointment smelled bad, and TR knew it. But the fact was he needed somebody in there who knew those corporate titans and knew how to put the squeeze on them. It was, as things turned out, an inspired decision. Cortelyou was a fabulously energetic fund-raiser. By November he had collected hundreds of thousands of dollars from big business: $250,000 from E. H. Harriman, $150,000 from Standard Oil executives, $166,000 from U.S. Steel, $50,000 from Henry Clay Frick, and $150,000 from J. P. Morgan. The donations, of course, were made in secret, but word of them leaked out, causing a firestorm of criticism. TR's advice to Cortelyou: Launch "the most savage counterattack possible."

After the election Roosevelt renewed the crackdown on trusts that he had suspended as the campaign heated up. Businessmen were shocked. Felt betrayed. They'd bought the sonofabitch and then he hadn't stayed bought. As a vice president with Standard Oil subsequently testified before a congressional hearing, "Darkest Abyssinia never saw anything like the course of treatment we experienced at the hands of the administration following Mr. Roosevelt's election in 1904."

TR had had no intention of selling out to the giants of American business. He'd simply wanted their money—and he got it.

He won in a landslide. The greatest in American history up to that time. Fifty-eight percent of the popular vote. More votes than even McKinley.

There was just one thing about that landslide. It only made TR president, leaving Congress in charge of legislating, and TR found that inconvenient. As he exclaimed uncontrollably one day, "Sometimes I wish I could be president and Congress too." What upset him in the months after the election was that Congress wouldn't give him the authority to clean up a mess in Santo Domingo* that desperately needed cleaning up. Although it was a small country, an insignificant country really, the backside of Haiti, the government had run up enormous debts to Europeans. The Europeans in turn had threatened to take over the country until their debts were repaid. It seemed like Venezuela all over again. Only it wasn't. In Venezuela TR had recommended arbitration. This time he wanted the United States to intervene directly, to take over the country's custom houses. Then the United States could repay the Europeans and, not incidentally, the U.S. companies which were also owed money.

It was an unprecedented scheme, potentially very disagreeable and complex. If a revolution broke out, as was definitely possible, the United States would have to decide which side to support, involving us in yet another Latin American country's internal affairs. It would be Panama again, only this time we wouldn't be getting a canal. But TR felt that intervention was necessary. If we failed to do something, the European powers would. TR explained all this to Congress, but Congress didn't want to be involved and turned down his request for authority to intervene. Become involved in Santo Domingo? Why? Let it be, let it be. But TR could not simply let it be. He wanted action. So he took it. After obtaining Santo Domingo's approval, he ordered the navy to take over the custom houses.

TR had a theory about the presidency. He called it the Stewardship Theory. It was very simple. If a president felt he needed to take action in the public interest, he could, as the steward of the people, "unless such action was forbidden by the Constitution or by the laws." If Congress didn't like what he had done, it could always stop him by passing a law. Or impeaching him. But of course Congress was to do no

---

*It is now known as the Dominican Republic.

such thing. Not in the wake of TR's resounding election victory. So nothing was done. For two years the United States administered the custom houses of Santo Domingo by presidential fiat. Then, finally, Congress approved a convention between the United States and Santo Domingo giving the president formal authority to intervene in the island country's financial arrangements. TR felt vindicated.

It had been a strange affair. Usually a president with a great personal following didn't have to abuse his powers. He could get done what he wanted done simply by appealing to popular opinion. That, anyway, was how things had usually worked in the past. But now things were different. For now a president was playing on the international stage. And there American popular opinion had little impact. What counted was raw power, military power. For a president to be able to control events, he had to be able to send in the navy or the army or the marines and to send them in quickly. Fail to act quickly, and others would act in his stead—and in their interests, not ours. So he had to act even if he lacked the requisite legal cover.

In his autobiography TR claimed that in his presidency, "I did not usurp power, but I did greatly broaden the use of executive power." But in the Santo Domingo affair at least, he did usurp power. After the Senate had expressly turned down his request to intervene in Santo Domingo, he had gone ahead and intervened anyway. He insisted it wasn't because he was power hungry but because intervention was necessary. But what was really going on was that he felt that in foreign affairs he should control things. And to obtain that control he was willing to do almost anything, even ignore Congress.

It had been TR's hope that he would have a great crisis while he was president, something perhaps on the order of the Civil War, a crisis that could show him to advantage, in which he could flex his muscles, really prove who he was and what he could do. But when the crisis finally came it was Woodrow Wilson who was in the White House. Teddy, on the sidelines, could watch but do nothing.

15. World Power: III

How Woodrow Wilson led the country into a war he claimed to be neutral about and then used scare tactics to persuade people to support the war and the peace

---

TR is followed by his hand-picked successor, William Howard Taft. Taft remains largely faithful to Roosevelt's vision but alienates TR by declining to consult him. Annoyed and anxious to return to power, TR decides to challenge Taft for the Republican nomination. Party bosses rally behind Taft, leading Roosevelt to form a third party, splitting the Republican Party in two, guaranteeing the election of Democrat Woodrow Wilson.

Woodrow Wilson, "Tommy" in his youth, was the son of a Presbyterian minister and often reminded people of one himself. But his grand dream, from the time he was a student at Princeton, was to be a statesman. He told one friend he hoped to be a governor. He told another he wanted to be a senator. After graduating from Princeton he became a lawyer. "The profession I chose was politics," he explained. But "the profession I entered was the law. I entered the one because I thought it would lead to the other."[1]

It was an excellent plan. For somebody else. Wilson, it turned out, couldn't bear the drudgery of lawyering, the studying of precedents, the filing of documents, the filling out of forms. He passed the bar in October 1882; in May 1883 he quit and decided to become a professor; a professor at least could spend his days working on ideas, and Wilson's imagination was always fired by ideas. Ideas could change the world. The following fall he entered Johns Hopkins University, eventually taking a degree in history.

He seems to have given up any hope that he might someday be elected to office. But he remained ambitious. In some ways he became more ambitious than ever. His new plan was to change the very way Americans governed themselves. Deeply influenced by the writings of British political scientists, he became convinced that the United States should adopt a parliamentary form of government. His first book, written while he was at Hopkins, explained that the switch was necessary because the way the government was currently organized, presidents couldn't lead; too many powerful committee chairmen in Congress prevented presidents from getting anything done. A radical notion, but at the time not an unreasonable one. Presidents in the 1880s, when the book appeared, *were* prevented from leading by powerful chairmen.

The book made a big splash. Wilson went on to become one of the most successful professors of his generation. Two years at Bryn Mawr were followed by two years at Wesleyan, which in turn were followed by an appointment to Princeton. There he became the most popular teacher in the school's history, regularly drawing to his lectures crowds of three hundred students and more. When he wasn't preparing his lectures he was writing. In ten years he wrote nine books, including a five-volume history of the United States. Most of the histories weren't very good, but they got his name out to a wider public, gave him a platform for his views. He wasn't a senator or a governor, but he was becoming influential. He was even being asked to lecture before public audiences.

It was all very satisfying, but to a man with Wilson's great drive, not nearly satisfying enough. After devoting himself to developing a career as a man of ideas, he found that ideas alone didn't seem to accomplish much. A turning point came in the course of his battle with the university administration over a plan to establish courses in sociology. Wilson argued eloquently in favor of sociology, declaring that it was essential to the development of a modern society. The administration responded that sociology posed a threat to religion and turned Wilson down.

Convinced of the futility of pushing ideas when he lacked power, Wilson went after power, becoming at age forty-six the new president of Princeton, the first in the history of the school who wasn't a minister. His was the most explosive tenure of any president in the school's history. He fought to abolish the school's entrenched elite "eating clubs" on the grounds that they undermined the principle of equality.

Next he fought to modernize the curriculum, establishing his much beloved sociology courses. He also managed to expand the school's staff, adding fifty professorships at one fell swoop.

Consumed by his work as an administrator, Wilson seems not to have given much thought to a political career. He was changing society by changing Princeton. But others, watching what he was doing at Princeton, were beginning to think of him as a possible political leader. George Harvey, a power in the Democratic Party, was the first to recognize Wilson's potential. In 1906, just four years after Wilson had taken over Princeton, Harvey began touting him as a presidential candidate. Harvey even had Wilson's face put on the cover of *Harper's Weekly*, which Harvey controlled.

In light of Wilson's subsequent reputation as a blazing liberal, Harvey's attraction to the Princeton president was ironic. Harvey liked Wilson because he was under the impression that Wilson was a conservative. Wilson himself seems to have thought of himself that way, and his writings left that distinct impression. He opposed labor unions, supported industrial monopolies, scorned Populism, condemned William Jennings Bryan, publicly praised J. P. Morgan, and ridiculed Teddy Roosevelt. If anybody could be more conservative than he was it was hard to see how.

Harvey slowly built up support for Wilson in the national press and then, in 1910, succeeded in getting the Democratic boss of New Jersey, James Smith, Jr., to back Wilson for governor. Up to this point Wilson had watched Harvey's activities somewhat in amazement, without taking them very seriously. But now that Smith offered to support him for governor he began to rethink the direction of his career. He bit.

Harvey and Smith, however, knew not what they wrought. Wilson, the much-heralded conservative, turned out to be extremely liberal. Stumping for the governorship he endorsed the direct election of senators, backed government regulation of utilities, and even supported a crackdown on monopolistic trusts.

Boss Smith worried that he'd been snookered. Wilson argued that he had simply undergone a genuine transformation in his views. There was, however, an element of political calculation in Wilson's change. For even after it became clear that he now had a liberal agenda, he continued to exploit his connections with the bosses. As he made his way around the state he was accompanied by machine politicians, who helpfully informed him about the dynamics of the local races. Then, from the stage

Wilson would denounce the bosses, going so far as to tell audiences he had no connections with them whatsoever—none. This even on occasions when Boss Smith was sitting in the audience directly in front of him.

Wilson was behaving as if he not only wanted to win but had to. And by now he actually did. For his presidency at Princeton was fast coming to a terrible conclusion. After eight years the trustees were tiring of Wilson and his fights, particularly a recent one over where to locate the new graduate school building. They were about to get rid of him.

He was, anyway, now deeply committed to politics. The more he talked the more he came to believe that he was finally fighting in the realm he should. He wasn't very good at the hail-fellow-well-met duties of the politician, but he was a superb speaker, as good as any the country had seen. And he had this magical ability to inspire people. In front of large groups, the bigger the better, he excelled. Directly as a result of his stirring performances, the Democrats swept to power in the state for the first time in a generation, putting Wilson in the governor's chair and Democrats in control of the lower house of the legislature. (The Republicans retained a slim majority in the senate.)

He proved to be a sensational governor. Totally devoted to politics, so devoted he complained he had no private life anymore, he rammed one bill after another through the legislature, putting in place the liberal agenda he had put before the voters, confirming Boss Smith's suspicions that Wilson really was a liberal. Over a six-month period he established direct primaries, effectively ending the boss's control of the presidential nominating process in the state; established a revamped and much more powerful public utility commission; and provided for workmen's compensation.

The bosses were appalled and declared war on him. But before they could exact their revenge Wilson, just a year after his election, was off and running for the Democratic nomination for president. At the Democratic Convention in 1912 the race came down to him and Champ Clark, the Speaker of the House of Representatives. Wilson won by branding Clark as the candidate of the bosses. The truth was that they both had the backing of various machines, without whose support winning the nomination was impossible. But Wilson succeeded in pulling the wool over the eyes of the delegates and ran off with the nomination.

That wasn't his only subterfuge. Wilson, who was fifty-six, had a history of serious health problems. As is now known, he had already suffered two strokes—his first at age forty—one of which resulted in blindness in his left eye. But this information was kept from the delegates and the public, as was the news that he suffered serious digestive ailments. Like Arthur and Cleveland before him, Wilson wasn't about to let his health get between him and the White House.

His first year as president, like his first year as governor, was, simply, spectacular. First he tackled the tariff, in three months winning a substantial reduction in the rates. Then he persuaded Congress to reform the banking system, providing for the establishment of the Federal Reserve. Then he went after the trusts. In his first year he called Congress into special session twice and addressed the members three times, a record. (Once was to deliver the State of the Union Address; it was the first time a president had delivered the address in person since John Adams.)

By his third year it wasn't domestic issues that preoccupied Wilson, however. It was the war that had broken out in 1914 in Europe, the so-called Great War. And it would be the war that would dominate the rest of his presidency.

Unlike TR, Wilson did not want a war in which to prove himself. And he certainly did not want world war. World war would change the United States, he predicted. It wouldn't be a little puny thing like the Spanish-American conflict. It would be all-consuming, terribly destructive of the American way of life, terribly destructive of civil liberties. And for the first few years of the war he worked hard to keep us out of it.

There was, however, no way actually to stay out. Not if the war lasted. The United States was too involved in the world by now to remain aloof from international affairs. Fifty years before, yes. Maybe even twenty-five years before. But not now. Now we had a colonial empire. Now Americans traveled routinely on the high seas. Now American farmers depended on sales to Europe to survive (and Europe depended on them). So although Wilson wanted to keep us out of war, he could not. As the war dragged on the United States found itself being pulled, against its will, into the conflict—and pulled in one direction. In the Allies' direction.

Wilson insisted that the United States was neutral, neutral in thought, neutral in deed. But the country wasn't, really. What the United States was was neutral in name. In truth Wilson wanted the Allies to triumph—and the foreign powers knew it. The British knew it, of course. When they illegally imposed an overly broad blockade on Germany, stopping ships on the high seas from proceeding to Germany, the Wilson administration barely managed to squeak out an objection, though the action contravened our rights as a neutral power under international law. And when they needed money to keep the war going—the Germans were blowing up so many British ships that Britain was going broke—the United States agreed to lend Britain billions. The Germans knew it, too. It wasn't just the loans, wasn't just Wilson's winking at the British blockade. It was Wilson's vigorous condemnation of German submarine warfare.

The Germans knew that it was upsetting, of course, that Americans were being killed on ships sunk by submarines. But there was an easy way the United States could solve the problem: Simply ban Americans from traveling on belligerent ships. Of course, this would have been a serious concession; as citizens of a neutral nation Americans had the right to travel however they pleased. But it would have been no greater a concession than had already been made to the British. And in fact many prominent Americans, among them Wilson's own secretary of state, William Jennings Bryan, actually advocated the passage of a law limiting the travel rights of Americans. But Wilson flat-out refused to support a ban, even after the sinking of the *Lusitania* in 1915, which cost the lives of 128 Americans. Bryan became so furious as a result that he resigned.*

So the British knew. The Germans knew. Wilson knew. But the American people in general did not. And in 1916 they went ahead and reelected Wilson in the belief that he really was neutral. They knew the facts, naturally, knew them as well as anybody. But because Wilson kept insisting we were neutral, people overlooked the facts, let Wilson define reality in the face of the facts. Presidents still had that power. While some presidents recently had taken some liberties with the truth—everybody, for instance, knew that TR had been responsible for the revolution in Panama even if he didn't admit it—people still were inclined to believe a president.

---

*In the 1930s FDR came to the conclusion that it would have been better for the American people if Bryan had remained. Like many Americans, Roosevelt had come to believe it was foolish to go to war to protect American neutrality. All the country should demand, he said, was that belligerents stay out of American territory. U.S. citizens, he said, should be prohibited from traveling on belligerent ships. Frank Freidel, *Franklin D. Roosevelt* (1990), p. 181.

And anyway, even if people doubted our neutrality, he *had* kept us out of war. That was an incontrovertible fact. And that was the only fact that really mattered. American boys weren't being sent to fight and die in Europe's war. But things weren't exactly that simple. While the country was not then at war, Wilson had put us in a position where war was likely. War wasn't even his decision anymore: It was the Germans'. If they resumed submarine warfare—which they had stopped at American insistence—the United States was almost pledged to go to war against them. So the slogan Wilson ran on—He Kept Us Out of War—was misleading. At any moment American boys might have to fight. If the Germans desired, they could force the United States into the war right in the middle of the campaign, making a mockery of the slogan Wilson was running on. The peace president suddenly would be the war president.

Wilson himself was quite aware of the emptiness of the slogan, confessed to friends that it was empty, but he also knew it had great popular appeal, that it was what people wanted to hear, wanted to believe. And he very much wanted to be reelected. So he did what no president before him ever had. He ran on a peace platform while privately preparing for war.

Two months after the election, on January 31, 1917, Germany announced it was resuming submarine warfare. In March three United States ships were sunk. In April the United States went to war.

There were, of course, complicating factors. A telegram had been intercepted that proved that Germany was preparing to strike an alliance with Mexico in the event of war with the United States, which put the war in the American backyard, inflaming public opinion. But it was the resumption of the sub attacks that was the key.

The German sub campaign was wildly successful, sinking in February alone a quarter of all British ships at sea. And because it was successful, it was frightening to Americans, frightening because it was so deadly and frightening because there seemed no way to defend against it. If it were possible to distinguish between civilized and uncivilized weapons of war, the sub seemed uncivilized.

The British blockade in actuality was far far more costly in human life. Historians estimate that the blockade, by denying Germans the food and medicines they needed, cost tens of thousands of civilian lives—far, far more civilian lives than submarine warfare cost. Wilson, however, did not draw attention to the human cost of the blockade. To win the war he had to demonize the Germans, had to convince Americans that Germans were inhumane.

Three things were new about the war the United States was now fighting, new and shocking. The first was that it was being fought in Europe. No president had ever before been to Europe during his term in office, and now millions of Americans were being sent over there. The second was that it was being fought mainly by men who were forced to. In all other U.S. wars, save for the Civil War, the armed forces had relied on volunteers. And even in the Civil War whole regiments had been drawn from volunteers. In World War I the government relied on the draft. During the conflict twenty-four million were registered, three million drafted. The third was that it led to unprecedented violations of civil liberties (worse in some ways than had occurred even in the Civil War). And all three factors were related. Because the war was taking place in Europe, because Americans weren't fighting for turf at home but for an idea abroad (the idea, Wilson told them in his war message, was to "make the world safe for democracy"), the government wasn't sure it could count on people's support. Millions of Americans didn't want to have anything to do with Europe and certainly didn't want to send their boys over there to die. That was the reason it was necessary to institute a draft. And it was also the reason civil liberties were violated so infamously. The government simply wasn't confident that the country would rally behind the war effort. The possibility of mass dissent was enormous.*

Wilson had been afraid of what the war would do to America and as things turned out he was right to be afraid. It wasn't just that the country would have to be put on a war footing, that the country would have to be militarized. It was that the country would have to be whipped into a frenzy—and in a frenzy people's civil liberties would be trampled. The forces unleashed would be so powerful that nobody, not even the

---

*After the war, in one of the first scientific polls ever, more than 60 percent said that they thought the war had been a mistake. But probably more opposed the war after it was over than before because of the revelations that came out in the 1920s about war profiteering, the duplicity of our Allies, and the general disillusionment that followed the failure of the Senate to approve the Treaty of Versailles. See Hazel Erskine, "The Polls: Is War a Mistake?" *Public Opinion Quarterly* (1970), pp. 135–38.

president, could control them. It would be 1798 all over again, only worse.* For with the modern tools of propaganda at the government's disposal, people could be controlled far more effectively than they had been in the eighteenth century.

And yet it was all necessary. To win the war Wilson had to win the hearts and minds of the people. And to do that he had to take desperate measures. To make sure that Americans understood sufficiently the great threat they faced, he had to scare hell out of them, which he did through the Committee on Public Information, which issued millions of inflammatory pamphlets containing stories about German butchery and German militarism and the threat the Germans posed to civilization.

To make sure Americans got only one view of the war, he authorized the postmaster to censor the mails for radical literature. Only the truly subversive stuff was supposed to be removed, but the postmaster got carried away, even confiscating at one point a whole issue of the *Nation* after the magazine published an editorial critical of labor leader Samuel Gompers, who was prowar. (Wilson in that one case overruled his postmaster, ending the confiscations of the magazine.) To stop radicals from giving speeches against the war, Wilson supported a law giving the government the right to jail anybody who said anything disruptive; the socialist Eugene V. Debs was given a ten-year sentence for saying that the United States was not a democracy.

The worst was the spying. Not the spying of German agents but the spying of Americans on Americans or Americans on aliens. This was carried out ostensibly by a private organization, the reactionary American Protective League, which quickly grew into a massive force of repression a quarter million strong. But it was actually in effect a branch of the government, Attorney General Mitchell Palmer using it as if it were his own private police force. Sometimes members even resorted to wiretapping. Wilson thought the group was "very dangerous" and disapproved of Palmer's decision to give it semiofficial status. But he never used his power to act against the league. Once the war had gotten underway and the propaganda machine had been put in place, it was a force unto itself. To gain control over the war effort Wilson had had to unleash a force over which he had little control.

---

*It was in 1798 that John Adams had so inflamed public opinion against the Jeffersonians that Congress passed the Alien and Sedition Acts.

It wasn't just the war that was responsible for the repression. It was also what the war had brought about, particularly in Russia. Americans had been genuinely frightened by the triumph of the Bolsheviks, frightened that communism might be coming to the United States, too. And they became very susceptible to anti-Communist hysteria, what was then known as the Red Scare. Of course, it was hard to know who was a Communist and who wasn't. Most didn't brag. But millions were sure they knew where to start—with the immigrants. So Palmer (preparing to make a run for the presidency) went after the immigrants, and in a spectacular series of raids lasting over a three-month period he rounded up four thousand of them, dragging them from their homes if need be, to cart them off to jail—and he did this without obtaining a single warrant. Some Americans protested, and one, Roger Baldwin, grew so incensed that he founded an organization to protect Americans from arbitrary government repression, the National Civil Liberties Union (later the American Civil Liberties Union). But there was no stopping the wave of abuse once it got going. It had to exhaust itself.[2]

Once the United States became involved in the war the fortunes of the Allies shifted quickly. In little over a year it was over, giving Wilson the opportunity he had been waiting for, the opportunity to use the peace process to change the world. Wilson, the minister's son, would play minister to Europe, lifting the moral sights of the victors, so that in the new world that emerged countries would base their foreign policies on morality, not self-interest.

In the fall of 1919, after working for months and months in Europe on the Versailles Treaty, Wilson came home to try to sell it to the American people. He didn't very much like the treaty. It was full of compromises. The British, the French, the Italians, the Japanese—all had used the conference to advance their own territorial claims. So things hadn't turned out at all as he'd hoped. But he had gotten one thing: The League of Nations. And the League, he believed, could solve all the problems the treaty itself had left to fester. Over time the League could contribute to world peace. Over time it could help the oppressed obtain self-determination. So he decided to work very hard to get the American people to give their support to the treaty.

To win them over to the cause he arranged to go on a punishing tour of the country. But when he got out to speak to people, he sensed very quickly that they weren't with him: They were tired of the war, tired of Europe, just plain tired. So he did what he had done at the outset to get people's attention. He tried to scare hell out of them. Again. Only this time the enemy wasn't Germany, it was Russia, communist Russia. Playing on people's genuine fears of communist revolution, he compared the opponents of the treaty to the Bolsheviks. "Opposition," he intoned, "is the specialty of those who are Bolshevistically inclined."

It was ugly and it was dirty and, worst of all, it was for naught. Despite his exertions—exertions so great he suffered a collapse on the road and had to be rushed back to Washington, where he suffered a second collapse, this time a life-threatening stroke—the treaty went down to defeat.

History had spun out of his control.[3]

Three Republican presidents followed Wilson: Warren Harding, Calvin Coolidge, Herbert Hoover. Each in his own way was ambitious.

Harding as a nineteen-year-old founded his own newspaper. Subsequently, notwithstanding his limited intellectual abilities, he shrewdly succeeded in getting himself elected to the Senate. He never planned on being president, but when the party bosses decided that he would make a golden candidate (whom they could control), he avidly accepted despite serious qualms about his own abilities.

Coolidge was more ambitious. That he'd even made a career in politics was proof. For he was extremely shy. As a boy he would run out the kitchen door anytime a stranger arrived.

Like Harding, he, too, never planned on being president. But the media turned him into a national star overnight. The Boston police, paid a paltry annual salary, had joined a union and in 1920 gone on strike. Coolidge, then governor of Massachusetts, bluntly declared that "there is no right to strike against the public safety by anybody, anywhere, any time." He quickly found himself, much to his satisfaction, on the Republican ticket as Harding's running mate.

Unlike Harding and Coolidge, Hoover was not a career politician. An orphan who pulled himself up by his own bootstraps, he became a skilled engineer and an amateur translator; in their spare time he and his wife translated into English a medieval Latin classic on mining. Venturing into business, he became wildly successful, becoming a millionaire by thirty. He got involved in government service during World War I, when he was appointed by Wilson to help feed the starving children of Belgium, which made him a hero. Harding subsequently appointed Hoover to the cabinet as secretary of commerce. Of the three presidents of the 1920s, he was the one who went after the presidency most aggressively. Highly self-confident, he was sure he could make a great president, and certainly a better president than either Harding or Coolidge.

In his 1928 race for the Republican nomination, a tough and difficult race, he had found out that Republican senator James Reed of Pennsylvania was planning to take to the road on a national speaking tour against him. Hoover had become alarmed, incensed, really, and had set out to crush Reed. Hiring a detective with money from his own pocket, Hoover apparently found damaging information against Reed—and blackmailed him with it. Reed abandoned his planned trip.[4]

But ambitious as all three presidents were, they had very little ambition to shape history. Harding and Coolidge especially believed in letting things take their natural course. "Four-fifths of all our troubles in this life would disappear," Coolidge philosophized, "if we would only sit down and keep still."

Coolidge might have tried to do more as president than he did. As governor of Massachusetts, after all, he showed energy in ending the Boston police strike. But after succeeding Harding he faced a personal tragedy. His son was out playing on the White House tennis court one day when he stubbed his toe. He was not wearing socks, his foot became infected, and he became gravely ill with blood poisoning. It seemed almost ridiculous, but Coolidge and the doctors could do nothing to help. Just a few days later the boy died. The president went into mourning and never seemed to come out of it.[5]

It seemed through most of the 1920s that presidents didn't *have* to do much. The international scene was calm and the economy was prosperous. Both Harding and Coolidge expressly announced that there was no reason to do much of anything but get out of the people's way. Americans, tired of Wilsonianism, tired of the regimentation of war, welcomed the return to normalcy, as Harding called it.

The United States was changing, to be sure. But the changes seemed promising. Wall Street was booming. The skylines of the cities were being transformed, skyscrapers rising one after another after another. And consumers were finding that they could buy things on the installment plan, an innovation of the twenties, which rapidly improved people's standard of living.

The Harding scandals diminished people's respect for him; never before had two cabinet secretaries been indicted. But he was dead by the time the worst allegations surfaced, and Coolidge was honest. So the good times rolled on.

*   *   *

Then, of course, came the crash.

Hoover wasn't as bad as the Democrats were to make him out later. When the stock market crashed, he energetically tried to rally the country, holding numerous press conferences and meetings. It was at his urging that the Congress established the Reconstruction Finance Corporation, with a budget of two billion dollars, a phenomenal amount (the budget of the entire government then was just five billion). But he projected, despite himself, an image of do-nothingism. Cold in person, he came across cold on the airwaves and in newsreels. And after championing standpatism during the twenties he was in no position to begin arguing for change in the thirties. It simply seemed too inconsistent with who he was and what he had stood for—laissez-faire.

As long as the economy had done well, as long as the country had been content with standpatism, the Republicans had done well. Since their ambition was to do nothing, they could easily achieve their goal. But now Hoover was actually trying to do something. And not succeeding. And it made him extremely frustrated. And when Hoover wanted something and couldn't get it—or thought he might not get it for some reason—he cheated a little, as he had in 1928 in hiring a detective to investigate Senator Reed.

In 1932, in the middle of his campaign and again sensing defeat, he once again stooped—and in a way no president before him ever had: He lied about government statistics. He couldn't change history, couldn't change the fact that Wall Street had crashed and the country had slipped into a profound depression. But he could change the statistics that were being used against him with such devastating effect by the Democrats. It wouldn't change reality, but it could possibly, just possibly, fool people into thinking that things weren't quite as bad as they thought. That unemployment wasn't quite as bad as they thought. That the Hawley-Smoot Tariff wasn't quite as bad. (The tariff, passed in 1930, was being blamed for deepening the depression.) So he manipulated the numbers a little and passed them off as real.

The way he jiggled with the tariff numbers was fascinating. The criticism of Hawley-Smoot was that it set the rates too high, on average about 60 percent, the highest in American history. This number was arrived at by averaging all the tariffs imposed on all imported goods. Hoover, however, averaged in all of the so-called "free goods," on which no tariff at all was imposed. Calculated in this manner—as nobody else had ever calculated the tariff before—the Hawley-Smoot rate came out to about 16 percent. It was, of course, a sham number, totally indefensible, and wildly misleading. But Hoover stuck by it.[6]

The Democrats had no trouble beating him, however. The times were against him. It didn't matter what numbers he used. The economy was terrible and people knew it.

On March 4, 1933, he and Franklin Roosevelt rode in the same car up Pennsylvania Avenue to attend Roosevelt's inauguration. FDR smiled and laughed and waved. Hoover sat rigidly still and mostly said nothing the entire way. Hoover didn't talk because he felt that FDR had betrayed the country in the previous month by refusing to work with the administration on a plan to save the banks. But what really was there to say anyway? He had lost, and FDR had won.

16. FDR: The Great Depression

How FDR triumphed during the Great Depression through charm, talent, and deviousness, several times using the IRS to go after his political enemies

The United States was being reshaped by two great forces of history when FDR took over. The obvious one was the depression. The other was radio. Both had worked against Hoover. Both worked for FDR.

When Roosevelt first took the oath as president it wasn't clear exactly what kind of president he would make. Walter Lippman had famously charged in a devastating column during the campaign that FDR was a lightweight, "a pleasant man who, without any important qualifications for the office, would very much like to be President." But Roosevelt quickly proved Lippmann and all those of his ilk wrong, Lippmann himself becoming a huge Roosevelt fan. Indeed it almost came to seem as if FDR was made for the presidency, and made for it at this particular moment in history. And he was. He possessed the right political skills, the right mixture of ambition, self-confidence, authority, idealism, realism, imagination, and—and this cannot be overstated—the right voice.

Probably FDR would have done well without radio, but with it he was, simply, untouchable. It was his good fortune to come to national prominence just as radio was coming into its own. Until FDR radio had hardly been used by presidents. It had been regarded almost as a curiosity. Wilson had been the first president to broadcast over a radio and Harding had been the first to own one, but neither he, Coolidge, nor Hoover ever mastered the medium. Harding bloviated, which didn't work well on radio. (William Gibbs McAdoo, Wilson's treasury secretary, said Harding's "speeches left the impression of an army of pompous phrases moving over the landscape in search of an idea. Sometimes these meandering words would actually capture a struggling thought and bear it triumphantly, a prisoner in their midst, until it died of servitude and overwork.") Coolidge had a high nasal voice which was simply unsuitable. And Hoover came across the way he did in person, which was none too endearingly.

Roosevelt, however, was made for radio—his deep rich voice, conversational style, use of homely metaphors, and exquisite sense of timing. All this made his famous "fireside chats" deeply memorable. It required the president to be a bit of an actor, of course. You couldn't just read a speech into the microphone; you had to rehearse it like a play, had to hear how the words sounded, testing out the cadence, which bothered FDR's advisers, who resented the time he spent preparing. But the fact was that FDR *was* an actor, loved acting, and was quite good at it. *There are two great actors in America,* he told Orson Welles once. *You are the other one.*

And when he got on the radio the United States was his, with sometimes as many as fifty million people tuning in to listen. And that gave him an enormous edge over everybody else. For in going to the people directly, talking with them in effect in their own homes, he could reach over the heads of Congress and over the heads of the newspaper publishers (the great majority of whom were Republicans). And that gave him the opportunity to shape the national debate. In time others would figure out how to use radio and try to join in the debate, reshaping it to *their* ends, men like Father Coughlin, a hate-spewing right-wing anti-Semite. But none would ever approach FDR's masterful use of the medium. He was so popular on the radio that industry executives implored him to go on the air more frequently. But FDR declined. He didn't want to wear people out, and he knew he could. In his twelve years in office he would deliver just thirty fireside chats.[1]

The other great force was the depression. It broke Hoover. It made Roosevelt.

It was not, as is sometimes supposed, the nation's first real depression—there had been sixteen before. And in 1933, when FDR took power it was, as yet, far from the longest—the four great depressions of the nineteenth century had all lasted longer.* But of all of them this was easily the most frightening. A quarter of the people were unemployed. A third of the railroads were bankrupt. And nearly all the banks were closed. To many it seemed as if the country itself might go under. A friend told Roosevelt that if he was able to fix the economy he'd go down as the greatest president in American history. *And if I don't,* FDR had retorted, *I'll be the last.*[2]

The first hundred days were the most productive of his entire presidency—and the easiest. Not easiest relative to the hours he had to put in, though: He probably worked harder than he ever had in his entire life, frequently toiling until midnight and beyond. But easiest in the exercise of power. Because he had Democratic majorities in both houses and because things were bad, so bad that Hoover himself had confessed his last night in the White House that he was at the end of his "string," FDR was given a free hand, the Congress virtually abdicating its constitutional responsibility to make laws.

In those first hundred days it was Roosevelt who made the laws. His people drafted the bills and Congress passed them, the members often not even bothering to read them through. The first week there was the emergency banking bill, which allowed banks to borrow money from the Reconstruction Finance Corporation against their own stock, ending the banking crisis. The second week there was the economy bill, which cut pensions, business subsidies, and defense spending. By the end of March, his first month in office,† Congress had established the Civilian Conservation Corps (CCC) and the Public Works Administration (PWA) and somehow had found time to reform securities laws and provide for the sale of weak beer.‡ In April there was a bill to save farmers from mortgage foreclosures and a bill to establish the Tennessee Valley

---

*At the time of Roosevelt's inauguration the depression had gone on for forty-two months. The depression of 1815–21 had lasted seventy-one; that of 1837–43, seventy-two; that of 1873–78, sixty-six; and that of 1893–97, forty-eight. Richard Morris, ed., *Encyclopedia of American History* (1970), p. 536.

†FDR was inaugurated March 4, 1933; not until the next inauguration would presidents begin their terms in January.

‡The amendment to repeal Prohibition had been passed but not yet gone into effect. The law to allow the sale of weak beer (that is, beer with an alcoholic content of 3.2 percent) was a stop-gap measure.

Authority (TVA). In May Congress passed the Agricultural Adjustment Act (AAA), which rewarded farmers for plowing their crops under, a radical idea that people had a hard time understanding at a time when so many were starving. But they went along with it anyway; FDR said it was needed to limit production and that that would be good for the economy. In June, just as the Congress was wrapping things up, it established the National Recovery Administration (NRA), the most astonishing New Deal innovation of all: an agency that would work with business to establish industrial codes to prevent price wars.[3]

FDR was a legend after those first hundred days, the man who could do anything and everything. And yet the reality was he really only had been all-powerful for the first two of the four months that the Congress was in session. By May he was having to compromise with both Republicans and Democrats. Republicans had forced him to shelve his plan to reorganize the railroads and had raised so many objections to the farm subsidy program, the AAA, that FDR had had to promise to appoint a Republican to implement it. Meanwhile Democrats had also grown increasingly difficult, demanding greater reforms than FDR was willing to support. He hadn't wanted legislation establishing federal bank insurance to protect depositors. The radicals in Congress had shoved it through anyway.

FDR hadn't squandered his powers. If anything he had exploited them brilliantly, demonstrating political skills no one had ever suspected he had. It was simply that it was impossible for anybody to maintain the control he had had those first few months. The paradox was that FDR himself had unwittingly undermined his own power by being so extraordinarily successful. The more confident people became that the economy would turn around and the more willing they were to dare to go back into the stock market, the less dire the crisis seemed. And as the perception grew that the crisis was receding, the necessity of following an all-powerful president seemed to recede as well. The dynamic had begun to shift in April, when the market suddenly regained 60 percent of its value, encouraging the hope that the worst was over. In fact, come summer the market would again go into a sharp decline, but the fear that had initially gripped the country—that "nameless, unreasoning, unjustified" fear—never returned. The emergency was over.

Roosevelt would never recover the dictatorial powers he had held those first few dramatic months, not even during World War II. Ever after he would face media criticism, partisan sniping, and intraparty squabbling, requiring him to fight to maintain his popularity, as other presidents had had to fight to maintain theirs.

But as Roosevelt came to understand, and understand so well, he did not need a crisis in order to govern successfully. All he needed was to keep alive the *memory* of the crisis. People did not need to be fearful, they only had to remember that once they had been. That was enough to keep them in line, to keep the unwieldy New Deal coalition together: the farmers, blacks, southerners, union laborers, and liberals. Keep reminding them how bad times had been under Hoover and they would remain loyal New Dealers. As FDR would advise other Democrats, *Just keep running against Hoover. Whoever the Republicans put up, run against Hoover.**

Moreover, the memory of the crisis gave him extraordinary freedom to reinvent the government, to give it vast new powers, more powers than anybody in American history had ever dreamed of giving to the government before except in times of war. In the past, as Roosevelt biographer Frank Freidel points out, the federal government had mainly been confined to the exercise of police powers and, when it had gone beyond that, to tax, spend, borrow, and regulate, it had usually done so gingerly. Now all that changed. Now the government acted boldly, intrusively in many ways, as officials began powerfully to direct the economy. And that helped FDR maintain his grip on the country.

In a way he was able to turn a negative, the Great Depression, into a positive, to bend this tremendous force to his own ends, an astonishing and almost unprecedented development. Usually in American history the great new forces had complicated governing. This force seemed to make governing easier. But that was because there had been a crisis. As Lincoln had discovered in the Civil War, in a crisis—a truly awful, terrible, fear-inspiring crisis—a capable president could rally people and accomplish great things as he could not in more normal times. A crisis made a president superpowerful, almost as powerful as George Washington.

The Great Depression, of course, was so great a force, that FDR could not end it—it was simply too vast, too deep, too powerful—World War II did. To end it he would have had to embrace budget deficits the likes of which nobody in world history had ever embraced before. And he and the public lacked the imagination to do that. He and they could barely understand the deficits in the billions that the New Deal was racking up.

---

*For a half century, Democrats were to continue running against him.

They could not grasp the reality of deficits in the tens of billions. And they certainly could not comprehend government budgets in the hundreds of billions. This was simply beyond people. And it was to remain beyond them until World War II changed their outlook. Then they would become used to these gigantic, nearly unfathomable figures. But only then.*

FDR, however, did not actually have to end the depression to remain popular, he only had to appear to be trying. And he had more than enough power at hand to do that. Anytime his popularity seemed to wane, as it did in the fall of 1933 as the economy slipped backward, all he had to do was spend more government money, set up more government programs. As Roosevelt aide Harry Hopkins was to remark, "We will spend and spend, tax and tax, and elect and elect." FDR couldn't spend like crazy, of course, and he didn't want to. To the end of his presidency he remained committed to balanced budgets and never himself became a Keynesian. To the end of his life he continued to believe that deficits are intrinsically bad. In many ways he was always his own worst enemy. Every time the economy picked up, he ordered the bureaucrats to cut back on spending, which inevitably brought about a slowdown in the economy, which in turn necessitated even bigger increases in spending. But he could always spend enough to keep hope alive.

If there were any doubts that FDR was connecting with people and doing so in a powerful way, the elections of 1934 laid those doubts to rest. Instead of losing seats as the party in power always had in midterm elections, the Democrats gained seats. When the new Congress assembled in 1935, three-fourths of the seats in both houses were held by Democrats, forming the greatest majority in history. Even FDR was astonished. And he immediately set out to exploit his new majority, setting in motion a second New Deal, sending up bills to establish the Social Security system, the Works Progress Administration (WPA), and finally the National Labor Relations Board (NLRB), which guaranteed labor's right to collective bargaining. Conservatives howled in protest, claimed the country was going to hell, Hoover crying that Roosevelt was

---

*The cost—some $560 billion—of World War II remains unbelievably expensive even by today's standards. (The Korean War cost $70 billion and Vietnam $121 billion.) *Dictionary of American History* (1976), p. 228.

socializing America, undermining American liberty. But FDR had the votes: The measures passed.

Despite the conservatives' noises, Roosevelt appeared to most Americans to be in control of things in 1935, but he was and he wasn't. While he dominated Congress as no president ever had save Jefferson and Wilson, he had plenty of reason to worry about his ability to maintain his dominance. While it had been fantastic to have the majorities he had, the very fact that they were so large made governing difficult. FDR afterward recalled that he had been so exhausted by the end of the session in 1935 that "I would have . . . gotten pleasure out of sticking pins into people and hurting them." The lesson of 1935 was that it was easier to keep control of the Democrats when they were smaller and when there was real opposition. You needed the opposition to keep your own forces in line.

None of the groups that made up the Democratic Party was as yet threatening to bolt, but the southern conservatives were antsy. While they loved the farm subsidies, there was the Eleanor Roosevelt problem: Eleanor was on a daring mission to bring blacks into the mainstream. Even though her efforts mainly affected blacks living in the North, helping blacks anywhere seemed wrong to southern whites, almost a challenge to their belief in white supremacy. And it made them suspicious. As the years went by they would become more and more suspicious, especially when the WPA began paying southern blacks raking leaves more than they could earn as farmhands. So FDR worried. For he needed both the blacks and the southern whites. He needed the blacks to win in several of the big northern cities where they were continuing to migrate in droves during the thirties, and he needed the southern whites to maintain his lock on the electoral college.

Looking back now, of course, it appears that FDR had nothing to worry about at all. But he didn't have the advantage we have. He didn't know then that he was going to be elected president four times, that he was going to get Democratic congresses throughout his years in office, that the people were going to stick with him. As president you just could never tell what might happen next. One minute you were on top of the world, and then . . . well, things changed and often they changed fast. *Just look what had happened to Wilson,* FDR would frequently point out. Six years on top and then he lost the Congress and the country.

Going into 1935 FDR's most immediate worry was Huey Long, "the Kingfish," the great southern demagogue of demagogues, the dictator of Louisiana, who controlled the state legislature as no governor had ever controlled a state legislature before. Of all the politicians in the country who might challenge FDR, Long was probably the ablest. Born poor, he had been ruthless in climbing to power, willing to say anything, do anything, to build up his political machine. He was very smart, very brash, and very imaginative. As a young politician he had taken a minor post as a state public service commissioner and turned it into a platform for social justice, excoriating the utilities and oil companies in a blatant bid for the populist vote. Like FDR he was a great promiser; Long himself said that FDR was worried because he knew Long could outpromise him. And he did. FDR in 1932 had promised labor unemployment insurance, promised farmers he'd limit production, promised conservatives he'd balance the budget, and promised business he'd expand the Reconstruction Finance Corporation. Long in 1934 and 1935 was promising to guarantee every American family a minimum annual income of $2,500, which in the 1930s was a kingly sum (Long's slogan was "Every Man A King"). He planned on paying for the handout by steeply taxing the rich, in effect, doing what nobody in American history had ever done, actually redistributing the wealth of the country.[4]

And Long was planning to run for president in 1936.

FDR was very frightened of Long. Long wasn't like a normal politician. He was a freak of history. A product of the crazy politics of Louisiana and of the desperate times of the 1930s. You could try to outmaneuver him as you would some other candidate, and FDR tried, moving left in 1935, to blunt the appeal of Long's redistribution plan. Abandoning caution, FDR embraced a proposal to raise the top bracket on personal income taxes from 59 to 75 percent. He also backed a plan to raise corporate income taxes from 13 $\frac{3}{4}$ to 16 percent.

But with a fellow like Long you had to do more, you had to destroy him—or he'd destroy you. So FDR decided to destroy Long. In late 1934 he had the Treasury Department begin a massive tax fraud investigation into Long's political machine. Fifty IRS agents were dispatched to Louisiana to dig up dirt on Long and his friends. It was a terrible abuse of federal power, and it was unprecedented. No other president in the seventy-two-year history of the IRS had ever dared use the agency against a political rival. But these were desperate times, and FDR was desperate to maintain control, desperate to make sure that a demagogue like Huey Long never gained power.*

Long tried to cut a deal with the administration, twice going to see the attorney general to get the investigation stopped. But FDR refused to negotiate. He wanted Long politically dead. And then suddenly Long *was* dead, killed by an erstwhile supporter as he was walking down a hallway in the statehouse in Baton Rouge.

FDR's other great concern in 1935 was the Supreme Court. On May 27—"Black Monday" to New Dealers—the Court struck down the law establishing the NRA, which immediately led to massive price wars. Which in turn led to lower profits. Which in turn led to layoffs, wage cutbacks, and a longer workweek as companies tried to compensate for the reduction in their profits. Within weeks a million clerks were affected. Clerks who had been working forty-eight hours a week suddenly were told they had to work sixty or seventy hours—and for less pay.

Presidents of course had seen the Court go against them before. The Court had gone against Thomas Jefferson, Andrew Jackson, and Grover Cleveland. But never before had the Court's rulings been so critical to the daily well-being of Americans, for never before had the government been so involved in their lives. When the Court acted now to strike down a law it often was, in effect, taking food out of peoples' mouths. The stakes were simply enormous.

---

*The first president to abuse the power of the IRS was apparently Hoover, who in 1931 ordered the agency to investigate the Navy League, an association of shipbuilders who'd begun criticizing the administration for cutbacks in the navy's shipbuilding budget. When FDR came to power the IRS went after Hoover's secretary of the treasury, millionaire Andrew Mellon. Years later Elmer Irey, the director of the Criminal Division of the IRS, wrote that "the Roosevelt Administration made me go after Andy Mellon." (He did not say if FDR personally was behind the probe.) The criminal case was eventually dismissed, but in civil court Mellon was found guilty of underpaying his taxes by half a million dollars. In 1940 the Treasury Department opened an investigation into the finances of Father Coughlin, the radio hate preacher. That same year FDR personally ordered an investigation of Rep. Hamilton Fish, a right-wing Republican isolationist who was allegedly consorting with Nazis. In 1944 FDR ordered the IRS to lay off Lyndon Johnson, then a Texas congressman, and Johnson's benefactor, Brown and Root, a gigantic Texas construction company; FDR needed the support of both to win Texas in the upcoming presidential election. (The state Democratic Party had been taken over by an anti-Roosevelt clique.) See David Burnham, *A Law Unto Itself: The IRS and the Abuse of Power* (1989), chap. 10.

And increasingly now the Court was striking down laws: the Railroad Retirement Act; a law to limit wasteful practices in oil fields; a law allowing the Treasury Department to eliminate the requirement that it pay off government bonds in gold. The Court even knocked down New York state's minimum wage law, which truly shocked. New Dealers had always worried that the Court might not allow the federal government to regulate wages, but not allow a state to do so? If a state couldn't, then nobody could. The law of the jungle was to prevail. Then—and this was the final blow—the Court invalidated the funding mechanism of the AAA. The law had provided that farmers who took land out of production were to be paid a subsidy funded by a tax on food-processing companies. The Court ruled that the tax was not a tax at all, but was really a method of manipulating farmers, and therefore illegal.

FDR was frustrated by the Court's rulings—and alarmed by them. Frustrated because if the Court continued handing down these kinds of decisions the New Deal would be eviscerated and people would lose hope. Alarmed because all the New Deal really had going for it *was* hope—he had been unable to end the depression after all—and if the hope vanished so would his chances to be reelected in 1936. Besides, by taking power as brazenly as he had, he had encouraged the idea that he could solve people's problems or at least ameliorate them. Before he had come along nobody had expected the president to help them if they lost their job. Presidents weren't held responsible for such things. Getting a job and keeping it was the individual's responsibility. But FDR had changed how people thought. Now they believed a president *was* responsible.

During the campaign a woman came up to FDR with a note requesting his assistance. She explained that she and her co-workers at a local garment factory in New Bedford, Massachusetts, had been earning eleven dollars a week. But then their pay was cut to four dollars. "You," she said to Roosevelt, "you are the only man that can do anything about it." And yet of course he couldn't, not if the Court kept handing down these negative decisions. "These people think that I have the power to restore things like minimum wages and maximum hours and the elimination of child labor," FDR complained, "and, of course, I haven't got any power to do it."

FDR remained the odds-on favorite in the election. But he wasn't a surefire winner. The newspapers were hysterical in their opposition. Business hated him. And he wasn't running against Hoover this time, much as he might try to persuade people he was, he was running against Alf Landon, the governor of Kansas, and Landon was progressive. And Landon was raising a lot more campaign money than he was, fourteen million to FDR's nine million. The Gallup poll consistently showed FDR ahead nationwide in the popular vote, but it also showed that he could only count on the Rocky Mountain states for sure. Everywhere else his support was a little soft.

Of course, of course, this was Franklin Roosevelt who was running, the president who had helped elect the greatest majority in the history of Congress, the president after whom people were naming their children, the president whose picture could now be found hanging in people's homes right next to the pictures of Grandma and Grandpa and Jesus. But then again . . . you could never tell in politics. A winner could easily become a loser overnight. And there hadn't been too many winners in the presidential history of the Democratic Party. In the second half of the nineteenth century there had been only one, Cleveland, and he had lost his first try at reelection. So there was plenty of precedent for losing.

But if there was one thing FDR did not plan on doing, it was losing.

It was under these circumstances that FDR decided to do what so many other presidents had done, and that was to use his patronage powers to build up a huge political army, in effect using the federal payroll to fund his campaign. There was, however, a difference. FDR had by far more jobs to hand out than any other previous president— thousands and thousands more jobs, WPA jobs. Harry Hopkins, the head of the WPA, tried hard to keep politics out of the agency but as the election came closer and closer he found that he couldn't. While anybody, Democrat or Republican, could get on the relief rolls, in many states only Democrats in good standing could get the managerial jobs, the ones that paid well. The politicians saw to that.

After the election, which he won handily, with 60 percent of the popular vote, more than anybody else had ever gotten, FDR set out to reform the Supreme Court, the Court that had given him so much trouble, the Court that had jeopardized so much that he and the Congress had done. It was an interesting step to take, a sign both of his power and his powerlessness. Because of his victory and its size he was of course in an extremely powerful political position and feeling confident and cocky. But the very reason that the Court loomed large in his thinking was that for four years he had been unable to end the depression—and there seemed no reason to think he could end it anytime soon, which made the survival of the New Deal extremely important; people had to be able to hope the government could offer relief, reform, and recovery. Which meant that the Court couldn't be allowed to keep on declaring New Deal laws unconstitutional. The fact was that events were out of his control. He couldn't get the economy going, and he couldn't change the Court's mind about the New Deal.

Out of desperation, he came up with about the worst possible plan to gain control over the Court. It was so bad it was almost laughable. People had to hear it two and three times before they got it. *The justices are overworked,* the president wrote in a message to Congress in early February 1937, *and need help.* So for every justice over the age of seventy he wanted the authority to appoint a new justice. As there were six justices over seventy, that meant six new justices.

People of course were appalled. It was one thing for FDR to reinvent the executive branch and dominate the legislative branch, quite another for him to tamper with the third branch. The Supreme Court was sacred. (Not for nothing did the new building just designed for the Court resemble a temple.) People indeed thought of the Court as a temple, thought of the justices as godlike—well, maybe not all of them. Willis Van Devanter, James McReynolds, these men did not seem so much like gods as devils. But people still did not approve of Roosevelt's plan. An attack on the independence of the judiciary smacked of fascism, just what the dictators in Europe were doing.

Worst of all, FDR wasn't straightforward about what he was up to. Everybody knew what he really wanted was to force the Court to approve the New Deal. But because he did not want to say so directly he came up with this harebrained idea about the justices needing help because they were old. FDR made much of the fact that it was essentially based on a plan that the now-seventy-five-year-old McReynolds had come up with when he was Wilson's attorney general a quarter century before (except that McReynolds's plan expressly excluded the Supreme Court). This was indeed an irony of sorts. But what impressed most Americans even more was the crudeness of FDR's power grab.

FDR worked hard to convince people he should have the power to name additional new justices. The opposition, however, worked harder. In July FDR asked his vice president, John Nance Garner, the former Speaker of the House, if he had a chance. *Do you want it with the bark on or off, captain?* Garner asked. *The rough way,* Roosevelt answered. *All right, you're beat.* And he had been "beat" in the worst possible way: He couldn't even get the bill out of committee.

In the end it didn't matter, for the Court was already changing. In January, before FDR had announced his plan, one of the justices, Owen Roberts, had decided to switch sides, which gave the liberals on the Court a fighting chance to prevail. Then the chief justice, Charles Evans Hughes, switched to the liberal side, apparently fearful that the Court was in danger of being marginalized as it had been during Reconstruction. In March the Court upheld state minimum-wage laws. In April the Court upheld the act providing for the establishment of the NLRB. Then, out of the blue, crusty old Van Devanter retired, giving FDR the chance to make his first appointment to the Court, the Alabama liberal Hugo Black. And then, in May, the Court upheld the Social Security Act.

So the New Deal *would* prevail. But FDR himself was damaged. Where once he had seemed invincible, now he didn't. It was almost the way it had been with Wilson. One minute you're up, the next you're down. Sure, he finally had the Court in his pocket, but now his enemies smelled blood and that made him vulnerable, almost human in people's eyes, ungodlike. The loss was intangible. You couldn't see it or touch it. But all FDR had had going for him anyway was the intangible emotion of hope. So the loss of invincibility counted. And it came at the worst possible time. For just then the economy was heading south again. By the fall the country was in a recession—the Roosevelt recession.*

It was a little ridiculous to be talking about a recession when the country had never really emerged from the depression, but it was particularly stinging.† The depression had been Hoover's fault. The recession was Roosevelt's. Because of his insistence on reining in government expenditures, the economy had again contracted. Ironically the new sys-

---

*A recession, by definition, occurs when the economy contracts over a period of two quarters.

†The stock market *had* staged a recovery, downtown offices were being rented again, cars were being sold, and unemployment was down from the 1933 high, when twelve million were jobless. But nearly eight million remained unemployed, 14 percent. *Historical Statistics of the United States* (1975), part 1, p. 135.

tem of Social Security had contributed to the contraction. In its first year of operation, officials had collected two billion dollars—and paid out nothing. Not until the second year would people begin receiving checks. The effect of the contraction was devastating. By 1938 two and a half million more people were unemployed, rocketing the unemployment rate to 19 percent.

FDR's response was to fight harder than he ever had before for the New Deal—and to fight in a way he never had before. Throughout the thirties he had maintained his links with the party's conservative southern Democrats, refusing to break with them even though they repeatedly tried to block him in Congress. Now he decided to purge their leaders, to run them out of the party by defeating them in the upcoming primaries and conventions in 1938.

Like his move against the Supreme Court, this one was both a sign of his power and his powerlessness. He dared make the move because he felt that the people were still with him. But he was forced into it because the conservatives in Congress had effectively taken control of the national agenda. It was they who had sabotaged his Supreme Court plan, they who had forced him to face political humiliation, they who thwarted him at every turn. He had been particularly incensed at their attempts to hinder the passage of the Fair Labor Standards Act, which provided for the establishment of national wages and hours. He finally had gotten the bill through, but the conservatives had materially weakened it, insisting on delaying the forty-hour workweek until 1942 and the forty-cent minimum wage standard until 1945. (In the interim, lower standards were established.)

The papers called it the great Roosevelt purge, which made it sound demonic. They played it big, turning each race FDR intervened in into a front-page story, which FDR relished. He always loved a good fight. It boosted his spirits. But in nearly all the fights he took on, he lost. In the end he was able to crow about having gotten rid of just one crusty conservative, the head of the House Ways and Means Committee. The rest were able to hang on by playing the race card. (A large group of blacks and poor whites supported FDR, but in the South they were frequently prevented from voting because they couldn't pay the poll tax.)

FDR had been damaged again, and in the general election it showed. In 1934, defying history, his party had picked up seats in both houses. In 1938 the party lost seats in both houses: seventy in the House of Representatives, seven in the Senate.

The thing about FDR was that despite setbacks, he never gave up.

He just never gave up. Once he had settled on a goal he'd let nothing get in his way. But he had this great gift. Although he was driven, he never let himself *look* driven. What people saw in his face was confidence, the famous Roosevelt grin. And it concealed who he really was.

Franklin Delano Roosevelt. The name told you a lot. On both sides of the family, the Delanos and the Roosevelts, he came from money and privilege.

It was because he was a Roosevelt that he dreamed of being president. Cousin Teddy after all had been president.

Teddy was a big influence. Teddy smiled all the time—so would Franklin. Teddy ran for the state legislature in his twenties—so would Franklin. Teddy became assistant secretary of the navy—so would Franklin. Teddy ran for vice president at age forty-one—Franklin would run at age thirty-eight. In college Franklin broke with the Democrats to vote for Teddy and even joined the local Republican club. When he ran for the state legislature he asked Teddy for his approval. Even in marriage he managed to involve Teddy: Eleanor was Teddy's niece. (When Franklin asked for her hand in marriage he told her he had big plans. The naive Eleanor asked, "Why me? I am plain. I have little to bring you.") At their wedding Teddy, then president of the United States, gave away the bride.

But he was as much a Delano as a Roosevelt. It was because he was a Delano that he was able to struggle on in the face of adversity. Delanos knew how to keep going. Warren Delano, his grandfather, had managed to rebuild his fortune, not once, but several times. He'd lose it, start over, win it back, then lose it again, and start over again.[5]

FDR hardly knew adversity for the first thirty-eight years of his life. It had all been up, up, up. Everything he strived for, he'd achieved: state legislator, assistant naval secretary, vice presidential nominee.

When he'd gotten into scrapes he showed he could get out of them. When he was running in 1920 for vice president, with James Cox, Republicans had questioned the willingness of Latin American countries to vote with the United States if the United States joined the League of Nations. During a speech FDR insisted they would, and then added, to prove that America's influence over Haiti was great, "The facts are that I wrote Haiti's constitution myself and, if I do say it, I think it a pretty good constitution." Actually, the facts were that he hadn't had anything to do with Haiti's constitution. He was lying. And the lying got him in trouble.

Haiti was insulted, and the Republicans made a big deal about it. Pressed afterward to explain himself, FDR lied again, this time denying he'd made the statement in the first place. Politicians lie all the time, of course. FDR usually lied better, more artfully. The controversy blew over.

The Cox-Roosevelt ticket happened to lose to Harding and Coolidge. But FDR's aim wasn't to win—nobody thought the Democrats could take the White House in 1920—it was to get his name around.

He had been a success at everything, one would have to say, except his marriage. His marriage was a sham. A few years earlier FDR, bored with Eleanor and tired of her constant moralistic carping, had got himself a mistress on the side—her social secretary, Lucy Mercer. In 1918 Eleanor discovered a packet of their love letters, which for her was an enormous turning point, the end of naïveté. But FDR was only concerned about the possible political consequences. If Eleanor sued for divorce, his political career would be over. Fortunately, Eleanor would not sue if Franklin stopped seeing Lucy. He agreed, and his career went on as before. (He would, in fact, continue seeing Lucy; he would invite her to all four of his inaugurations. Later he would have an affair with *his* social secretary.)

Then, in 1921, at age thirty-nine, FDR faced catastrophe. Polio. Now he would need to prove that he really was a Delano.

It struck during the summer when he was on vacation off the coast of Maine. At first he didn't know what it was and even managed to keep on with his activities. One day, despite the ache in his legs, he went out and helped fight a forest fire. But within two weeks he was totally immobilized.

His mother put pressure on him to give up politics and return to Hyde Park, where he could live out his life as a country squire. But Roosevelt ignored her. The year before he had run for vice president of the United States. Someday he would run for president. So he kept up with his political correspondence—not even at the beginning of the illness, when he was in the most pain, did he stop writing—and kept dreaming and working. It took him a year or so to be able to move a single toe, another year to get out of bed and move around, and it took a full seven years for him to learn how to walk again, though it wasn't really walking but more like shuffling. By resting one arm on the shoulder of one of his sons and leaning on a cane with the other, he could thrust himself forward, shifting his weight back and forth from one leg to the other. The legs were utterly withered; in a rare photo they look about as big around as a child's arm. But with the help of fourteen-pound steel braces he could stabilize himself well enough to stand.[6]

Most Americans never realized he was a paraplegic. They didn't know because FDR made sure they didn't. Outside the family hardly anybody knew. When friends early on asked him what on earth had happened to him, he lied. Said he was suffering from a bit of lameness. Eventually the truth came out that he had polio. But FDR managed to leave the impression he had pretty much overcome it. The public never saw his braces; they were painted black to match his pants, which were cut long as a further precaution. And the public never saw him in a wheelchair: Photographers were forbidden to take his picture in one. There are just two pictures of FDR in a wheelchair; both were taken by a family member while he was at Hyde Park. Neither was ever distributed. It would be impossible today to do what FDR did then. The press now wouldn't put up with the strictures. But in the thirties, press people lived by a different code. When FDR shouted out, "No movies of me getting out the machine, boys," as he climbed out of his car, they obediently shut off their cameras.

By 1928 he was personally ready to run for office again, but the timing seemed terrible. The economy was booming, the Republicans reaping the political benefits. So when supporters talked up the possibility of his running for the party's presidential nomination he quickly objected. The nomination wouldn't be worth having in 1928. Instead, he supported the candidacy of Al Smith, the incumbent governor of New York, and a very close friend.

Smith ran and won the nomination, which gave FDR a shot at the governor's chair. (Smith's term was up; he couldn't run for reelection as governor while also running for president.) Again FDR was reluctant because 1928 didn't seem to be a Democratic year. But after hesitating he finally changed his mind during the state Democratic convention and was nominated by acclamation.

In November he barely won. Out of the four million votes cast, he defeated his opponent by a mere twenty-five thousand. But his timing was excellent. The following year the market crashed, putting the Republicans on the defensive. In 1932, as governor of New York, FDR would be in a perfect position to run for the presidency.

He thought himself completely hidden. *No man knows me,* he told his daughter Anna once. *Not even my own family.* But his actions repeatedly betrayed him. What he was was ambitious—deeply, deeply ambitious. He didn't fall into the presidency by accident. He worked for it for decades. He had been coy with Eleanor about his dreams, only admitting while he courted her that he expected to achieve something really big someday. But to a friend around the same time he came right out with it: He wanted to be president. He was barely twenty-one.

Going into 1939 Roosevelt began thinking about what he would do after his second term was up. His big plan was to establish a giant library at Hyde Park to house his presidential papers. Nobody before FDR had ever had a presidential library, but Roosevelt was very aware of history and knew that there would be great interest in his presidency. Before, there had been nothing like the Great Depression, nothing like the New Deal. And of course FDR wanted people to remember that. A library would help. He even got a cousin to begin going through his stuff.

So as the year began there was every indication that Roosevelt was going to return to private life; a year and a half earlier he'd told Jack Garner, his vice president, that he'd never again run for public office. FDR even began trying to line up a successor. He all but anointed Postmaster General Jim Farley, though he let several others think they might really get his support. (FDR was too cagey ever to let anybody know what he really would do.)

Nobody thought FDR himself would run again. No president had ever been elected to a third term, and none had ever made a serious try for one directly after their first two.* Since George Washington, Americans had considered two terms enough for any president.

The German invasion of Poland on September 1, 1939, changed everything.

---

*Grant campaigned for a third term three years after he had left the White House. TR, of course, let Taft succeed him after his first two terms and only then ran again (TR considered the three and a half years he served after McKinley died as his first term). Wilson made noises about running for a third term but never in fact did.

17. FDR: World War II

How FDR waged war without asking Congress to declare it

It had been the thing that the people of the old Allied countries feared the most during the late 1930s: a second European war. England and France had feared war so much that they had been willing to do anything to prevent it. It would be conveniently forgotten later, after Hitler turned out to be a megalomaniac, that both countries enthusiastically applauded British prime minister Neville Chamberlain's appeasement of Hitler at Munich in 1938.

But now war was a reality. Hitler was not going to be satisfied with Austria or Czechoslovakia, as the appeasers had claimed at Munich. He wanted it all, all of Europe, just as he had indicated in *Mein Kampf* two decades earlier. And he was willing to go to war to get it. And that altered Americans' outlook—and FDR's.

The effect on the country was to give new vitality to the internationalists, who until then had almost had to hide how they felt. As the polls showed, Americans were resolutely isolationist. But now the internationalists could begin arguing openly that isolationism wouldn't protect the United States from war, that we had to begin building up our defenses, that there was a moral difference between Germany and Britain, and that in our own self-interest we would have to take Britain's side—again.

The effect on FDR was equally energizing. Ever since the debacle over the Supreme Court, he had been on the defensive, his administration seeming almost to drift, the New Deal running out of steam. But now he had a clear new crisis to handle—and he seemed to welcome it. It focused his energy again as nothing else had in some time.

* * *

World War II would bring many changes to the United States. The first was to give us our first third-term presidency, a momentous turning point. After Washington had turned down a third term, Americans had come to believe that all presidents should. So not even Jefferson or Jackson went for a third term, though each easily could have convinced himself—and the country—that it was his duty to do so. Democratic leaders were eager to see Roosevelt run again. He had loomed so large for so long that nobody else in the party seemed comparable and, most importantly, nobody else seemed likely to be able to win. But Americans generally were far far less willing to smash tradition, much as they loved FDR. A three-term president? It seemed a bit much.

FDR got the Democratic nomination easily enough—he never even formally announced he was a candidate; the first time Democrats were officially informed that he would be available to run again was at the Chicago convention—but it was far from clear that he would be able to win the election. And in fact the Gallup poll in August 1940 showed the Republican nominee, Wendell Willkie, was ahead—not by much, but ahead nonetheless. FDR would be in the race of his life.

At the outset there was a semblance of civility to the election campaign, Willkie and Roosevelt agreeing (through intermediaries) not to play politics with one of the hottest issues dividing the country: Roosevelt's deal with Winston Churchill in the late spring to trade American destroyers in exchange for British bases. But as the campaign moved forward there was enormous pressure on both sides to play politics with foreign policy, and Willkie reneged.

Willkie really had little choice but to exploit foreign policy. There weren't any other issues on which he could make a credible fight. Though he had fought the New Deal for years, he and the Republicans now had no alternative but to downplay their opposition (except when preaching to the Republican choir). The public had showed in election after election that it approved of the New Deal. So, inevitably, it was foreign policy about which Willkie and Roosevelt argued.

And that brought out the worst in both of them. Willkie, sensing Americans' resistance to involvement in yet another European war, came out as an isolationist, boldly proclaiming that he would never send American boys to fight Europe's wars. Which was blatantly political on three grounds. One, Willkie himself until then had been a bona fide internationalist, as much an internationalist as FDR, so claiming that he was really an isolationist was rather shameless. Two, no president could make the promise he was making and still sleep soundly at night. For there simply was no way to escape from Europe's wars anymore. The Great War had proven that. The world was getting smaller and smaller. And three, by conspicuously making the claim that he was *against* sending American boys to fight over in Europe, he was implying that Roosevelt was *for* sending them. As the campaign heated up, Willkie came right out and said so. Thus Willkie played politics with the issue, but so did FDR.

Although Roosevelt was convinced that the United States had to support Britain (as he had demonstrated in striking the destroyer/base deal), he now held that he was as opposed to getting involved in a new war in Europe as Willkie. Isolationism was simply too great a force to resist, he believed. In the summer of 1939 when he had asked Congress—a Democratic Congress—to repeal the neutrality laws so that the U.S. could side with Britain, he had been turned down (this was *after* Munich, *after* Charles Lindbergh had reported that Germany possessed more planes than the rest of Europe combined, *after* Hitler had laughed in public at FDR's request that he promise not to invade any country for ten years). Not until November 1939, more than two months after Germany invaded Poland, did Congress finally revise the country's neutrality laws and endorse the sale of arms and weapons to the Allies. So he had to fudge his differences with Willkie, make people think both of them would keep America out of war. Only gingerly was he even willing to push for preparedness. When his army advisers told him a draft would be needed, he let Congress pass it without his open support. (The bill became law in September, in the middle of the election campaign.)

The need for preparedness was obvious. Troops were having to train with fake rifles made out of leftover wood, the artillery was a generation old, and there weren't any tanks. When FDR had traveled to upstate New York in August to watch soldiers training, the scene had had a surreal quality. Soldiers firing wooden guns at trucks covered in canvas tops with the word "TANK" written on them, a telephone pole stuck on top symbolizing the tank's barrel.

But FDR felt strongly that he couldn't get out in front of the people, not too far out in front, at least. And certainly not in an election year. It was an irresponsible position to take given the way events were going, and would hurt us immensely when war came; it would take the country more than two full years after Pearl Harbor to gear up for the invasion of Normandy. And it was probably one of the most irresponsible positions any American president had taken in history. To know a war is coming and not prepare for it because you're in the middle of an election cycle? And yet it was also entirely understandable.

A leader *could* only lead so far. And no president in the past had ever faced such a situation. None before had ever been in the position of facing a madman capable of plunging the whole world into war. Woodrow Wilson had probably been in the closest similar position. He knew during the election of 1916 that war with Germany was likely, given the way things were going. But the odds in favor of a second war with Germany in 1940 were far, far greater, as Roosevelt himself privately acknowledged.

War wasn't a certainty, of course. Nothing was ever certain. And FDR had some hope that if the United States built up its defense forces, Germany (and Japan, for that matter) might back down. But that was a forlorn hope, and since the United States was *not* building up its forces—not building them up enough to wage war *and* meet the armaments needs of the Allies—almost a foolish hope.*

Given what had already taken place, prudence dictated preparedness. Germany, in defiance of the Versailles treaty, had rearmed; forced a union with Austria; seized Czechoslovakia; made a nonaggression pact with its eastern neighbor Russia (in order to be able to focus on its western neighbors); overrun Poland; invaded Denmark and Norway; taken over Holland, Belgium, and Luxembourg; invaded France; driven British forces across the Channel after humiliatingly trapping them at Dunkirk, defeated France; and then launched the Battle of Britain, raining bombs down on London nightly.

Meanwhile Japan had also been on the move, making war on China and threatening war on the Dutch East Indies (Malaysia). And Italy

---

*In the spring of 1940 FDR had asked Congress to approve a five-billion-dollar increase in the defense budget and had called for the production of fifty thousand planes. But given Great Britain's needs, the increase was not nearly enough, as FDR knew. Worse, the Selective Service Act had been delayed so long that not until after the election would anybody be drafted. As late as the summer of 1941, just 128,000 men would be enrolled in military training. Army leaders such as George Marshall warned the president that the country was unprepared to fight even a one-ocean war, and certainly not a two-ocean war.

had taken over Ethiopia. So there was more than enough reason in the fall of 1940 to believe that the United States would find itself drawn into war.

And yet FDR refused to tell the American people the truth. They didn't want to hear the truth. Because he wanted their votes, he gave them what they wanted: false hope that the United States could remain out of the war.

By the end of the campaign the contest had turned into a virtual street brawl, both sides playing dirty. Not in new ways, but in the same old ugly ways American politicians had been playing dirty for decades. Willkie's camp went after Roosevelt's boys, claiming they'd been given privileged command positions in the armed forces. (Roosevelt's son Elliott, for instance, had been made a captain in the air corps.) FDR, for his part, authorized a good old-fashioned smear campaign against Willkie.

Roosevelt had learned that the Republicans had got hold of letters that Henry Wallace, the Democrats' vice presidential candidate, had written to a mystic, which made Wallace look like a nut. Frightened that the letters might just discredit the entire Democratic ticket and put his campaign on the defensive, FDR privately threatened to sue the Republicans if they published the letters. In the event the letters somehow got out, Roosevelt told his advisers they should leak information about an affair that Willkie was having with a society woman. "Spread it as a word-of-mouth thing," he told his campaign staff in a conversation captured by a secret Oval Office tape machine. "We can't have any of our principal speakers refer to it, but the people down the line can get it out."[1]

By election day FDR was back out in front—far out in front. He won by five million votes. It was, however, the smallest of his three presidential victories. (He had won by seven million in 1932, by eleven million in 1936.)

The election changed FDR's political position dramatically. It gave him the freedom to move the country in the direction of war without having to worry that his policy might lead to his electoral defeat. So he was in a much more powerful, much more comfortable place. Willkie, switching back to internationalism, even agreed to help out. But the president still could not simply plunge the nation into war. The army wasn't prepared and neither was the country. So FDR had to take baby steps toward war.

The most important of these steps was Lend-Lease, which he dreamed up while on vacation in the Caribbean in December, just prior to his inauguration. It was a bold plan, without precedent, daring really. Instead of selling the Allies an arsenal of weapons, the United States would loan them what was needed, the way neighbors loan each other a fire hose in an emergency. It was mildly deceptive, of course. As the Republican isolationists like Robert Taft pointed out, loaning weapons was like loaning chewing gum: "You don't want it back."* But it was a necessary deception. Britain was out of money, couldn't afford to buy weapons without a loan, and Americans refused to make a loan. (The Allies still hadn't paid back the loans from World War I, which annoyed Americans tremendously; Coolidge had won a lot of points in the twenties when he refused to forgive the loans. As he effectively had put it, "They hired the money, didn't they.")

If events had played into FDR's hands he wouldn't have needed to resort to deception, but the fact was that they did not. There was just one event that would have helped, and that was an event no president could welcome, an armed attack. So FDR just kept nudging people forward, toward a war they would have to fight even though they didn't want to.

Only the nudging didn't work. While Congress approved Lend-Lease, and the British began to get the arms and weapons they needed, going into the summer of 1941 they needed more. They needed American muscle. As Churchill told Roosevelt at a secret meeting off the coast of Newfoundland, Britain could not hold out on its own much longer.

FDR, pressed as he had never quite been pressed before, chose to act in a characteristically deceptive way. He would wage war without declaring it. For months he had refused Churchill's requests to allow American destroyers to accompany British convoys across the Atlantic. Too risky, FDR told Churchill, much too risky. For what would happen when a German sub approached? The destroyer would probably have to destroy it or risk being destroyed. But now, faced with the prospect that the British were indeed on the precipice, he changed policies, agreeing to allow the destroyers to begin patrolling. It was an astonishing decision, blatantly deceptive, wholly unprecedented. But as FDR would say later,

---

*And in fact few of the weapons ever were given back. Of fifty billion dollars in equipment loaned under Lend-Lease, just ten billion dollars came back at the end of the war. Clive Ponting, *1940* (1990); A. J. P. Taylor, *Politicians, Socialism, and Historians* (1982), pp. 230–35.

"I am perfectly willing to mislead and tell untruths if it will help win the war."

Unlike President Polk, who had created an incident hoping to provoke an attack which he could use to obtain a declaration of war, FDR was hoping to wage war without a declaration at all, a first in American history. Down the road, after the country had built up its defenses, he could openly ask for war, using as a justification any of the numerous incidents which were sure to take place on the open seas. But *he* would have the option of deciding precisely when he chose to go to the Congress with a request for a declaration. In effect, he would gain control over a situation that until now he had been unable to control. A brilliant maneuver, though decidedly manipulative and undemocratic. (Hitler at any time might, of course, declare war on us, but military leaders believed he probably would not. He had his hands full already.)

The destroyer patrols began in August. By early September he had his first incident. The USS *Greer*, traveling to Iceland, received a radio report from a British plane that a German sub was in the vicinity. The *Greer* then located the sub and hounded it for nine hours. In accordance with orders, the Americans refused a British request to bomb the sub, but the British flew by and dropped several depth charges. The Germans, not knowing who had bombed them—the Americans or the British—fired a torpedo at the *Greer* (but missed). In a fireside chat FDR promptly used the incident to rouse American ire, misleadingly telling his radio audience that the attack had been entirely unprovoked. He declined to ask for a declaration of war. He didn't want war just yet. He needed more time to prepare.

Over the next few months there would be several more incidents. In mid-October the *Kearny* was attacked, with eleven deaths. In late October, the *Reuben James* with more than a hundred. Still FDR refused to go to the country to demand war. He still wasn't ready. Still didn't feel the country was militarily ready, still not psychologically ready. Probably he was about the last person in the administration to feel this way (excepting notably Secretary of State Cordell Hull).

There are some who believe he might never have been willing to use an incident he himself had provoked to go to the Congress for a declaration of war. But then, suddenly, he no longer had to decide. The Japanese decided the issue for him at Pearl Harbor, sending much of the American Pacific fleet to the bottom of the ocean on that "day that will live in infamy." A few days later Hitler declared war against the United States.

* * *

The war greatly simplified things for FDR. As commander in chief he would be able to do just about everything he needed to do to win the war. Congress could continue to block him occasionally on domestic legislation—and did. In 1943 legislators would override his veto of an antilabor bill. In 1944 the Senate would reject his demand for an increase in Social Security taxes. But in foreign affairs power remained almost exclusively in his hands. A commander in chief could do as he pleased.

In the fall of 1942 FDR wanted to lower farm prices. They were simply too high, he felt. It was, ironically, the government that was keeping them high; the Price Act required the executive to maintain prices at an inflationary level. But he didn't want to wait for the Congress to act. The Congress would take too long. He wanted action now. So he went to the attorney general to ask if there wasn't something he could do. Sure, he was told. Under his war powers he could just repeal the law, repeal a law made by Congress—as commander in chief he could even do *that*.

Still FDR was a little worried about going that far. That sounded extreme. So he asked Hugo Black for advice. Black by then was on the Supreme Court and shouldn't have agreed to give the president advice. He might someday have to rule on the matter. But it was war and the old rules didn't seem to be as important so he went ahead and gave the advice. Told the president that if he unilaterally repealed a law the courts would get involved and things would get ugly. Probably he'd lose.

But there was a way. He could tell Congress he intended to repeal the law by a certain date if he hadn't heard from legislators in the interim. It wasn't much of a concession, but the courts likely would look far more favorably on his action. For he would seem to be leaving the power to decide the issue with Congress. So on September 7 FDR gave Congress three weeks to pass a law basically saying that legislators wanted the old law they passed before to remain in effect. When they didn't act he used his war powers to lower farm prices. It was another unprecedented move, another rung up the ladder of the imperial presidency. But hardly anybody complained. This was after all war and FDR was the American leader in the war. And in a war a leader can do almost anything.

As the election of 1944 approached, he again had to take politics into consideration, which always limited his options. It meant he had to think about the effect of his policies on the groups that made up the Democratic majority. He wouldn't necessarily have to alter his policies, but he might have to keep them secret. Which is in fact just what he did at Teheran, where he met with Churchill and (for the first time) Stalin.

There Stalin revealed that he had territorial ambitions in the war, he wanted a chunk of Poland and the control of certain Baltic ports. FDR said he was perfectly willing to accommodate his friend Joe, but he had a favor to request. He wanted the concession kept secret for the time being. If the Poles in Chicago found out it would complicate his reelection efforts. Stalin said he understood. So they struck a deal and then kept quiet about it.

It wasn't right. Not right at all. Not respectful of democratic tradition. Not respectful of the war itself, which supposedly was being fought in the name of self-determination (as the Atlantic Charter said). But one always had to make compromises, and in wartime sometimes you had to make big compromises. Even in a war a president had to compromise.

FDR should not have run again in 1944. He was too ill. Or he should, at least, have leveled with the American people about the likelihood that he would die in office. But he refused to acknowledge he was ill, even to family members, though Eleanor said afterwards he knew—of course, *he knew*. Many days he could not get out of bed, so he just stayed in bed and worked there. The last year of his life he only rarely tried standing; the braces simply hurt too much now, they were too heavy and his muscles too weak.

But he insisted on running, wouldn't even consider not running. After three terms in office the fire that had roared in him his whole life, that had got him through polio and then through the ordeals of the Great Depression, still roared. Now that peace seemed so close he wanted to be there to enjoy it. Be there to see the war through to its end. Be a part of history, not just watch it from the sidelines. It was the power of the presidency that nourished him during these years, and surrendering that power was inconceivable. It was, after all, the only thing that kept people from seeing, when they looked into his cragged, weathered old face, the face of an old man. Instead, what they saw was the face of the president of the United States. FDR liked that.

The Republicans picked Thomas Dewey, a fiery prosecutor from New York, who ran a hard race. He ran harder than FDR. But FDR was—FDR. And by now there was very little doubt that the American people loved him. So Dewey really didn't have much of a chance.

Roosevelt hardly made an effort at all to campaign. It was enough to show people that he was leading the war effort. So he made a big show of being president, of being commander in chief. That would be far more effective than running around the country giving campaign speeches. As part of his strategy he made a trip during the summer of 1944 to Hawaii. It was to review the troops, reassure Americans he hadn't forgotten about the war in the Pacific, and to have his photograph taken with his Pacific commanders, particularly the charismatic pipe smoker, Douglas MacArthur. It was a barely concealed campaign stunt, and it annoyed MacArthur. "The humiliation of forcing me to leave my command," he complained bitterly afterward, "for a picture-taking junket." But it went over well with people, just as FDR had expected.

There was still the question of his health, though, which was appearing more fragile by the day; on the way back from Hawaii he had stopped at the naval shipyard in Bremerton in Washington State. Just moments into his speech before ten thousand sailors he suffered an angina attack and nearly crumpled over. It lasted fifteen minutes. Still FDR refused to stop talking, refused even to cut short his address, which was being carried by radio to a nationwide audience. So he kept on talking—and talking and talking, for a full thirty-five minutes.

To prove he was alive and well, to demonstrate his physical fitness, he arranged for several campaign appearances near election day. The first was to take him to New York City, where he would go on a fifty-mile swing around several of the boroughs in an open car on the morning of October 21. On any day this would have been a difficult journey for a man in FDR's condition. He was suffering now from hypertension, cardiac disease, a swollen heart, a chronic cough, and weight loss; he'd lost so much weight a friend said you could get your fist between his neck and his collar. And of course it was October and New York in October can be cold, sometimes very, very cold.

As it happened this October day was not only very, very cold, it was rainy and it was windy. At times the winds reached up to fifty miles an hour. But FDR wouldn't cancel and he couldn't very well ride around with the top up. That would raise enormous questions. So he went through with the ride, traveling in that open car with the cold and the rain and the wind beating against his tired old face for four long hours, though it damn well near killed him.[2]

Woodrow Wilson's widow happened to be in attendance at his inauguration. She took one look at FDR and knew. "He looks exactly as my husband did when he went into his decline." Hearing this Frances Perkins, the secretary of labor (and the first woman cabinet officer), told her, "Don't say that to another soul. He has a great and terrible job to do and he's got to do it even if it kills him."

His death three months later, on the very eve of the war's end, shocked the country and his family. It seemed impossible that FDR had not been able to cheat death. He himself may have thought he could.

The presidents who came after lived in his shadow. It made governing for all of them more difficult. None could compare.[3]

18. The Cold War

How the Cold War forced Harry Truman and Dwight Eisenhower to take new and extraordinary steps to gain power and to keep it

Harry Truman took over from FDR just as one world was ending and a new one beginning. No one had any confidence that he was up to the task. Not even Harry Truman.

His was the usual midwestern farm boy's story in a way. Raised on a small Missouri farm, he learned good sturdy values, became exceptionally responsible, and worked hard at whatever he was doing. And like many farm boys, he dreamed of getting away from the farm, moving to the city to escape.[1]

But his story was different, too. Because he had poor eyesight, because he loved to read, because he wasn't physically imposing, he always felt different from his peers. And he *was* different. His dreams were different. While he, like them, dreamed of opening a store or lighting out West to make a quick buck as a land speculator, he also dreamed of becoming a politician. As a boy he became fascinated by Woodrow Wilson. In his thirties he began expressing a desire to someday be elected governor or president. It was just in jest, he assured his friends. But it wasn't. He really dreamed of someday being elected governor, being elected president.

There didn't seem at the outset much chance of either of those developments taking place. For Truman wasn't a very self-confident young man. He was shy. He was reluctant to leave the family farm. And not until his father finally died, when Truman was thirty, did he dare. But his father's death was a turning point. The following year Truman campaigned to be named postmaster in Grandview, where the family farm was located. After winning the job he then got himself named a county road supervisor, where he oversaw the building of new roads. Two minor posts, but at least he finally was moving out of the family orbit. Over the next couple of years he was to try for every opportunity that came his way. One month he went down to Texas to see about buying some land. Another he considered becoming a homesteader in Montana. On still another occasion he became involved in a mining operation.

He never managed to turn any of these ventures into a successful career, but what was interesting about Truman was that he kept trying. There simply was no getting him down. He was driven. Somehow, some way, he was going to succeed.

There was a romantic reason for trying so hard. Truman had fallen in love with Bess Wallace, but when he had asked for her hand in marriage, she turned him down. Truman figured it was because he hadn't made anything of himself. So he kept on working at being a success and also kept working on Bess. Week in, week out, he'd trudge over to her place at 219 Delaware Street in Independence to court her.

He finally found a career at which he could excel in 1917, when he turned thirty-three: It was war. In the army he was given charge of a notoriously difficult regiment, Kansas City's Battery D, which already had gone through two captains. Truman quickly won over the two-hundred-man force and whipped them into shape.

He emerged from the army a far more self-confident person than he had been when he went in. His first step was to open a haberdashery with a friend. His second was to ask Bess again to marry him. Now that he had prospects, she did. It was a measure of both his love for her and his intense drive that he had kept asking. She had first turned him down in 1911. It was now 1919.

The haberdashery failed after two years when the economy slipped into a brief recession, loading Truman down with thousands of dollars in debts. But it was a fortuitous turn of events. It gave him the opportunity to get into politics. Just as the store was going bankrupt, a friend affiliated with the local machine came in one day to see if Truman wanted to run for county commissioner (formally, county judge). Truman immediately accepted.

Truman didn't really fit into the prevailing culture of local Missouri politics. He was honest, and politics was corrupt. But Truman wasn't so honest that he was unwilling to cooperate with the machine. As long as the machine didn't ask him to break the law, he and his friends in the machine got along fine. He was always willing to compromise his standards when it came to his political friends. You couldn't survive in Missouri politics if you didn't show a little moral flexibility. In Missouri the price of power was having friends who might one day end up in jail.

Boss Tom Pendergast could afford to let Truman be honest because he had another commissioner in his pocket who was willing to dole out corrupt contracts to the machine's friends. (Each of the county's three commissioners held a different portfolio.) But because Truman *was* honest his political prospects weren't very bright. Pendergast didn't want Truman in any powerful positions. And for ten years Truman never had the opportunity to rise any higher in the government than county commissioner. His chance finally came in 1934, however, when Pendergast couldn't get any of his favorite politicians to run for a seat in the U.S. Senate. Five turned him down. Truman applied and was accepted. With the machine's backing, he won easily.

He spent the next decade in the Senate, but it wasn't until he was well into his sixth year, in 1941, that he finally came into his own as the powerful head of the Truman Committee. This was a committee, largely of Truman's own creation, whose job was to ferret out corrupt defense contracts. It was an ironic role for Truman. At the same time that he was vigorously exposing corruption in Washington, his old pals in Missouri, the same pals who got him elected, the same pals he was regularly helping with small favors—getting a boy into Annapolis, helping a company gain access to the bureaucracy—were busily stealing the local government blind. Recently, Boss Pendergast himself had been hauled into court for tax evasion. But Truman performed masterfully and honestly, eventually saving taxpayers, at least according to his own press statements, more than fifteen billion dollars.

In early 1944 he had the chance to be named Franklin Roosevelt's running mate. He passed. Being vice president seemed a quick path to nowhere. In the Senate he had power. Vice presidents then had no power. He'd just be another of Franklin Roosevelt's flunkies. He didn't realize that FDR was near death. He'd only seen FDR in person a few times. And nobody who knew of FDR's precarious physical condition bothered to inform Truman. Besides, Bess didn't approve of the idea of his getting onto the national ticket. Being vice president would put him—and the family—in the media spotlight.

Key party leaders pressured Truman repeatedly over the six month period leading up to the national convention to accept the post. For six months he begged off. Just a few days before the delegates were scheduled to vote on the nomination, FDR himself finally called. Until then

FDR had left a confusing trail of signals as to who he wanted as a running mate in order not to alienate any of the leaders who were lobbying for the post. It had never been clear to Truman therefore that the president really wanted him. It seemed that it was just the bosses who wanted him. Now that he heard directly from Roosevelt, he immediately accepted. In 1944, at the height of the war, it was simply impossible to turn down a direct request by the president of the United States.

When FDR suddenly died in April 1945, just a few months into his fourth term, Harry turned to Eleanor to ask what he could do to help. *Mr. President,* Eleanor replied, *it's you who need the help now.* Harry confessed that he felt like the sun, the moon, and the stars had fallen on him. *If you've ever prayed for anybody, boys,* he implored some reporters, *pray for me now.*

At first he showed that he was obviously in over his head. While he had vast experience in domestic politics and knew domestic issues as well as anybody, he was ignorant of foreign affairs. FDR hadn't even bothered to tell him about the Manhattan Project. Compounding his difficulties, he was impulsive, overly prickly, obviously insecure. Tired and tense, he dashed off a letter to a music critic who had found fault with his daughter, Margaret's piano playing. Truman indecorously wrote that if he weren't president he'd punch the fellow in the nose.

Determined to prove he was up to the job, he bragged about his decisiveness, but he was often a little *too* decisive. An undersecretary of state remembered bringing him fourteen separate questions that needed to be decided, most relating to serious matters. Truman took fifteen minutes to dispose of them all. As historian Daniel Yergin has pointed out, it was an example of decisiveness, but it meant that the president had disposed of each problem in sixty-four seconds and "64 seconds per problem is not necessarily the way Americans would want important matters decided."

In over his head or not, he had to make the big decisions. As he loved to say, the buck stopped with him. And there were many many big decisions to be made, some just awfully hard.

The first of these was whether to drop the bomb on Hiroshima. Truman afterward claimed he simply made the decision and that was that: *Never lost a single night's sleep over it.* But it was a complicated decision, morally complicated, strategically complicated. If—and it was a big if—the bomb worked, and if—and this too was a big if—it was a fact that *only* the bomb would bring about a speedy end to the war, it was still morally unclear that the president should agree to use the weapon. It would after all destroy an entire city filled with civilians. Already during the war the Allies had firebombed many cities, including Dresden, which had cost the lives of hundreds of thousands of civilians, so the moral line had been crossed. But the decision to kill civilians still remained awful and difficult. And it wasn't even clear that the bomb was needed.

While some military advisers said it was and estimated that it would save the lives of tens of thousands of American GIs—the soldiers who probably would have to die in the series of battles needed to storm and take the Japanese home islands—these were after all just estimates. It was always possible, as a few maintained (and as the U.S. Strategic Bombing Survey was to conclude after the war), that the Japanese would be willing to surrender without making a last-ditch fight. There was some reason to believe that if the United States agreed in advance to retain the Japanese emperor, Hirohito, the Japanese might very well consent to a negotiated settlement (as diplomatic intermediaries were suggesting and as the United States itself concluded on the basis of intercepted Japanese messages in July, several weeks *before* Hiroshima).

Of course a negotiated settlement was not what the country expected. FDR had promised that the United States would accept nothing less than unconditional surrender. But there was nothing to stop the United States from indicating through diplomatic back channels that it would agree to the retention of Hirohito, as in fact it subsequently agreed to do. And anyway, was honoring FDR's promise worth the lives of hundreds of thousands of Japanese civilians?

As Truman weighed these factors he also had to consider one other: that the bomb was needed not only to win the war against the Japanese but to intimidate the Soviets, who were already showing signs that they intended to make an aggressive grab for several of the countries of Eastern Europe. As Truman confessed in his diary on May 15, 1945, as he was making his decision, the bomb would be America's "master card" in the struggle with the Communists for dominance in the post-war world.

The bomb did what it was supposed to. Stopped the war and frightened the Communists. It also did one other thing: It transformed the presidency.

With the power of the bomb at his fingertips the president of the United States instantly became the most powerful person in the world, more powerful than anybody in human history had ever been. More powerful than the Founding Fathers had ever imagined that a president should be.

No one but a handful of scientists understood the bomb, knew how it worked or what it really was all about, but its power was undoubted. When Hiroshima was bombed, Truman, trying to make sense of it for the American people, said that the "bomb had more power than 20,000 tons of TNT . . . more than two thousand times the blast power of the British 'Grand Slam' which is the largest bomb ever used in the history of warfare."

It gave Truman enormous control over events, more control than probably any president had ever had over anything; what president had, after all, been able to end a world war instantly? But deciding whether to use it involved terribly difficult moral decisions, decisions no American president had ever been called on to make. In effect, it gave Truman the power to play God.

Most Americans were persuaded that Truman on this occasion played God well, but that he was playing God was dismaying. Presidents were supposed to make their decisions on the basis of an informed public debate. Here was an instance in which a president made a profoundly important decision without any public debate at all. Given the circumstances, it couldn't be avoided. But it showed just how vastly more complicated the world was becoming.[2]

The second great decision of his presidency was what to do about the Soviet Union. This one too was complicated. It was quite clear by 1947, 1948 that the Soviets intended to use the presence of their army in Eastern Europe to control the governments there. Although in 1945 they had allowed free elections in several countries—in Hungary, Bulgaria, Czechoslovakia, and Austria—they subsequently began to clamp down and clamp down hard, imposing Communist governments and routing democrats, systematically killing off anti-Communist leaders. In 1947 the Soviets had subsequently made moves on Turkey and Greece, prompting Truman to issue his famous promise that the United States would come to the rescue of any European country threatened by Communist subversion—the Truman Doctrine. By 1948 the wartime alliance between the Soviets and the United States had completely broken down.

But it did not appear to most government analysts that war was likely; the Soviets simply did not seem strong enough to stage a war. That didn't mean that they wouldn't keep pushing their interests and pushing them hard, just short of war. As George F. Kennan, the State

Department's senior policy-planning analyst, explained, the West could expect the Soviets to be extremely aggressive, to press and prod wherever they spotted weakness. To beat the Soviets at this game, Kennan said, the West—and the United States in particular—would have to vigorously contain the Soviets, which would probably engage the United States in a struggle that could last decades and decades.

Kennan privately had doubts that Americans could keep up the struggle; as he well knew democracies usually lose their focus after a short while. And anyway Americans by tradition weren't very interested in foreign policy. And they certainly had little interest in maintaining the kind of gigantic military establishment that would be needed for containment. As they had after World War I, they had pressed the Congress to demobilize and demobilize quickly as soon as the war ended, and Congress had, cutting back on the military budget and ending the draft. To be sure, because of what had happened after the First World War, Americans were willing to embrace the idea of collective security and enthusiastically backed the creation of the United Nations. But they remained deeply apprehensive about taking a leading role in world affairs, preferring to focus on things domestic: getting a good education via the GI Bill of Rights, getting a good job, moving to the suburbs.

So Truman faced a difficult task.

There really was only one way for Harry Truman to persuade Americans to take on the burden he felt they had to and that was to scare the hell out of them. To persuade them the commies were as bad as Nazis, worse even, because at least the Nazis didn't try to pretend they were your friends as Commies did. All around the world the Commies were telling people, average people, they were on *their* side. The Commies put on such a convincing demonstration that even some Americans fell for their line and they did this even knowing, as by now people were beginning to know, how utterly ruthless Stalin was.

Truman wasn't the first president to feel he had to scare hell out of the country. Woodrow Wilson, after all, had felt *he* had to after World War I in order to get the League of Nations approved. But Wilson had only needed to create a big-enough scare to get a single piece of legislation passed, legislation that was unlikely to affect people in any direct way. Truman, by contrast, needed to instill enough fear in Americans for them to approve a gigantic new commitment to internationalism, a commitment totally at odds with the country's deepest traditions. We wouldn't merely be joining an international debate club, as Wilson had wanted.

We would be taking charge of the Free World's security, which would entail huge financial costs, and risk transforming the United States into a giant militaristic society.

FDR, to be sure, had had to confront the forces of isolationism, too. But he hadn't had to manufacture the threat the enemy posed. It was obvious. In Europe Hitler was taking over one country after another after another by military force. In Asia the Japanese were conquering huge parts of China by force.

Truman, by contrast, didn't have nearly so easy a task. For while the Soviets were clearly undermining the elected governments in Eastern Europe, they weren't conquering them in the blatant way Hitler had. And they certainly hadn't yet shown any interest in rolling tanks into France or dropping bombs on Great Britain.

Complicating matters, his poll numbers were terrible, just 35 percent of Americans approving of his presidency in the spring of 1948. And Congress was in the hands of the Republicans; in the off-year elections of 1946 they had swept out the Democrats in both houses, the first time in a generation that the Republicans had taken over.

So he had no clout. No capital to draw on. Hell, he hadn't even been elected to the job. His own mother-in-law felt he wasn't up to it and went around telling people so.

If Truman was going to succeed, he would really have to do a number on the country. It would be downright ugly, requiring the government, in effect, to practice the demonic arts of propaganda. To defeat the enemy we would have to become like him in a way.

But what choice did Truman have? As he saw it, none at all. Not if he were to remain in control of events. And he very much wanted to. Not just for the noble reason that he believed what he was doing was for the good of the country. And he very clearly believed it was. But for a personal political reason. In November he'd be up for election. Like Teddy Roosevelt before him, he very much wanted to win the presidency in his own right. Also like Teddy, he felt he needed to show that he was in charge and that he had accomplished something. Something big. Something to make people realize he wasn't a little man but a big one.

But how *do* you scare hell out of the country? Truman decided that going all out was the only way to catch people's attention. So he went all out: Told the people that the country was facing a terrible, terrible crisis—maybe even facing war. "War" was the word to use, he knew. As White House adviser Clark Clifford had written to Truman in a long, influential memo the year before, in 1947, "the worse matters get, up to a fairly certain point—real danger of imminent war—the more is there a sense of crisis. In times of crisis, the American citizen tends to back up his president."

So there was to be a war scare. It came in March 1948, and it was totally contrived. There *was* no threat of war and no evidence that the Soviets were planning on it. No evidence of dangerous Soviet troop movements. No evidence that the Soviets were even thinking of waging war. The only new developments were in Finland and Czechoslovakia, and neither was shocking.

Finland had signed an agreement with the Soviets that guaranteed its independence—if anything, a positive development. Of course, any agreement with the Soviets could be regarded as suspicious, but this one happened not to be; by signing, Finland was not indicating in any way that it was suddenly becoming a Soviet satellite (and few in the U.S. government thought that it was).

In Czechoslovakia the Soviets were taking over, but there was really little new in this. As the U.S. ambassador in Prague explained in cables to Washington, the Soviets were not suddenly beginning to exercise control over the country—they already *had* de facto control of almost all of it: By the beginning of 1948 Soviet-backed Communists held nearly every portfolio in the government.

If there were to be a war, Gen. Lucius Clay, the U.S. commander of military forces in Europe and the military governor of Germany, was probably in the most vulnerable position in Europe. If the Soviets were willing to take substantial risks, they could overrun his forces quickly. If anyone was to be worried about war it should have been him. And he wasn't. On March 1 he wrote Sen. Henry Cabot Lodge, Jr., that things were peaceful: "I believe American personnel are as secure here [in Berlin] as they would be at home. . . . Probably no occupation force ever lived under as secure conditions and with greater freedom from serious incidents than do the American forces living in Germany."

Curiously, however, Clay wrote another letter (in the form of a telegram) that same day, which seemed to leave an entirely different impression. This letter was sent to Gen. Stephen Chamberlin, the director of army intelligence. "For many months, based on logical analysis," General Clay began, "I have felt and held that war was unlikely for at least ten years. Within the last few weeks, I have felt a subtle change in Soviet attitude which I cannot define but which now gives me a feeling that it may come with dramatic suddenness." He admitted that while he could not point to any data to support his belief, there was "a feeling of a new tenseness in every Soviet individual with whom we have official relations."

Clay was telling the truth in the first letter, lying in the second. He lied because he had been asked to lie, as he virtually admitted years later. In February, General Chamberlin had come to see him in Berlin. "He told me that the Army was having trouble getting the draft reinstituted, and they needed a strong message from me that they could use in congressional testimony. So I wrote out this cable. I sent it directly to Chamberlin and told him to use it as he saw fit."

Actually there was more to the affair than Clay knew. The military wanted more than the draft, it wanted the defense budget increased, wanted the airline industry saved (because of defense cutbacks the industry was in sharp decline, many companies on the verge of bankruptcy). And it wasn't just the military that was involved. So was the State Department. State wanted the pressure of the threat of war to force Congress to pass the Marshall Plan, which had been proposed in 1947 and still hadn't been approved.

President Truman had concerns of his own. In the fall he'd be up for election. And as of the moment he had little to go to the country to brag about. Inflation was getting bad. Strikes were rampant. And he still seemed to many Americans to be too small for the job he was in, too much like them. So he badly needed *something*. Something to rally people around the administration. And as Clark Clifford had explained in his memo, getting into a dogfight with the Commies could help. "There is considerable political advantage to the Administration," Clifford wrote, "in its battle with the Kremlin."

It was a dicey move, to ring the bell of war when there was no real risk of war. Ring it too loudly and you could, by raising Communist fears, actually bring on an actual war. And if the Congress and the country really began to believe that war was likely, they might conclude there wasn't enough money in the budget to finance the Marshall Plan *and* the military so the Marshall Plan, Truman's pet project, would probably be cut back or scrapped completely.

And of course it was another of those morally sticky situations. While a lot of good reasons could be given to defend the war scare, it was in the end a ploy, an extremely manipulative ploy—and wholly unprecedented. Never before had a president rung the bell of war in the absence of war. It was, of course, an extremely effective way to take control of events. But once the administration had begun to ring the bell there was the danger that others would start ringing it, too, and then keep ringing it for their own reasons. And there might be no way to stop it from ringing. Even if war itself did not come about, other events associated with war might, such as the repression of civil liberties.

And soon civil liberties *were* being repressed. The war scare could not be held solely responsible for the repression that was to come, for McCarthyism, for the hearings held by the House Un-American Activities Committee, for the Red-baiting that was to help Richard Nixon win a seat in the United States Senate in 1950. These were the result of the general Cold War crisis, not the war scare of 1948. In March 1947, one full year earlier, President Truman had signed the nation's first Loyalty Act, which authorized investigators to crack down on suspected subversives working for the federal government. But the war scare would contribute to the repression, further legitimize it.

The climax of the war scare came on Saint Patrick's Day, 1948, when President Truman addressed a joint session of Congress. To thunderous applause he demanded action to defend the United States from unnamed countries that, by their actions, had showed that they were utterly untrustworthy. The president did not say war was imminent; he didn't have to. Merely to address a joint session of Congress was to underline the seriousness of the situation. And of course the speech went over beautifully. By shrewdly wrapping himself in the flag, his cause became the American cause. To be against Truman was almost to be against the country.

Having raised the specter of war the administration quickly backpedaled, fearful that the country might actually think war *was* imminent. But officials handled things just right. By the end of the year the Selective Service System had been reinstituted, the military budget had been increased dramatically (the budget for airplane production alone upped by a third), and the Marshall Plan was approved. A stunning string of successes, especially given Truman's low standing in the polls and the general hostility of the Republican Congress to Truman and the Democrats.

In the fall Truman won his bid to be elected president in his own right. It wasn't only because of the war scare or even mainly because of it that he won. There were many many factors: the Cold War crisis atmosphere in general, Truman's feisty performance on the campaign trail, the willingness of the Democrats to run against Herbert Hoover yet again, the lackluster performance of the overconfident Republican candidate, Thomas Dewey. But the war scare helped.

Truman would face, in his remaining years in office, many crises. In 1949 the Soviets would explode their first atomic bomb, three years earlier than American experts had predicted, which was a shock, and led to a new round of Red-baiting. The government was not, as it happened, filled with spies, but there *were* spies, and they did in fact help the Soviets steal America's technical plans for the bomb. Late in 1949 China turned Communist. Then in 1950 North Korea invaded South Korea.

As events crashed down on his head, one after another, like avalanches, Truman struggled to avoid being buried and was barely able to. The Soviets said history was on their side, and sometimes it almost seemed that it was. Which prompted Truman, as ambitious as ever to control events, to take extraordinary measures. To back the use of Mafia and Corsican thugs to break Communist-led strikes in France. To allow former Nazis into the United States to undertake rocket research. To wage war in Korea without first obtaining a declaration of war from Congress—all in all an astounding record, but from Truman's perspective, totally defensible.[3]

The Cold War presidents who followed Harry Truman faced situations every bit as difficult as he did. They weren't the same situations. For the nature of the conflict with the Communists changed as people slowly realized that Communism wasn't monolithic, that the Soviets and the Chinese did not coordinate their military strategy, that both were often driven more by nationalism than ideology. And it particularly changed after the Soviets began building bombs at a rapid clip. Then the balance of power shifted considerably; once both sides had nuclear weapons, using them became unthinkable. If we tried to blow them up, they could blow us up, which in a way contributed to international stability. But the situations remained similar and drove presidents to use extraordinary means—just as Truman had—to gain control.

Dwight Eisenhower arrived on the national political stage a far more powerful figure than Truman did. He after all had orchestrated the victory in Europe, including the most sensational of all our battlefield victories, the invasion of Normandy. Even *he* had to take extraordinary measures to gain control.

Like Truman, Ike was a midwestern farm boy. And like Truman, he flowered in the army, using appointment to West Point to work his way quickly up the military ladder. He didn't crush his competitors on the way up. He didn't need to. He was so bright, so quick, so easygoing, that his superiors naturally advanced him quickly, promoting Ike to captain. Then, however, his career stalled. Because there weren't any wars to fight, the army stopped handing out promotions. So Ike returned to school, enrolling in the Army General Staff School in Leavenworth, Kansas, to get the advantage of an advanced degree in the expectation that that might somehow help bring him attention. He graduated first in his class.

He then used his personal charm to secure a string of appointments with the most powerful officers in the army, early on becoming a top aide to Douglas MacArthur. (Ike's specialty: speech writing, which was ironic in light of his own celebrated inarticulateness. Actually, Ike could be highly articulate when he wanted to. It's just that it often suited his purposes to be inarticulate, as on occasions when he wanted to obfuscate his position.)

When the United States went to war against Hitler, George Marshall was expected to take over American operations in Europe. But Roosevelt insisted on keeping Marshall in Washington. And Marshall picked Ike.

It turned out to be one of the best decisions Marshall ever made. For Eisenhower not only had a strong strategic sense but enormous political skills. He alone among American commanders was capable of controlling the cocky British field marshal Bernard Montgomery. "I have his personal equation," Ike told Marshall, "and have no lack of confidence in my ability to handle him."

It was a measure of his enormous ambition that he was willing to put up with Montgomery's foolishness, his incessant squabbling. For Ike, despite his charming demeanor, had an enormous temper. Friends referred to it as a Bessemer furnace, one aide reporting that Ike blew up every fifteen minutes or so. But losing his temper with Montgomery would have been counterproductive, turning a potential ally into a certain enemy. So Ike simply made sure he never lost his temper with Montgomery.

After the war nearly everybody* wanted him to run for president, both the Democrats and the Republicans. Ike was deluded into thinking he might get both parties' support. When it became clear he wouldn't, he became a Republican, naively expecting in 1952 to be nominated by acclamation. Unfortunately, between him and his goal stood Robert Taft. And Taft was a huge presence in the party. Son of a president and a man who himself expected to be president. And a hero of the conservatives.

---

*Everybody, that is, except for his son and a brother. They were against his running. Ike ignored their objections.

Ike took him on, portraying Taft to the world as a political dinosaur.

The general's most effective blow came as the convention was getting underway. Taft's people, worried that Taft didn't come across well on television, demanded that television coverage of the convention be limited. Ike let out a public wail that got the attention of the whole country. *Bob Taft,* said the general who had defended freedom in Europe in World War II, *is limiting the freedom of Americans here at home. That isn't right. The cameras should be allowed in to broadcast everything so that the American people can see the candidates for themselves and decide who they favor.* It wasn't a fair attack. It overstated things. But Ike went with it.

During the high-pressure days that followed Ike authorized his managers to offer key Republicans the control of patronage in their states in exchange for their support. He even promised to name the governor of Pennsylvania to the cabinet if the governor helped firm up support in the state's delegation. Taft also played hard, also made patronage promises. But Ike played harder. Ike won.

The biggest headache Eisenhower faced in the general election was what to do about Joseph McCarthy, the Republican senator from Wisconsin. Ike detested McCarthy. In his diary, in an entry on March 13, 1951, he had noted that McCarthy reportedly was digging up "dirt with which to smear me if I run for president." But in 1952 McCarthy was at the height of his power, and Ike didn't want to alienate him. Nothing McCarthy did could get Ike to change his mind, not even when the senator took on George Marshall, questioning Marshall's patriotism. Ike let the attack on his mentor go unchallenged. A short time afterward Eisenhower even agreed to campaign with McCarthy. It was a morally offensive thing to do and disappointed a lot of people. But Ike wanted to win. If he had to hold hands with Joe McCarthy, he'd do *that* to win.[4]

The marvelous thing about Eisenhower was that he could get away with it because he had been a hero and because he had that great, winning smile. The hero status, the big grin, hid who he really was, hid his ambition, hid his drive. And yet it was his drive that was his most important quality.

Three months after he became president came a day on which he demonstrated his drive amply. It was Thursday, April 16, 1953, and Ike was in Georgia on a little golfing vacation. But he had to interrupt his respite to give a speech and appear at some events. It was to be a terribly trying day. First he had to fly up to Washington, then give his speech, then take part in a limousine caravan, then throw out the first ball at the season-opening game of the American League, then fly down to Charlotte, North Carolina, then make an eighty-mile drive to Salisbury to attend a county celebration and make another speech, then finally fly back to Georgia.

That morning, however, he woke up in terrible pain. Nothing the doctor did could relieve the pain. On the flight up north it got worse and worse. But Ike went through with his schedule pretending nothing was wrong, even though when he gave his first speech he had to hold the lectern to keep from falling over. At the end of the day, back in Georgia, he collapsed, his wife, Mamie worrying that he "really was sick—we were really concerned about him." She was right to be concerned. It turned out he'd probably suffered a heart attack that morning.

Why hadn't he simply stayed put once he'd realized he was ill? He had a variety of reasons. He didn't want people thinking he was too old to be president. He didn't want people speculating about his ability to finish his term. And he very much wanted to give that first speech. It was a big speech, to be broadcast live nationally, and he was going to use it to settle once and for all a nasty turf battle between himself and Secretary of State John Foster Dulles over the control and direction of the government's foreign policy. Ike wanted to commit the United States to a policy of world disarmament, negotiation, and peace, and Dulles did not. Just a few days earlier Ike had exploded in rage at Dulles's opposition. "All right, then," he had said. "If Mr. Dulles and all his sophisticated advisers really mean they can *not* talk peace seriously, then I am in the wrong pew. For if it's *war* we should be talking about, I *know* the people to give me advice on that—and they're not in the State Department. Now either we cut out all this fooling around and make a serious bid for peace—or we forget the whole thing."

So he gave the speech and went on with all the other events as well. But he was wretched all day long. In his memoirs he recalled that that day was "one of the most miserable" of his entire life.

At the end of the day his press spokesman told the media that Ike was suffering from food poisoning. Nobody really knew what he was suffering from. The heart attack diagnosis wasn't made until years later. But the very fact that Ike sent out his spokesman to downplay his health problems was telling. Ike was perfectly willing to mislead people about his health if need be. In 1949 he'd suffered a heart attack and never revealed it; he knew it might very well prevent his being elected president.[5]

He succeeded in quickly ending the Korean War, but the Cold War was a source of continuing crises. And just as Truman had, Ike stretched his presidential authority in an attempt to gain control over them, perhaps most spectacularly in 1958. That year, without congressional authorization, he rushed fourteen thousand marines to Lebanon to forestall a Communist takeover. Later, when Fidel Castro came to power in Cuba, the administration had the CIA consider ways to assassinate him.

After Eisenhower come the presidents on our side of the line that divides American history, that imaginary and misleading line that supposedly divides the unambitious presidents from the ambitious ones. Because the media have so relentlessly worked over the ambitiousness of these leaders it is unnecessary to do so here. But it may be worthwhile to sketch in some of the larger forces at work in the society, which shaped the way these men exercised power.

First and most important of these forces was, until the time of Bill Clinton, the Cold War.

John Kennedy was committed to fighting the Cold War ruthlessly. It was he who created the Green Berets in an attempt to deter the Communist revolutions then breaking out throughout the Third World. Repeatedly during his brief three years in office he resorted to whatever means he thought necessary to defeat Communism. In 1961 he secretly authorized a rogue force of Cuban exiles trained by the American military to invade Cuba and topple Fidel Castro. In 1962 he engaged in a high-stakes poker game with the Soviet Union in Cuba that could have ended in nuclear war; in the end he happened to choose the right policy, but for thirteen days, thirteen frightening days for those who knew what was going on, Kennedy played God as no president before him had ever had to, not even Truman, deciding on a policy that could have ended in nuclear holocaust without so much as a single town meeting to let people know how their fate was being decided. Like Truman, Kennedy played God well, it was generally believed, but who wanted a president playing God at all?

It wasn't always clear that the presidents themselves knew what their own administrations were up to, but that was part of the legacy of the Cold War. To win it the government had to become so big, the military-industrial complex so . . . complex, that presidents no longer could know all that was being done in their name. To this day it is still disputed whether Kennedy knew about the army and CIA assassination plots against Castro—the use of Mafia hit men, poison pens, even a poisoned scuba suit (which was to be given to Castro as a gift). Probably he knew something of these plans; recently declassified documents show that in the spring of 1962 his own Joint Chiefs of Staff endorsed as "suitable for planning purposes" a broad range of assassination schemes in addition to other "dirty tricks" intended to undermine the Cuban government. (One particularly heinous plot, nicknamed Operation Bingo, involved staging a fake attack on the American forces stationed at Guantanamo Bay to give the United States an excuse for launching a military assault on Havana.) It's likely that if he did not know, it was because he did not want to know. Presidents during the Cold War were careful to maintain "plausible deniability," so they wouldn't have to lie blatantly about their involvement in nefarious undertakings in the event that these ever became public.[6]

Lyndon Johnson by nature was untroubled by lying. As his biographer Robert Caro has amply demonstrated, he almost enjoyed it, and he'd engaged in prevarication for decades; in college he'd been known as "Bullshit Johnson" because of his lies. If it suited his purposes, suited his ambitions, he lied. So he probably would have lied as president even without Vietnam. But Vietnam turned him into a master liar, the greatest liar in American presidential history. Unable to win the war, unable to say for sure it could be won, but damned eager to persuade Americans that he had adopted the right policy, the prudent policy, he began to lie about our prospects, and to lie so often and to such an extent that eventually no one knew when he was telling the truth.

When he and the military leaders proclaimed the Tet Offensive in 1968 a defeat for the Vietnamese Communists, even though the Vietcong had actually been able to penetrate Saigon's defenses and had even broken through to the American Embassy there, he lost all credibility. It might have been true, as some historians insist today, that Tet was a defeat for the Vietcong, but by then nobody could trust Lyndon

Johnson to tell the truth even if he were telling it. The "credibility gap" was simply too vast to be bridged.[7]

Just as Vietnam was a result of the Cold War, so was Watergate in a way. It wasn't that Vietnam and Watergate were inevitable. They weren't. To a very great extent they reflected the weaknesses of the particular men who happened to be serving as president at the time. Without a Johnson, there might not have been a Vietnam; without a Nixon, there almost certainly wouldn't have been a Watergate. But both disasters were directly linked to the Cold War. Without the Cold War neither Vietnam nor Watergate would have been possible.

Nixon, like Johnson, of course, was seriously flawed, a politician who always looked like a cartoonist's idea of a politician—shifty eyes, perspiring lip, insincere smile. An introvert in an extrovert's profession, but so ambitious to be president that he was willing to reinvent himself, to pretend to be an extrovert. He was who he was, calculating and shrewd and untrustworthy, willing to fire off the most sensational accusations if it would help his career: that Jerry Voorhis, his opponent in his first race for Congress, voted the Communist line; that Helen Gahagan Douglas, his opponent in his race for the Senate, was the "pink lady" of American politics; that Truman was a traitor to his party.[8]

So Nixon was disposed to political deviltry. In a way, it was his deviltry that made him almost inevitable in the Cold War. He seemed to reflect all the evil that the war brought out in Americans, the snooping into people's private lives, the leveling of vicious smears, the mindless belief that America was always right and everybody else always wrong. Without the Cold War a Richard Nixon would have been impossible. He lived off people's fears, knew how to exploit them as no other politician in the era did save Joe McCarthy (who lacked Nixon's intelligence and restraint).

It was considered remarkable that he was able to stage a comeback after losing both the presidency and the governorship of California. But the times favored him. After Vietnam went sour and the country began to come apart, demonstrators taking over the universities and even in a way the White House itself, shouting so loud and so often—*Hey, hey, hey, LBJ, how many kids did you kill today*—that Johnson, sitting in the Oval Office, could literally hear their chants, driving him nuts, Nixon suddenly became desirable, a symbol of harshness and order. He pretended to be smooth and almost soft, chattering endlessly about the need to bring the country together, but everybody knew what Nixon was: He was the man who would stop the nonsense, the tough son of a bitch who would throw the demonstrators in jail and take America back. Just as in the fifties he had been the respectable man's Joe McCarthy, in the sixties he was the respectable man's George Wallace.

The fact was the country *was* falling apart. Nixon was right about that. It wasn't just Vietnam. It was also the civil rights movement, which inspired blacks to demand power, which put a further strain on the system. More than at any other time in American history save for the Civil War, things were out of control. Nixon was determined to retake the streets, to retake control of America, and to do it any way he could. Which led to the plan to break into the office of the psychiatrist of Daniel Ellsberg, the antiwar intellectual who leaked the Pentagon Papers to the *New York Times*. Which led to the creation of the Plumbers, who were needed to plug damaging leaks in the government. Which led to the plan to firebomb the Brookings Institution, a bastion of liberalism. By the third year of the administration Nixon was on the defensive, convinced that people were out to get him: the Jews in the IRS, the liberals in the media, the Harvards. Which led to the "enemies list."

Nixon's response to the events unfolding in the country was, of course, morally offensive. But it was, given who he was and what was happening, not altogether incomprehensible. And given his ambitious drive to keep power, it was as inevitable as anything in history ever is.

That Nixon's men, in the summer of 1972, would think nothing of ordering a break-in at the Democratic national headquarters at Watergate was, given all this, hardly surprising or shocking. It was almost the natural outgrowth of all that had gone before.

Two years later, as the Nixon administration began falling apart along with the country, people began wondering why he simply didn't write off the Watergate burglars, letting them twist slowly in the wind, in John Ehrlichman's choice phrase. The answer was simple. He couldn't without worrying that all the other crimes would become public, the crimes that had been inspired by Vietnam, inspired by the Cold War. For some of the very same people who were caught breaking into Watergate had been used in the break-in of Ellsberg's psychiatrist's office. And the high officials who had authorized the Watergate break-in earlier had authorized the Ellsberg break-in. And they were the same people who had considered firebombing the Brookings Institution. So it was all connected. If one or two people went down, the whole ship would sink.

So Nixon covered up and covered up and covered up some more.

He tried every trick in the book to hang onto power, even tried using television's great force to hang on as he had used it in 1952 in the "Checkers speech." But television no longer was his friend. As Lyndon Johnson had discovered, it could work against you as easily as it could work for you. While it made the president human, brought him closer to the people, it also shone an unrelenting spotlight on his weaknesses. Once upon a time a president's weaknesses could be hidden. Now they no longer could. Now because of all the lies and tricks and deceitfulness journalists no longer trusted presidents. And after the presidents lost the journalists, they lost the people, too.

Things did not necessarily have to turn out this way. If there had been no Cold War, no bomb. If governing hadn't gotten so damned complicated that presidents felt driven to take extraordinary measures to maintain American security. If the world had just remained the simple place it once had been—then there wouldn't have been a Vietnam, wouldn't have been a Watergate. Presidents, ambitious as they were, wouldn't have done some of the outrageous things they had done. But there *was* a Cold War, there *was* the bomb. And they did do those things. And the country was never the same again.

Once the Cold War ended, controlling things did become easier for presidents. But that didn't necessarily mean they stopped being manipulative. Politics is a school for scandal, each generation of politicians passing onto the next the tricks of the trade. Once a trick was learned it was never forgotten.

Conclusion

At the outset of our era there had been no warning of what was to come. As John Kennedy looked out over the crowd that gathered for his inauguration on that bright clear cold day in January 1961, summoning citizens to ask not what their country could do for them but what they could do for their country, Americans wondered in awe at their great good fortune in their selection of presidents. Then, in short order, Lyndon Johnson lied about the Vietnam War, Richard Nixon covered up Watergate, and Kennedy was posthumously exposed as a philandering liar who'd possibly used mob connections to achieve victory in Illinois.

They and their successors had an astonishingly rich variety of flaws. But one failing in particular seemed common to nearly all of them: They were too damn ambitious, too willing to do anything to gain power, to keep power, and to get things done. If they had to overpromise to win, they overpromised. If they had to lie, most of them lied. If they had to manipulate, they manipulated. Democrat, Republican, liberal, conservative—it didn't matter. They all seemed to behave about the same way. By the 1990s Americans were fed up.

Cynicism did not come naturally to Americans. Americans are believers. Even after Vietnam and Watergate they desperately wanted to continue believing. When Ronald Reagan promised he could cut taxes, increase defense spending, and balance the budget simultaneously, they believed. David Stockman, Reagan's own budget wizard, told people it couldn't be done. They still believed. In 1984, after four years of Reagan, the facts were in: It could not be done. Still believing, they reelected Reagan anyway. Four years after that they elected George Bush, who promised, "Read my lips. No new taxes"—and believed *him*.

In the 1990s Americans stopped believing. It was a decisive turning point.

George Bush promised he wouldn't raise taxes. He raised them. Bill Clinton promised he would reform welfare without hurting the poor. He signed the Republican welfare bill even though his own people admitted it *would* hurt the poor.* Bush and Clinton hadn't gone beyond the others. Lyndon Johnson and Richard Nixon had both told bigger lies. But a generation of lying—and getting caught at lying by the media—had finally taken its toll. The jig was up.

Clearly something had gone wrong, but what, exactly? Some pundits believed that the moral character of the presidents had declined. This, surprisingly, was the view of Richard Nixon, whose own character was so often questioned. "You know," he commented to his assistant Monica Crowley in 1993, "politics has really gone down the tubes since 1960, and frankly it's not much fun to watch. Maybe it's because the characters involved are weak on either the personal or political side—or both. I don't know. I ran in '68 and '72, so I guess that applies to me too. It just seems that good candidates aren't running anymore. It's better not to run anyone at all than a turkey. The top offices in this country require better."[1]

Others took a broader view of things. It wasn't the moral character of the presidents that was declining, it was society. America was becoming corrupt, the feeling of community dying. In its place a selfish individualism was taking hold. In ruthlessly pursuing their own individual ambitions, the presidents were merely reflecting the rampaging individualism of the society at large.

When the historians were asked whether there really was a decline, they almost always said that there wasn't, that you could find examples of presidential duplicity and manipulativeness going back to the first days of the Republic. On the op-ed pages they reminded Americans of their presidents' not-so-noble history: how John Adams had used the Alien and Sedition Acts to go after his political enemies; how James K. Polk had lied about the circumstances leading up to the

---

*Dick Morris, Clinton's erstwhile media guru, told him if he signed the Republican welfare bill he'd win the 1996 election by fifteen points; if he refused, he'd lose by three. Hillary Clinton explained: "We have to do what we have to do, and hope our friends understand." Dick Morris, *Behind the Oval Office* (1997), p. 298.

Mexican War in order to fool the country into thinking the Mexicans had made an unprovoked attack; how Warren Harding had concealed malfeasance during the Teapot Dome scandal. All of which was true. And all of which pointed to ambitiousness as a continuing problem in American history. It seemed that some politicians had always been too ambitious for the country's good. If they wanted something badly enough, they lied or cheated to get it.

But the historians misled. Politics really *had* changed. Things weren't about the same as always. They *had* gotten worse. The fact was they had been getting worse all the time. If you looked carefully at American history you could see a clear pattern of decline. Instead of things getting better and better over time, as Americans liked to fantasize, they had gotten worse and worse, almost as if a reverse law of progress were in operation.

In the late 1790s presidents had begun playing politics with national security; in the early 1800s compromising their principles; in the 1820s bargaining for the presidency; in the 1830s packaging themselves like soap; in the 1840s lying; in the 1850s countenancing bribery; in the 1870s tolerating corruption; in the 1880s pandering to immigrants and stealing votes; in the 1890s manipulating the press; in the early 1900s exploiting their families; in the 1910s exploiting the fear of communism; in the 1930s abusing the Internal Revenue Service to go after their political enemies; in the 1940s creating war scares; in the 1950s regularly sending troops into war zones without congressional approval; in the 1960s plotting the assassination of foreign heads of state; in the 1970s concealing their own malfeasance; and in the 1980s manipulating public opinion through elaborately staged events for television.

To be sure, the pattern of decline wasn't relentlessly downward. Corrupt bosses no longer controlled the political process. Wall Street no longer dominated the parties. Vote stealing was now rare. And from time to time a truly moral figure had become president and conducted himself in a truly moral way, leading many to conclude that the moral tone of the presidency depended mainly on the character of the person who happened to be holding the office, a Nixon conducting himself badly, a Carter reasonably well. And to a certain extent, that had been the case.

But by focusing on the individual alone people had missed the underlying pattern of reverse progress evident in the way most presidents played politics—missed the fact that the system over time had become more and more politically promiscuous, ever more tolerant of a wider and wider range of unseemly presidential behavior.

What had driven politics downward? It wasn't that the character of the presidents had declined. It was more impersonal and complicated than that. It was history that was responsible, change—the stupendous transformation of the United States from the insignificant, small, largely rural, largely eastern-coastal country that once was, into the bulging colossus of fifty states in the throes of post-industrialism that had come to be. As the country became more complex, governing and politics also became more complex, making it harder for national politicians to acquire and maintain control.

It simply was harder to control a country of more than a quarter billion than one of four million, harder to control a nation made up of people of many backgrounds than those of mainly one, harder to control a nation that was urban than one that was rural. And the harder maintaining their control became, the more desperate presidents became, forcing them to take increasingly desperate measures to reassert their control.

First of the great changes to transform the nation was the birth of the two-party system. Then, in dizzying succession, came the suffrage explosion, manifest destiny, the crisis over slavery, industrial capitalism, machine politics, immigration, the revolution in mass communications, the arrival of the United States on the world stage, the Great Depression, the Cold War, the bomb, and television, among others. With each new change a new low was often reached, a threshold crossed that previous presidents wouldn't have dreamed of crossing, circumstances drawing them further and further away from the high standards set by George Washington.

We are taught to be suspicious of people who are ambitious for power. But without their ambition presidents could never face the forces of history head-on, wrestling with them for control, eventually adapting to them in certain ways, and sometimes modifying and influencing them. So ambition, while dangerous, is also essential.

We like to pretend that normal people should be elected president. People, that is, with a normal amount of ambition. But normal people don't have what it takes—that extraordinary drive to succeed.

Table of Presidents

1.	George Washington	1789–1797	no party
2.	John Adams	1797–1801	Federalist
3.	Thomas Jefferson	1801–1809	Democratic-Republican
4.	James Madison	1809–1817	Democratic-Republican
5.	James Monroe	1817–1825	Democratic-Republican
6.	John Quincy Adams	1825–1829	Democratic-Republican
7.	Andrew Jackson	1829–1837	Democrat
8.	Martin Van Buren	1837–1841	Democrat
9.	William Henry Harrison	1841	Whig
10.	John Tyler	1841–1845	Whig
11.	James K. Polk	1845–1849	Democrat
12.	Zachary Taylor	1849–1850	Whig
13.	Millard Fillmore	1850–1853	Whig
14.	Franklin Pierce	1853–1857	Democrat
15.	James Buchanan	1857–1861	Democrat
16.	Abraham Lincoln	1861–1865	Republican
17.	Andrew Johnson	1865–1869	Democrat (ran as Republican)
18.	U. S. Grant	1869–1877	Republican
19.	Rutherford B. Hayes	1877–1881	Republican
20.	James A. Garfield	1881	Republican
21.	Chester Arthur	1881–1885	Republican
22.	Grover Cleveland	1885–1889	Democrat

23.	Benjamin Harrison	1889–1893	Republican
24.	Grover Cleveland	1893–1897	Democrat
25.	William McKinley	1897–1901	Republican
26.	Theodore Roosevelt	1901–1909	Republican
27.	William Howard Taft	1909–1913	Republican
28.	Woodrow Wilson	1913–1921	Democrat
29.	Warren Harding	1921–1923	Republican
30.	Calvin Coolidge	1923–1929	Republican
31.	Herbert Hoover	1929–1933	Republican
32.	Franklin Roosevelt	1933–1945	Democrat
33.	Harry Truman	1945–1953	Democrat
34.	Dwight Eisenhower	1953–1961	Republican
35.	John F. Kennedy	1961–1963	Democrat
36.	Lyndon Johnson	1963–1969	Democrat
37.	Richard Nixon	1969–1974	Republican
38.	Gerald Ford	1974–1977	Republican
39.	Jimmy Carter	1977–1981	Democrat
40.	Ronald Reagan	1981–1989	Republican
41.	George Bush	1989–1993	Republican
42.	William Clinton	1993–	Democrat

Acknowledgments

If I had to say who was most responsible for me writing this book I'd have to say it was my grandparents, particularly Bess and Nat. For the two of them, two poor Jewish immigrants who came over on the boat in the late nineteenth century, the only history that mattered was American history. And after FDR became president, it was the American presidents who mattered most. And because Bess and Nat loved the presidents, my mother loved the presidents. And I, too. As a child I wanted to learn everything I could about them. From the time I was a junior in high school a giant framed picture of George Washington hung over my bed.

My friends also contributed. Bill Rorabaugh suggested that I focus on ambition. Scott Rensberger, a superb storyteller, so impressed upon me the importance of storytelling that I decided to build the book around stories. Jackson biographer Robert Remini kindly reviewed several chapters, helping save me from several embarrassing mistakes. Debbie Horne helped me think through the thousand and one knotty problems that cropped up during the years I worked on the book, patiently letting me talk on and on about them until finally things became clear.

Historian Bernie Weisberger, my college mentor, patiently went through the manuscript in detail, supplying me with page after page of helpful suggestions. When I finally went through them all, thinking that was the end of it, more arrived. And then, when I finished with those, pages and pages more. On the eve of a vacation in California he took the time to send me his final pages of notes.

Two editors worked with me on the book, each affecting it profoundly. Cynthia Barrett's genius is her ability to give an author just the right word of encouragement at just the right time. It was because of her kind praise that I agreed to take on what seemed like an overwhelmingly large topic. Simply put, without her help, there would have been no book.

Paul McCarthy, like Bernie, supplied me with reams of notes. Pages and pages of single-spaced, detailed notes. After critiquing a chapter as a whole, he would move on to critique each individual page, then each paragraph, then each sentence, then individual words. Often, he wasn't satisfied until he finally had got down to the nub, the syllable. Because of his thoroughness, because of his keen insights, this is a much better book than it otherwise would have been.

Most copy editors do not earn a mention in an author's acknowledgments. But Susan Llewellyn merits inclusion. Her work considerably improved this book.

Finally, I want to thank Bill McClure. For several years he has had to listen to me talk about the presidents, talk about them so much and so often that he now knows far more about the presidents than anybody should have to know except the historians. By helpfully reminding me that there is a world beyond the presidents, he saved me from the mistake of becoming obsessed with them (though I'm sure there were times when I did seem obsessed).

Notes

The chapters on Washington, Jackson, and Lincoln reflect my own research over the years in their personal and official papers. I am particularly familiar with Old Hickory's correspondence; in the 1970s I spend a pleasant year working as a low-level assistant researcher on the Jackson Papers Project, located at the Hermitage in Nashville. It was my duty to try to decipher Jackson's handwriting, which was a bad as his spelling. The chapters on the other presidents are based mainly on scholarly biographies and histories written over the last half century. But I also found myself turning frequently to James D. Richardson's *Messages and Papers of the Presidents* (1897–1917), an indispensable collection of the official papers of the presidents from Washington through Wilson. Of particular usefulness is the two-volume index, which amounts to an almanac of American history. Finally, I want to acknowledge the tremendous benefit I derived from reading the American Presidency Series published by the University Press of Kansas, which includes one-volume biographies of about two dozen presidents.

Prelude

1. Forrest McDonald, *The Presidency of George Washington* (1974), p. 27.
2. James Thomas Flexner, *George Washington and the New Nation* (1970), pp. 212, 245, 268; Ralph Adams Brown, *The Presidency of John Adams* (1975), p. 134.
3. Noble E. Cunningham, Jr., *The Presidency of James Monroe* (1996), pp. 22–23.
4. Flexner, *George Washington and the New Nation*, pp. 195–98, 203.

Chapter 1. In the Garden of Eden

1. James D. Richardson, ed., *Messages and Papers of the Presidents* (1897–1917), vol. 1, pp. 34–35.
2. Bernhard Knollenberg, *George Washington: The Virginia Period, 1732–1775* (1964), chap. 14.

3. Douglass Adair, *Fame and the Founding Fathers* (1974), chap. 1.

4. Garry Wills, *Cincinnatus* (1984), especially chap. 1.

5. Forrest McDonald, *The Presidency of George Washington* (1974), pp. 82–88.

6. McDonald, *Washington*, pp. 177, 184.

7. All believed that parties were evil. But some, including Madison, believed it was impossible to do away with them altogether. See *Congressional Quarterly, Selecting the President* (1996), p. 3.

8. On Washington's use of power see Wills, *Cincinnatus.*

9. Carl Anthony, *First Ladies* (1990), vol. 1, pp. 128, 147.

Chapter 2. The Birth of the Two-Party System

1. Ralph Adams Brown, *The Presidency of John Adams* (1975), pp. 14–19.

2. Federalists, who had been tipped off to the contents of the XYZ correspondence, were eager to see it opened for public perusal. "We wish much for the papers if they can with propriety be made public," wrote Boston Federalist Jonathan Mason. "The Jacobins want them. And in the name of God let them be gratified; it is not the first time they have wished for the means of their destruction." Quoted in Page Smith, *John Adams* (1962), vol. 2, p. 959.

3. Brown, *Adams*, pp. 157–58.

4. James Morton Smith, *Freedom's Fetters* (1956), p. 270; Brown, *Adams*, pp. 123ff.

5. Brown, *Adams*, p. 199.

6. Nathan Schachner, *Thomas Jefferson* (1951), vol. 2, chap. 46; Merrill D. Peterson, *Thomas Jefferson and the New Nation* (1970), pp. 651ff.

7. Wilfred E. Binkley, *The Man in the White House* (1958), pp. 100–101.

8. Ibid., pp. 99–102.

9. C. Vann Woodward, ed., *Responses of the Presidents to Charges of Misconduct* (1974), pp. 39–40.

10. Leonard Levy, *Jefferson and Civil Liberties* (1963), chap. 3.

11. Henry Adams, *History of the United States of America* (1889–91), vol. 1, chap. 14; vol. 2, chap. 2.

12. Peterson, *Thomas Jefferson and the New Nation,* pp. 748–86.

13. Woodward, *Responses of the Presidents,* pp. 37–39.

14. Irving G. Williams, *The Rise of the Vice Presidency* (1956), chap. 4.

Chapter 3. The Revolution in the Suffrage

1. Sam B. Smith, ed. *The Papers of Andrew Jackson* (1980), vol. 1, p. 74.

2. Thomas A. Bailey, *Presidential Greatness* (1966), p. 163.

3. Noble E. Cunningham, Jr., *The Presidency of James Monroe* (1996), pp. 31–35.

4. Andrew Jackson, letter to Willie Blount, January 25, 1812, National

Archives, Washington D.C.; James Parton, *Life of Andrew Jackson*, vol. 1, pp. 349–60.

5. Robert Remini, *Andrew Jackson and the Course of American Freedom, 1822–1832* (1981), pp. 88–94; Samuel Flagg Bemis, *John Quincy Adams and the Union* (1956), vol. 2, p. 43.

6. Bemis, *Adams*, vol. 2, pp. 18–19; Albert Somit, "Andrew Jackson: Legend and Reality," *Tennessee Historical Quarterly* (Dec. 1948), pp. 291–313; John William Ward, *Andrew Jackson: Symbol for an Age* (1955), pp. 83–86; Irving G. Williams, *The Rise of the Vice Presidency* (1956), p. 38.

7. Lucius Wilmerding, Jr., *James Monroe: Public Claimant* (1960), esp. pp. 82–84.

8. Bemis, *Adams*, vol. 2, p. 77; Remini, *Jackson and the Course of American Freedom*, p. 110.

9. Remini, *Jackson and the Course of American Freedom*, p. 132.

10. C. Vann Woodward, ed., *Responses of the Presidents to Charges of Misconduct* (1974), p. 58; Remini, *Jackson and the Course of American Freedom*, pp. 133–34; Bemis, *Adams*, vol. 2, p. 141.

11. Remini, *Jackson and the Course of American Freedom*, p. 7.

12. Woodward, *Responses of the Presidents*, pp. 61–62.

13. Donald B. Cole, *Martin Van Buren and the American Political System* (1984).

14. Williams, *Rise of the Vice Presidency*, p. 43.

15. Cole, *Martin Van Buren*, p. 353.

16. Robert Allen Rutland, *The Presidency of James Madison* (1990), pp. 134–35; Oliver Chitwood, *John Tyler* (1939).

17. Holman Hamilton, *White House Images and Reality* (1958), p. 63.

18. Gil Troy, *See How They Ran* (1996), pp. 22–25.

Chapter 4. Manifest Destiny

1. Charles G. Sellers, *James K. Polk: Jacksonian* (1957); Sellers, *James K. Polk: Continentalist* (1966); Thomas A. Bailey, *Presidential Saints and Sinners* (1981), pp. 64–68; Frederick Merk, *Manifest Destiny and Mission in American History* (1966); Albert K. Weinberg, *Manifest Destiny* (1935).

Chapter 5. The Story of Franklin Pierce

1. This chapter is based mainly on Roy Franklin Nichols, *Franklin Pierce: Young Hickory of the Granite Hills* (1958). I also consulted the Pierce Papers at the Library of Congress.

Chapter 6. The Slavery Crisis

1. The main facts about Buchanan's life are drawn chiefly from: Philip S.

Klein, *President James Buchanan* (1962); George Ticknor Curtis, *Life of James Buchanan* (1883), vols. 1 and 2.

2. Charles G. Sellers, *James K. Polk: Continentalist* (1966), p. 34.

3. Elbert Smith, *The Presidency of James Buchanan* (1975), chap. 4.

4. These statistics are drawn from the charts published in Richard Morris, ed., *The Encyclopedia of American History* (1970); Richard Hofstadter, *The Age of Reform* (1956), p. 23.

5. David E. Meerse, "Buchanan, Corruption and the Election of 1860," *Civil War History* 12 (June 1960), pp. 116-31.

6. Allan Nevins, *The Emergence of Lincoln* (1950), vols. 1 and 2.

7. C. Vann Woodward, ed. *Responses of the Presidents to Charges of Misconduct*, (1974), p. 99.

8. U.S. House, *Covode Investigation*, H. Report No. 648, 35th Congress, 1st sess., June 16, 1860.

9. On Polk, see Woodward, *Responses of the Presidents*, pp. 83–84.

Chapter 7. The Story of Abraham Lincoln

1. In preparing this chapter I have relied primarily on the source documents included in Roy Basler, ed., *The Collected Works of Abraham Lincoln* (1953–55), especially vols. 1 and 2. I have also made extensive use of Benjamin Thomas, ed., *Lincoln: 1847–1853: Being the Day-by-Day Activities* (1936). Three biographies proved essential: David Donald, *Lincoln* (1995), Stephen Oates, *With Malice Toward None* (1977), Benjamin Thomas, *Abraham Lincoln* (1952). On Lincoln's early years as a boy and young politician I consulted Benjamin Thomas, *Lincoln's New Salem* (1954), Donald Riddle, *Congressman Abraham Lincoln* (1957), Paul Simon, *Lincoln's Preparation for Greatness* (1965). On his life as a mature politician I consulted: Don Fehrenbacher, *Prelude to Greatness* (1962), Harry Carman and Reinhard Luthin, *Lincoln and the Patronage* (1943). Three books were helpful in dissecting the myths about Lincoln: Richard Current, *The Lincoln Nobody Knows* (1958), Stephen Oates, *Abraham Lincoln: The Man Behind the Myths* (1984), David Donald, *Lincoln Reconsidered* (1961).

2. Arthur Schlesinger, Jr., *The Imperial Presidency* (1974), p. 68.

Chapter 8. The Birth of Industrial Capitalism

1. These statistics are drawn from the government's *Historical Statistics of the United States* (1975), vols. 1 and 2.

2. On the Monroe scandals see Robert Remini, *Andrew Jackson and the Course of American Freedom* (1981), pp. 19, 399; on the rest see C. Vann Woodward, ed., *Responses of the Presidents to Charges of Misconduct* (1974).

3. Government salary statistics are conveniently listed in the index (under salaries) in James D. Richardson, ed., *Messages and Papers of the Presidents* (1917).

4. Michael Medved, *The Shadow Presidents* (1979), pp. 39–52.

5. Geoffrey Perret maintains that Grant disallowed grants of immunity because he saw no reason why the "small fry" should be allowed to go free. I believe this is a naive view. See Perret's *Ulysses S. Grant: Soldier and President* (1997), chap. 30. For a more realistic view of Grant's role in the investigation, see William S. McFeely, *Grant: A Biography* (1981), chap. 24. McFeely does not say what he believes Grant's motive was in withdrawing grants of immunity, but is convinced that in the affair Grant acted to protect Babcock.

After I finished writing this section of the book I happened to read Leonard Garment's *Crazy Rhythm* (1997). On page 259 he explains that in 1973, when Richard Nixon was trying to undermine John Dean, John Ehrlichman suggested that Nixon should allow White House staffers to testify before the Watergate grand jury only on the condition that none would be eligible to receive immunity. Nixon immediately agreed. "If Dean did not have immunity," Garment writes, "Nixon and Ehrlichman reasoned, he would realize that sooner or later he faced criminal sanctions; his fate would then depend on Nixon's willingness to use presidential powers to lighten his sentence. Presumably Dean, frightened by this prospect, would be less likely to testify against Nixon to the grand jury."

Grant was not in the same position as Nixon. Grant wasn't worried that *his* aide could implicate *him* in a crime. But he was worried that Babcock's friends could implicate *him*. It was for that reason that Grant insisted that nobody receive immunity.

Dean, incidentally, testified in full without immunity anyway. When he found out from Garment that Nixon had personally insisted on eliminating immunity, "Dean reacted with a spasm of rage. . . . In short order he accelerated his conversations with the Watergate prosecutors and started spilling the beans."

6. Thomas C. Reeves, *Gentleman Boss: The Life of Chester Alan Arthur* (1975), pp. 51–56.

7. Richard Hofstadter, *The American Political Tradition* (1948), pp. 170–71.

8. Allan Peskin, *Garfield* (1978), pp. 362–63.

Chapter 9. The Birth of Machine Politics

1. This chapter is mainly based on Harry Barnard, *Rutherford B. Hayes* (1954); T. Harry Williams, ed., *Hayes: The Diary of a President: 1875–1881* (1964); Claude Bowers, *The Tragic Era* (1929); C. Vann Woodward, *Reunion and Reaction* (1991).

Chapter 10. The Story of Chester A. Arthur

1. This chapter is mainly based on Thomas C. Reeves, *Gentleman Boss: The Life of Chester Alan Arthur* (1975); Leonard D. White, *The Republican Era: 1869–1901* (1958), pp. 188–220.

Chapter 11. The Arrival of the Immigrants

1. Allan Nevins, *Grover Cleveland* (1932); James D. Richardson, ed., *Messages and Papers of the Presidents* (1897–1917), pp. 5142–43.

2. Charles G. Sellers, *James K. Polk: Continentalist* (1966), p. 158.

3. Thomas A. Bailey, *Presidential Saints and Sinners* (1981), pp. 269–70.

Chapter 12. The Media

1. Allan Nevins, *Grover Cleveland* (1932), pp. 523–48; Robert Ferrell, *Ill-Advised: Presidential Health and Public Trust* (1992), pp. 3–11; Michael Medved, *The Shadow Presidents* (1979), pp. 85–86; James D. Richardson, ed., *Messages and Papers of the Presidents* (1897-1917), p. 5833.

2. The following discussion of the media is based, in part, on James Pollard, *The Presidents and the Press* (1947), and Edwin Emery and Michael Emery, *The Press and America* (1984).

3. Lewis L. Gould, *The Presidency of William McKinley* (1980), chaps. 1, 3.

4. Though the transformation was gradual, there actually was a specific date on which the change became official. It was April 24, 1897, the day the *Washington Star*'s William Price became the first designated White House reporter.

5. Glyndon G. Van Deusen, *Horace Greeley* (1953), p. 420.

6. Albert Bigelow Paine, *Th. Nast: His Period and His Pictures* (1904), pp. 318, 336.

7. Lincoln Steffens, *The Autobiography of Lincoln Steffens* (1931), pp. 285–91; Melville Landon, *Eli Perkins* (1891), p. 24.

8. David S. Barry, "News-Getting at the Capital," *Chautauquan Monthly* 26 (Dec. 1897), pp. 282–86.

9. Thomas C. Reeves, *Gentleman Boss: The Life of Chester Alan Arthur* (1975), passim.

Chapter 13. World Power: I

1. The facts about McKinley's life are mainly based on Margaret Leech, *In the Days of McKinley* (1959).

2. Albert K. Weinberg, *Manifest Destiny* (1935); Theodore P. Greene, ed., *American Imperialism in 1898* (1955).

3. Allan Nevins, *Grover Cleveland* (1932), chap. 30.

4. Thomas A. Bailey, ed., *The American Spirit* (2nd ed.; 1968), chap. 31.

5. It used to be taught that Spain, on the eve of the war with the United States, had indeed given in to all of McKinley's demands. But historians now reject that viewpoint. See Lewis Gould, *The Presidency of William McKinley* (1980); H. Wayne Morgan, *America's Road to Empire* (1968).

6. Arthur Schlesinger, Jr., *The Imperial Presidency* (1974), chaps. 1–4.

Chapter 14. World Power: II

1. David McCullough, *Mornings on Horseback* (1981).

2. Edmund Morris, *The Rise of Theodore Roosevelt* (1979); Morton Keller, ed., *Theodore Roosevelt: A Profile* (1967); Henry Pringle, *Theodore Roosevelt* (1956).

3. Lewis Gould, *The Presidency of Theodore Roosevelt* (1991).

4. Irving G. Williams, *The Rise of the Vice Presidency* (1956), pp. 84–85.

5. Gil Troy, *See How They Ran* (1996), pp. 114, 121–22; Forrest McDonald, *The American Presidency* (1994), p. 435; Gould, *Presidency of Theodore Roosevelt*, pp. 20, 153–54.

Chapter 15. World Power: III

1. The basic facts on Wilson's life are drawn from: August Heckscher, *Woodrow Wilson* (1991); Kenneth Clements, *The Presidency of Woodrow Wilson* (1992); Arthur S. Link, ed., *Woodrow Wilson: A Profile* (1968); John A. Garraty, *Woodrow Wilson: A Great Life in Brief* (1956).

2. David Kennedy, *Over Here* (1980).

3. Thomas A. Bailey, *Woodrow Wilson and the Great Betrayal* (1945).

4. Michael Medved, *The Shadow Presidents* (1979), pp. 185–93.

5. Robert Gilbert, *The Mortal Presidency* (1992), chap. 2.

6. Thomas A. Bailey, *Presidential Saints and Sinners* (1981), pp. 203–4.

Chapter 16. FDR: The Great Depression

1. David Halberstam, *The Powers That Be* (1979), pp. 24–29.

2. Gene Smith, *The Shattered Dream: Herbert Hoover and the Great Depression* (1970).

3. The basic facts in this and the following chapter are drawn from Frank Freidel, *Franklin D. Roosevelt: A Rendezvous with Destiny* (1990).

4. T. Harry Williams, *Huey Long* (1970).

5. Geoffrey Ward, *Before the Trumpet: Young Franklin Roosevelt* (1985).

6. Robert Ferrell, *Ill-Advised: Presidential Health and Public Trust* (1992), p. 40.

Chapter 17. FDR: World War II

1. Robert C. Butow, "The FDR Tapes," *American Heritage* (Feb.-March 1982), pp. 18–19.

2. Robert Ferrell, *Ill-Advised: Presidential Health and Public Trust* (1992), p. 46.

3. William Leuchtenburg, *In the Shadow of FDR* (1983).

Chapter 18. The Cold War

1. The facts in this chapter on Truman's background are mainly drawn from Alonzo Hamby, *Man of the People* (1995).

2. David McCullough, *Truman* (1992); Daniel Yergin, "Harry Truman—Revived and Revised," *New York Times Magazine* (Oct. 24, 1976), p. 83.

3. Frank Kofsky, *Harry S. Truman and the War Scare of 1948: A Successful Campaign to Deceive the Nation* (1995); Walter LaFeber, *America, Russia and the Cold War* (1985); Daniel Yergin, *Shattered Peace: The Origins of the Cold War and the National Security State* (1977). I should note that while I remain impressed with the facts in Kofsky's book, I do not share his perspective on the Cold War.

4. Herbert Parmet, *Eisenhower and the American Crusades* (1972), pp. 130–33; Fred Greenstein, *The Hidden-Hand Presidency* (1982), chap. 5.

5. Dwight Eisenhower, *Mandate for Change: 1953–55* (1963), pp. 147–48; Robert Ferrell, *Ill-Advised: Presidential Health and Public Trust* (1992), pp. 53–72; John Emmet Hughes, *The Ordeal of Power* (1963), pp. 107–15; Parmet, *Eisenhower*, p. 278; *New York Times* (Apr. 17–22, 1953); Robert Gilbert, *The Mortal Presidency* (1992). I should note that there is considerable disagreement about whether Ike actually had heart attacks in 1949 and 1953. Ike's physician always insisted that he did not, and for years no one rose to dispute him. Recently Ike's cardiologist in the White House, Dr. Tom Mattingly, reviewed Eisenhower's heart records and came to the conclusion that he probably had had a heart attack in 1949 and almost certainly did in 1953. Ferrell cites Mattingly approvingly, as does Gilbert, though Gilbert refers to the 1949 and 1953 episodes as "mystery" illnesses.

6. *New York Times*, November 19, 1997, p. A25.

7. Robert Caro, *The Years of Lyndon Johnson: The Path to Power* (1982), chap. 8.

8. Fawn Brodie, *Richard Nixon: The Shaping of His Character* (1981).

Conclusion

1. Monica Crowley, *Nixon Off the Record* (1996), pp. 8–9.

Index

RICHARD SHENKMAN is the *New York Times* bestselling author of five books, including *Legends, Lies & Cherished Myths of American History*. He is an Emmy Award–winning investigative reporter and the former managing editor of the news department at the CBS-TV affiliate in Seattle. He appeared in his own history show on the Learning Channel and can be regularly seen on Fox News and MSNBC. He is currently the managing editor of TomPaine.com.